The R Reference Manual
Base Package
Volume 2

The R Development Core Team

Version 1.8.1

A catalogue record for this book is available from the British Library.

First printing, January 2004 (24/1/2004).

Published by Network Theory Limited.

15 Royal Park
Bristol
BS8 3AL
United Kingdom

Email: info@network-theory.co.uk

ISBN 0-9546120-1-9

Further information about this book is available from
http://www.network-theory.co.uk/R/reference/

Copyright © 1999–2003 R Development Core Team.

Permission is granted to make and distribute verbatim copies of this manual provided the copyright notice and this permission notice are preserved on all copies.
Permission is granted to copy and distribute modified versions of this manual under the conditions for verbatim copying, provided that the entire resulting derived work is distributed under the terms of a permission notice identical to this one.
Permission is granted to copy and distribute translations of this manual into another language, under the above conditions for modified versions, except that this permission notice may be stated in a translation approved by the R Development Core Team.

Contents

Publisher's Preface	**1**
1 Base package — graphics	**5**
abline	6
arrows	8
assocplot	10
axis	12
axis.POSIXct	15
axTicks	17
barplot	19
box	24
boxplot	25
boxplot.stats	29
bxp	31
chull	34
col2rgb	36
colors	39
contour	40
coplot	44
curve	48
dev.xxx	50
dev2	52
dev2bitmap	55
Devices	57
dotchart	59
filled.contour	61
fourfoldplot	64
frame	67
Gnome	68
gray	70
grid	71

gtk	73
Hershey	74
hist	78
hsv	82
identify	84
image	86
interaction.plot	89
Japanese	92
jitter	93
layout	95
legend	98
lines	104
locator	106
matplot	108
mosaicplot	112
mtext	116
n2mfrow	119
pairs	120
palette	123
Palettes	125
panel.smooth	127
par	128
pdf	137
persp	139
pictex	143
pie	145
plot	147
plot.data.frame	149
plot.default	150
plot.density	154
plot.design	155
plot.factor	158
plot.formula	159
plot.histogram	161
plot.lm	163
plot.table	166
plot.ts	168
plot.window	170
plot.xy	172
plotmath	173
png	177
points	180
polygon	183

CONTENTS iii

```
postscript . . . . . . . . . . . . . . . . . . . . . . . . 186
ppoints . . . . . . . . . . . . . . . . . . . . . . . . . 192
preplot . . . . . . . . . . . . . . . . . . . . . . . . . 193
pretty . . . . . . . . . . . . . . . . . . . . . . . . . 194
qqnorm . . . . . . . . . . . . . . . . . . . . . . . . 196
quartz . . . . . . . . . . . . . . . . . . . . . . . . . 198
recordPlot . . . . . . . . . . . . . . . . . . . . . . . 199
rect . . . . . . . . . . . . . . . . . . . . . . . . . . 200
rgb . . . . . . . . . . . . . . . . . . . . . . . . . . . 203
rug . . . . . . . . . . . . . . . . . . . . . . . . . . . 204
screen . . . . . . . . . . . . . . . . . . . . . . . . . 206
segments . . . . . . . . . . . . . . . . . . . . . . . . 209
stars . . . . . . . . . . . . . . . . . . . . . . . . . . 211
stripchart . . . . . . . . . . . . . . . . . . . . . . . 216
strwidth . . . . . . . . . . . . . . . . . . . . . . . . 218
sunflowerplot . . . . . . . . . . . . . . . . . . . . . 220
symbols . . . . . . . . . . . . . . . . . . . . . . . . 223
termplot . . . . . . . . . . . . . . . . . . . . . . . . 226
text . . . . . . . . . . . . . . . . . . . . . . . . . . 228
title . . . . . . . . . . . . . . . . . . . . . . . . . . 231
units . . . . . . . . . . . . . . . . . . . . . . . . . . 233
x11 . . . . . . . . . . . . . . . . . . . . . . . . . . . 234
xfig . . . . . . . . . . . . . . . . . . . . . . . . . . . 236
xy.coords . . . . . . . . . . . . . . . . . . . . . . . 238
xyz.coords . . . . . . . . . . . . . . . . . . . . . . . 240
```

2 Base package — math   243

```
abs . . . . . . . . . . . . . . . . . . . . . . . . . . . 244
all.equal . . . . . . . . . . . . . . . . . . . . . . . . 245
approxfun . . . . . . . . . . . . . . . . . . . . . . . 247
Arithmetic . . . . . . . . . . . . . . . . . . . . . . . 250
backsolve . . . . . . . . . . . . . . . . . . . . . . . 252
Bessel . . . . . . . . . . . . . . . . . . . . . . . . . 254
chol . . . . . . . . . . . . . . . . . . . . . . . . . . 257
chol2inv . . . . . . . . . . . . . . . . . . . . . . . . 260
colSums . . . . . . . . . . . . . . . . . . . . . . . . 262
convolve . . . . . . . . . . . . . . . . . . . . . . . . 264
crossprod . . . . . . . . . . . . . . . . . . . . . . . 266
cumsum . . . . . . . . . . . . . . . . . . . . . . . . 267
deriv . . . . . . . . . . . . . . . . . . . . . . . . . . 268
eigen . . . . . . . . . . . . . . . . . . . . . . . . . . 271
Extremes . . . . . . . . . . . . . . . . . . . . . . . 274
fft . . . . . . . . . . . . . . . . . . . . . . . . . . . 276
```

findInterval		278
gl		280
Hyperbolic		281
integrate		282
kappa		285
log		287
matmult		289
matrix		290
nextn		292
poly		293
polyroot		295
prod		297
qr		298
QR.Auxiliaries		301
range		303
Round		305
sign		307
solve		308
sort		310
Special		313
splinefun		315
sum		318
svd		319
tabulate		321
Trig		322
3	**Base package — distributions and random numbers**	**323**
	bandwidth	324
	Beta	326
	Binomial	328
	birthday	330
	Cauchy	332
	Chisquare	334
	density	337
	Exponential	342
	FDist	344
	GammaDist	346
	Geometric	348
	Hypergeometric	350
	Logistic	352
	Lognormal	354
	Multinomial	356
	NegBinomial	358

Normal	361
Poisson	363
r2dtable	365
Random	367
Random.user	372
sample	374
SignRank	376
TDist	378
Tukey	380
Uniform	382
Weibull	384
Wilcoxon	386

4 Base package — models 389

add1	390
AIC	393
alias	395
anova	398
anova.glm	399
anova.lm	401
aov	403
AsIs	405
C	407
case/variable.names	409
coef	411
confint	412
constrOptim	414
contrast	417
contrasts	419
deviance	421
df.residual	422
dummy.coef	423
eff.aovlist	425
effects	427
expand.grid	429
expand.model.frame	430
extractAIC	432
factor.scope	434
family	436
fitted	439
formula	440
glm	443
glm.control	449

glm.summaries	451
influence.measures	453
is.empty.model	457
labels	458
lm	459
lm.fit	463
lm.influence	465
lm.summaries	468
logLik	470
logLik.glm	472
logLik.lm	473
loglin	475
ls.diag	478
ls.print	480
lsfit	481
make.link	483
makepredictcall	484
manova	486
model.extract	487
model.frame	489
model.matrix	491
model.tables	493
naprint	495
naresid	496
nlm	497
offset	501
optim	502
optimize	509
power	512
predict.glm	513
predict.lm	515
profile	517
proj	518
relevel	521
replications	522
residuals	524
se.contrast	525
stat.anova	527
step	529
summary.aov	532
summary.glm	534
summary.lm	537
summary.manova	540

terms	542
terms.formula	543
terms.object	545
TukeyHSD	547
uniroot	549
update	551
update.formula	553
vcov	554
weighted.residuals	555

5 Base package — dates, time and time-series 557

as.POSIX*	558
cut.POSIXt	560
DateTimeClasses	562
diff	565
difftime	567
hist.POSIXt	569
print.ts	571
rep	572
round.POSIXt	575
seq.POSIXt	576
start	578
strptime	579
Sys.time	583
time	584
ts	586
ts-methods	589
tsp	590
weekdays	591
window	593

6 Base package — datasets 595

airmiles	596
airquality	597
anscombe	599
attenu	601
attitude	603
cars	605
chickwts	607
co2	608
data	609
discoveries	612
esoph	613

euro	615
eurodist	617
faithful	618
Formaldehyde	620
freeny	621
HairEyeColor	622
infert	624
InsectSprays	626
iris	627
islands	629
LifeCycleSavings	630
longley	632
morley	634
mtcars	636
nhtemp	637
OrchardSprays	638
phones	640
PlantGrowth	641
precip	642
presidents	643
pressure	644
quakes	645
randu	646
rivers	647
sleep	648
stackloss	649
state	651
sunspots	653
swiss	654
Titanic	656
ToothGrowth	658
trees	659
UCBAdmissions	660
USArrests	662
USJudgeRatings	663
USPersonalExpenditure	664
uspop	665
VADeaths	666
volcano	668
warpbreaks	669
women	671

Index   **675**

Publisher's Preface

This reference manual documents the use of R, an environment for statistical computing and graphics.

R is *free software*. The term "free software" refers to your freedom to run, copy, distribute, study, change and improve the software. With R you have all these freedoms.

R is part of the GNU Project. The GNU Project was launched in 1984 to develop a complete Unix-like operating system which is free software: the GNU system. It was conceived as a way of bringing back the cooperative spirit that prevailed in the computing community in earlier days, by removing the obstacles to cooperation imposed by the owners of proprietary software.

You can support the GNU Project by becoming an associate member of the Free Software Foundation. The Free Software Foundation is a tax-exempt charity dedicated to promoting computer users' right to use, study, copy, modify, and redistribute computer programs. It also helps to spread awareness of the ethical and political issues of freedom in the use of software. For more information visit the website www.fsf.org.

The development of R itself is guided by the R Foundation, a not for profit organization working in the public interest. Individuals and organisations using R can support its continued development by becoming members of the R Foundation. Further information is available at the website www.r-project.org.

Brian Gough
Publisher
November 2003

Introduction

This is the second volume of the R Reference Manual. This volume documents the commands for graphics, mathematical functions, random distributions, models, date-time calculations, time-series and example datasets in the *base* package of R. For ease of use, the commands have be grouped into chapters by topic, and then sorted alphabetically within each chapter.

The documentation for the other base package commands can be found in the first volume of this series. The first volume documents the core commands: programming constructs, fundamental numerical routines and system functions. The base package commands are automatically available when the R environment is started. Documentation for additional packages is available in further volumes of this series.

R is available from many commercial distributors, including the Free Software Foundation on their source code CD-ROMs. Information about R itself can be found online at www.r-project.org.

To start the program once it is installed simply use the command R on Unix-like systems,

```
$ R
R : Copyright 2003, The R Development Core Team
Type 'demo()' for some demos, 'help()' for on-line help,
or 'help.start()' for a HTML browser interface to help.
Type 'q()' to quit R.
>
```

Commands can then be typed at the R prompt (>). The commands given in this manual should generally be entered with arguments in parentheses, e.g. `help()`, as shown in the examples. Typing the name of the function without parentheses, such as `help`, displays its internal definition and does not execute the command.

To obtain online help for any command type `help(command)`. A tutorial for new users of R is available in the book *"An Introduction to R"* (ISBN 0-9541617-4-2).

Chapter 1

Base package — graphics

abline *Add a Straight Line to a Plot*

Description

This function adds one or more straight lines through the current plot.

Usage

```
abline(a, b, untf = FALSE, ...)
abline(h=, untf = FALSE, ...)
abline(v=, untf = FALSE, ...)
abline(coef=, untf = FALSE, ...)
abline(reg=, untf = FALSE, ...)
```

Arguments

a,b	the intercept and slope.
untf	logical asking to *untransform*. See Details.
h	the y-value for a horizontal line.
v	the x-value for a vertical line.
coef	a vector of length two giving the intercept and slope.
reg	an object with a `coef` component. See Details.
...	graphical parameters.

Details

The first form specifies the line in intercept/slope form (alternatively a can be specified on its own and is taken to contain the slope and intercept in vector form).

The h= and v= forms draw horizontal and vertical lines at the specified coordinates.

The `coef` form specifies the line by a vector containing the slope and intercept.

`reg` is a regression object which contains `reg$coef`. If it is of length 1 then the value is taken to be the slope of a line through the origin, otherwise, the first 2 values are taken to be the intercept and slope.

If `untf` is true, and one or both axes are log-transformed, then a curve is drawn corresponding to a line in original coordinates, otherwise a line is drawn in the transformed coordinate system. The h and v parameters always refer to original coordinates.

The graphical parameters col and lty can be specified as arguments to abline; see par for details.

References

Becker, R. A., Chambers, J. M. and Wilks, A. R. (1988) *The New S Language.* Wadsworth & Brooks/Cole.

See Also

lines and segments for connected and arbitrary lines given by their *endpoints*. par.

Examples

```
data(cars)
z <- lm(dist ~ speed, data = cars)
plot(cars)
abline(z)
```

arrows   *Add Arrows to a Plot*

Description

Draw arrows between pairs of points.

Usage

```
arrows(x0, y0, x1, y1, length = 0.25, angle = 30, code = 2,
       col = par("fg"), lty = NULL, lwd = par("lwd"),
       xpd = NULL)
```

Arguments

x0, y0   coordinates of points **from** which to draw.

x1, y1   coordinates of points **to** which to draw.

length   length of the edges of the arrow head (in inches).

angle   angle from the shaft of the arrow to the edge of the arrow head.

code   integer code, determining *kind* of arrows to be drawn.

col, lty, lwd, xpd
                  usual graphical parameters as in `par`.

Details

For each `i`, an arrow is drawn between the point (x0[i], y0[i]) and the point (x1[i],y1[i]).

If `code=2` an arrowhead is drawn at (x0[i],y0[i]) and if `code=1` an arrowhead is drawn at (x1[i],y1[i]). If `code=3` a head is drawn at both ends of the arrow. Unless `length = 0`, when no head is drawn.

The graphical parameters `col` and `lty` can be used to specify a color and line texture for the line segments which make up the arrows (`col` may be a vector).

The direction of a zero-length arrow is indeterminate, and hence so is the direction of the arrowheads. To allow for rounding error, arrowheads are omitted (with a warning) on any arrow of length less than 1/1000 inch.

References

Becker, R. A., Chambers, J. M. and Wilks, A. R. (1988) *The New S Language.* Wadsworth & Brooks/Cole.

See Also

segments to draw segments.

Examples

```
x <- runif(12); y <- rnorm(12)
i <- order(x,y); x <- x[i]; y <- y[i]
plot(x,y, main="arrows(.) and segments(.)")
## draw arrows from point to point :
s <- seq(length(x)-1) # one shorter than data
arrows(x[s], y[s], x[s+1], y[s+1], col= 1:3)
s <- s[-length(s)]
segments(x[s], y[s], x[s+2], y[s+2], col= 'pink')
```

assocplot   *Association Plots*

Description

Produce a Cohen-Friendly association plot indicating deviations from independence of rows and columns in a 2-dimensional contingency table.

Usage

```
assocplot(x, col = c("black", "red"), space = 0.3,
          main = NULL, xlab = NULL, ylab = NULL)
```

Arguments

x	a two-dimensional contingency table in matrix form.
col	a character vector of length two giving the colors used for drawing positive and negative Pearson residuals, respectively.
space	the amount of space (as a fraction of the average rectangle width and height) left between each rectangle.
main	overall title for the plot.
xlab	a label for the x axis. Defaults to the name of the row variable in x if non-NULL.
ylab	a label for the y axis. Defaults to the column names of the column variable in x if non-NULL.

Details

For a two-way contingency table, the signed contribution to Pearson's χ^2 for cell i, j is $d_{ij} = (f_{ij} - e_{ij})/\sqrt{e_{ij}}$, where f_{ij} and e_{ij} are the observed and expected counts corresponding to the cell. In the Cohen-Friendly association plot, each cell is represented by a rectangle that has (signed) height proportional to d_{ij} and width proportional to $\sqrt{e_{ij}}$, so that the area of the box is proportional to the difference in observed and expected frequencies. The rectangles in each row are positioned relative to a baseline indicating independence ($d_{ij} = 0$). If the observed frequency of a cell is greater than the expected one, the box rises above the baseline and is shaded in the color specified by the first element of col, which defaults to black; otherwise, the box falls below the baseline and is shaded in the color specified by the second element of col, which defaults to red.

References

Cohen, A. (1980), On the graphical display of the significant components in a two-way contingency table. *Communications in Statistics—Theory and Methods*, **A9**, 1025–1041.

Friendly, M. (1992), Graphical methods for categorical data. *SAS User Group International Conference Proceedings*, **17**, 190–200. http://www.math.yorku.ca/SCS/sugi/sugi17-paper.html

See Also

mosaicplot; chisq.test.

Examples

```
data(HairEyeColor)
## Aggregate over sex:
x <- margin.table(HairEyeColor, c(1, 2))
x
assocplot(x, main = "Relation between hair and eye color")
```

axis   *Add an Axis to a Plot*

Description

Adds an axis to the current plot, allowing the specification of the side, position, labels, and other options.

Usage

```
axis(side, at = NULL, labels = TRUE, tick = TRUE,
    line = NA, pos = NA, outer = FALSE, font = NA,
    vfont = NULL, lty = "solid", lwd = 1, col = NULL, ...)
```

Arguments

- **side**   an integer specifying which side of the plot the axis is to be drawn on. The axis is placed as follows: 1=below, 2=left, 3=above and 4=right.
- **at**   the points at which tick-marks are to be drawn. Non-finite (infinite, NaN or NA) values are omitted. By default, when NULL, tickmark locations are computed, see Details below.
- **labels**   this can either be a logical value specifying whether (numerical) annotations are to be made at the tickmarks, or a vector of character strings to be placed at the tickpoints.
- **tick**   a logical value specifying whether tickmarks should be drawn
- **line**   the number of lines into the margin which the axis will be drawn. This overrides the value of the graphical parameter mgp[3]. The relative placing of tickmarks and tick labels is unchanged.
- **pos**   the coordinate at which the axis line is to be drawn. this overrides the value of both line and mgp[3].
- **outer**   a logical value indicating whether the axis should be drawn in the outer plot margin, rather than the standard plot margin.
- **font**   font for text.
- **vfont**   vector font for text.

lty, lwd   line type, width for the axis line and the tick marks.

col   color for the axis line and the tick marks. The default NULL means to use par("fg").

...   other graphical parameters may also be passed as arguments to this function, e.g., las for vertical/horizontal label orientation, or fg instead of col, see par on these.

Details

The axis line is drawn from the lowest to the highest value of at, but will be clipped at the plot region. Only ticks which are drawn from points within the plot region (up to a tolerance for rounding error) are plotted, but the ticks and their labels may well extend outside the plot region.

When at = NULL, pretty tick mark locations are computed internally, the same axTicks(side) would, from par("usr","lab"), and par("xlog") (or ylog respectively).

Value

This function is invoked for its side effect, which is to add an axis to an already existing plot.

References

Becker, R. A., Chambers, J. M. and Wilks, A. R. (1988) *The New S Language.* Wadsworth & Brooks/Cole.

See Also

axTicks returns the axis tick locations corresponding to at=NULL; pretty is more flexible for computing pretty tick coordinates and does *not* depend on (nor adapt to) the coordinate system in use.

Examples

```
plot(1:4, rnorm(4), axes=FALSE)
axis(1, 1:4, LETTERS[1:4])
axis(2)
box() # to make it look "as usual"

plot(1:7, rnorm(7), main = "axis() examples",
```

```
        type = "s", xaxt="n", frame = FALSE, col = "red")
axis(1, 1:7, LETTERS[1:7], col.axis = "blue")
# unusual options:
axis(4, col = "violet", col.axis="dark violet",lwd = 2)
axis(3, col = "gold", lty = 2, lwd = 0.5)
```

axis.POSIXct *Date-time Plotting Functions*

Description

Functions to plot objects of classes "POSIX1t" and "POSIXct" representing calendar dates and times.

Usage

```
axis.POSIXct(side, x, at, format, ...)

## S3 method for class 'POSIXct':
plot(x, y, xlab = "", ...)

## S3 method for class 'POSIX1t':
plot(x, y, xlab = "", ...)
```

Arguments

x, at	A date-time object.
y	numeric values to be plotted against x.
xlab	a character string giving the label for the x axis.
side	See axis.
format	See strptime.
...	Further arguments to be passed from or to other methods, typically graphical parameters or arguments of plot.default.

Details

The functions plot against an x-axis of date-times. axis.POSIXct works quite hard to choose suitable time units (years, months, days, hours, minutes or seconds) and a sensible output format, but this can be overridden by supplying a format specification.

If at is supplied for axis.POSIXct it specifies the locations of the ticks and labels: if x is specified a suitable grid of labels is chosen.

See Also

DateTimeClasses for details of the classes.

Examples

```
res <- try(data(beav1, package = "MASS"))
if(!inherits(res, "try-error")) {
attach(beav1)
time <- strptime(
  paste(1990, day, time %/% 100,time %% 100),
  "%Y %j %H %M"
)
plot(time, temp, type="l") # axis at 4-hour intervals.
# now label every hour on the time axis
plot(time, temp, type="l", xaxt="n")
r <- as.POSIXct(round(range(time), "hours"))
axis.POSIXct(1, at=seq(r[1], r[2], by="hour"), format="%H")
rm(time)
detach(beav1)
}

plot(.leap.seconds, 1:22, type="n", yaxt="n",
     xlab="leap seconds", ylab="", bty="n")
rug(.leap.seconds)
```

axTicks   *Compute Axis Tickmark Locations*

Description

Compute tickmark locations, the same way as R does internally. This is only non-trivial when **log** coordinates are active. By default, gives the **at** values which **axis(side)** would use.

Usage

```
axTicks(side, axp = NULL, usr = NULL, log = NULL)
```

Arguments

side
: integer in 1:4, as for **axis**.

axp
: numeric vector of length three, defaulting to par("Zaxp") where "Z" is "x" or "y" depending on the **side** argument.

usr
: numeric vector of length four, defaulting to par("usr") giving horizontal ('x') and vertical ('y') user coordinate limits.

log
: logical indicating if log coordinates are active; defaults to par("Zlog") where 'Z' is as for the **axp** argument above.

Details

The **axp**, **usr**, and **log** arguments must be consistent as their default values (the par(..) results) are. Note that the meaning of **axp** alters very much when **log** is TRUE, see the documentation on par(xaxp=.).

axTicks() can be regarded as an R implementation of the C function CreateAtVector() in '..../src/main/graphics.c' which is called by axis(side,*) when no argument **at** is specified.

Value

numeric vector of coordinate values at which axis tickmarks can be drawn. By default, when only the first argument is specified, these values should be identical to those that axis(side) would use or has used.

See Also

axis, par. pretty uses the same algorithm but is independent of the graphics environment and has more options.

Examples

```
plot(1:7, 10*21:27)
axTicks(1)
axTicks(2)
stopifnot(identical(axTicks(1), axTicks(3)),
          identical(axTicks(2), axTicks(4)))

## Show how axTicks() and axis() correspond :
op <- par(mfrow = c(3,1))
for(x in 9999*c(1,2,8)) {
    plot(x,9, log = "x")
    cat(formatC(par("xaxp"),wid=5),";",T<-axTicks(1),"\n")
    rug(T, col="red")
}
par(op)
```

barplot *Bar Plots*

Description

Creates a bar plot with vertical or horizontal bars.

Usage

```
## Default S3 method:
barplot(height, width = 1, space = NULL,
  names.arg = NULL, legend.text = NULL, beside = FALSE,
  horiz = FALSE, density = NULL, angle = 45,
  col = heat.colors(NR), border = par("fg"),
  main = NULL, sub = NULL, xlab = NULL, ylab = NULL,
  xlim = NULL, ylim = NULL, xpd = TRUE,
  axes = TRUE, axisnames = TRUE,
  cex.axis = par("cex.axis"), cex.names = par("cex.axis"),
  inside = TRUE, plot = TRUE, axis.lty = 0, ...)
```

Arguments

height
: either a vector or matrix of values describing the bars which make up the plot. If height is a vector, the plot consists of a sequence of rectangular bars with heights given by the values in the vector. If height is a matrix and beside is FALSE then each bar of the plot corresponds to a column of height, with the values in the column giving the heights of stacked "sub-bars" making up the bar. If height is a matrix and beside is TRUE, then the values in each column are juxtaposed rather than stacked.

width
: optional vector of bar widths. Re-cycled to length the number of bars drawn. Specifying a single value will no visible effect unless xlim is specified.

space
: the amount of space (as a fraction of the average bar width) left before each bar. May be given as a single number or one number per bar. If height is a matrix and beside is TRUE, space may be specified by two numbers, where the first is the space between bars in the same group, and the second the space between the groups. If not given explicitly, it defaults to c(0,1)

	if `height` is a matrix and `beside` is TRUE, and to 0.2 otherwise.
`names.arg`	a vector of names to be plotted below each bar or group of bars. If this argument is omitted, then the names are taken from the `names` attribute of `height` if this is a vector, or the column names if it is a matrix.
`legend.text`	a vector of text used to construct a legend for the plot, or a logical indicating whether a legend should be included. This is only useful when `height` is a matrix. In that case given legend labels should correspond to the rows of `height`; if `legend.text` is true, the row names of `height` will be used as labels if they are non-null.
`beside`	a logical value. If FALSE, the columns of `height` are portrayed as stacked bars, and if TRUE the columns are portrayed as juxtaposed bars.
`horiz`	a logical value. If FALSE, the bars are drawn vertically with the first bar to the left. If TRUE, the bars are drawn horizontally with the first at the bottom.
`density`	a vector giving the density of shading lines, in lines per inch, for the bars or bar components. The default value of NULL means that no shading lines are drawn. Non-positive values of `density` also inhibit the drawing of shading lines.
`angle`	the slope of shading lines, given as an angle in degrees (counter-clockwise), for the bars or bar components.
`col`	a vector of colors for the bars or bar components.
`border`	the color to be used for the border of the bars.
`main,sub`	overall and sub title for the plot.
`xlab`	a label for the x axis.
`ylab`	a label for the y axis.
`xlim`	limits for the x axis.
`ylim`	limits for the y axis.
`xpd`	logical. Should bars be allowed to go outside region?
`axes`	logical. If TRUE, a vertical (or horizontal, if `horiz` is true) axis is drawn.
`axisnames`	logical. If TRUE, and if there are `names.arg` (see above), the other axis is drawn (with lty=0) and labeled.

`cex.axis`	expansion factor for numeric axis labels.
`cex.names`	expansion factor for axis names (bar labels).
`inside`	logical. If `TRUE`, the lines which divide adjacent (non-stacked!) bars will be drawn. Only applies when `space = 0` (which it partly is when `beside = TRUE`).
`plot`	logical. If `FALSE`, nothing is plotted.
`axis.lty`	the graphics parameter `lty` applied to the axis and tick marks of the categorical (default horizontal) axis. Note that by default the axis is suppressed.
`...`	further graphical parameters (`par`) are passed to `plot.window()`, `title()` and `axis`.

Details

This is a generic function, it currently only has a default method. A formula interface may be added eventually.

Value

A numeric vector (or matrix, when `beside = TRUE`), say mp, giving the coordinates of *all* the bar midpoints drawn, useful for adding to the graph.

If `beside` is true, use `colMeans(mp)` for the midpoints of each *group* of bars, see example.

Note

Prior to R 1.6.0, `barplot` behaved as if `axis.lty = 1`, unintentionally.

References

Becker, R. A., Chambers, J. M. and Wilks, A. R. (1988) *The New S Language.* Wadsworth & Brooks/Cole.

See Also

`plot(..., type="h")`, `dotchart`, `hist`.

Examples

```
tN <- table(Ni <- rpois(100, lambda=5))
r <- barplot(tN, col='gray')
# type = "h" plotting is 'bar'plot
lines(r, tN, type='h', col='red', lwd=2)

barplot(tN, space = 1.5, axisnames=FALSE,
   sub = "barplot(..., space= 1.5, axisnames = FALSE)")

data(VADeaths, package = "base")
barplot(VADeaths, plot = FALSE)
barplot(VADeaths, plot = FALSE, beside = TRUE)

mp <- barplot(VADeaths) # default
tot <- colMeans(VADeaths)
text(mp, tot + 3, format(tot), xpd = TRUE, col = "blue")
barplot(VADeaths, beside = TRUE,
        col = c("lightblue", "mistyrose", "lightcyan",
                "lavender", "cornsilk"),
        legend = rownames(VADeaths), ylim = c(0, 100))
title(main = "Death Rates in Virginia", font.main = 4)

hh <- t(VADeaths)[, 5:1]
mybarcol <- "gray20"
mp <- barplot(hh, beside = TRUE,
  col = c("lightblue","mistyrose","lightcyan","lavender"),
  legend = colnames(VADeaths), ylim= c(0,100),
  main = "Death Rates in Virginia", font.main = 4,
  sub = "Faked upper 2*sigma error bars",
  col.sub = mybarcol,
  cex.names = 1.5)
segments(mp, hh, mp, hh + 2*sqrt(1000*hh/100),
         col = mybarcol, lwd = 1.5)
stopifnot(dim(mp) == dim(hh)) # corresponding matrices
mtext(side = 1, at = colMeans(mp), line = -2,
   text = paste("Mean", formatC(colMeans(hh))), col = "red")

# Bar shading example
barplot(VADeaths, angle = 15+10*1:5, density = 20,
        col = "black", legend = rownames(VADeaths))
title(main = list("Death Rates in Virginia", font = 4))
```

```
# border :
barplot(VADeaths, border = "dark blue")
```

| box | *Draw a Box around a Plot* |

Description

This function draws a box around the current plot in the given color and linetype. The `bty` parameter determines the type of box drawn. See `par` for details.

Usage

```
box(which="plot", lty="solid", ...)
```

Arguments

which
: character, one of "plot", "figure", "inner" and "outer".

lty
: line type of the box.

...
: further graphical parameters, such as `bty`, `col`, or `lwd`, see `par`.

References

Becker, R. A., Chambers, J. M. and Wilks, A. R. (1988) *The New S Language*. Wadsworth & Brooks/Cole.

See Also

`rect` for drawing of arbitrary rectangles.

Examples

```
plot(1:7,abs(rnorm(7)), type='h', axes = FALSE)
axis(1, labels = letters[1:7])
box(lty='1373', col = 'red')
```

boxplot — *Box Plots*

Description

Produce box-and-whisker plot(s) of the given (grouped) values.

Usage

```
boxplot(x, ...)

## S3 method for class 'formula':
boxplot(formula, data = NULL, ..., subset)

## Default S3 method:
boxplot(x, ..., range=1.5, width=NULL, varwidth=FALSE,
   notch=FALSE, outline=TRUE, names, boxwex=0.8, plot=TRUE,
   border=par("fg"), col=NULL, log="", pars=NULL,
   horizontal=FALSE, add=FALSE, at=NULL)
```

Arguments

formula
: a formula, such as y ~ x.

data
: a data.frame (or list) from which the variables in formula should be taken.

subset
: an optional vector specifying a subset of observations to be used for plotting.

x
: for specifying data from which the boxplots are to be produced as well as for giving graphical parameters. Additional unnamed arguments specify further data, either as separate vectors (each corresponding to a component boxplot) or as a single list containing such vectors. NAs are allowed in the data.

...
: For the formula method, arguments to the default method and graphical parameters.

 For the default method, unnamed arguments are additional data vectors, and named arguments are graphical parameters in addition to the ones given by argument pars.

range
: this determines how far the plot whiskers extend out from the box. If range is positive, the whiskers extend

	to the most extreme data point which is no more than **range** times the interquartile range from the box. A value of zero causes the whiskers to extend to the data extremes.
width	a vector giving the relative widths of the boxes making up the plot.
varwidth	if **varwidth** is TRUE, the boxes are drawn with widths proportional to the square-roots of the number of observations in the groups.
notch	if **notch** is TRUE, a notch is drawn in each side of the boxes. If the notches of two plots do not overlap then the medians are significantly different at the 5 percent level.
outline	if **outline** is not true, the boxplot lines are not drawn.
names	group labels which will be printed under each boxplot.
boxwex	a scale factor to be applied to all boxes. When there are only a few groups, the appearance of the plot can be improved by making the boxes narrower.
plot	if TRUE (the default) then a boxplot is produced. If not, the summaries which the boxplots are based on are returned.
border	an optional vector of colors for the outlines of the boxplots. The values in **border** are recycled if the length of **border** is less than the number of plots.
col	if **col** is non-null it is assumed to contain colors to be used to color the bodies of the box plots.
log	character indicating if x or y or both coordinates should be plotted in log scale.
pars	a list of graphical parameters; these are passed to **bxp** (if **plot** is true).
horizontal	logical indicating if the boxplots should be horizontal; default FALSE means vertical boxes.
add	logical, if true *add* boxplot to current plot.
at	numeric vector giving the locations where the boxplots should be drawn, particularly when **add** = TRUE; defaults to 1:n where n is the number of boxes.

Details

The generic function **boxplot** currently has a default method (**boxplot.default**) and a formula interface (**boxplot.formula**).

Value

List with the following components:

stats   a matrix, each column contains the extreme of the lower whisker, the lower hinge, the median, the upper hinge and the extreme of the upper whisker for one group/plot.

n   a vector with the number of observations in each group.

conf   a matrix where each column contains the lower and upper extremes of the notch.

out   the values of any data points which lie beyond the extremes of the whiskers.

group   a vector of the same length as out whose elements indicate which group the outlier belongs to

names   a vector of names for the groups

References

Becker, R. A., Chambers, J. M. and Wilks, A. R. (1988) *The New S Language.* Wadsworth & Brooks/Cole.

See also boxplot.stats.

See Also

boxplot.stats which does the computation, bxp for the plotting; and stripchart for an alternative (with small data sets).

Examples

```
## boxplot on a formula:
data(InsectSprays)
boxplot(count ~ spray, data = InsectSprays,
        col = "lightgray")
# add notches (somewhat funny here):
boxplot(count ~ spray, data = InsectSprays,
        notch = TRUE, add = TRUE, col = "blue")

data(OrchardSprays)
boxplot(decrease ~ treatment, data = OrchardSprays,
        log = "y", col="bisque")
```

```
rb <- boxplot(decrease ~ treatment, data = OrchardSprays,
              col="bisque")
title("Comparing boxplot()s and non-robust mean +/- SD")

mn.t <- tapply(OrchardSprays$decrease,
               OrchardSprays$treatment, mean)
sd.t <- tapply(OrchardSprays$decrease,
               OrchardSprays$treatment, sd)
xi <- 0.3 + seq(rb$n)
points(xi, mn.t, col = "orange", pch = 18)
arrows(xi, mn.t - sd.t, xi, mn.t + sd.t,
       code = 3, col = "pink", angle = 75, length = .1)

## boxplot on a matrix:
mat <- cbind(Uni05 = (1:100)/21, Norm = rnorm(100),
  T5 = rt(100, df = 5), Gam2 = rgamma(100, shape = 2))
boxplot(data.frame(mat),
        main = "boxplot(data.frame(mat), main = ...)")
par(las=1) # all axis labels horizontal
boxplot(data.frame(mat),
        main = "boxplot(*, horizontal = TRUE)",
        horizontal = TRUE)

## Using 'at = ' and adding boxplots -- example idea by
## Roger Bivand :
data(ToothGrowth)
boxplot(len ~ dose, data = ToothGrowth,
        boxwex = 0.25, at = 1:3 - 0.2,
        subset= supp == "VC", col="yellow",
        main="Guinea Pigs' Tooth Growth",
        xlab="Vitamin C dose mg",
        ylab="tooth length", ylim=c(0,35))
boxplot(len ~ dose, data = ToothGrowth, add = TRUE,
        boxwex = 0.25, at = 1:3 + 0.2,
        subset= supp == "OJ", col="orange")
legend(2, 9, c("Ascorbic acid", "Orange juice"),
       fill = c("yellow", "orange"))
```

boxplot.stats *Box Plot Statistics*

Description

This function is typically called by boxplot to gather the statistics necessary for producing box plots, but may be invoked separately.

Usage

```
boxplot.stats(x, coef = 1.5, do.conf=TRUE, do.out=TRUE)
```

Arguments

x
: a numeric vector for which the boxplot will be constructed (NAs and NaNs are allowed and omitted).

coef
: this determines how far the plot "whiskers" extend out from the box. If coef is positive, the whiskers extend to the most extreme data point which is no more than coef times the length of the box away from the box. A value of zero causes the whiskers to extend to the data extremes (with this setting no outliers will be returned).

do.conf, do.out
: logicals; if FALSE, the conf or out component respectively will be empty in the result.

Details

The two "hinges" are versions of the first and third quartile, i.e., close to quantile(x, c(1,3)/4). The hinges equal the quartiles for odd n (where n <- length(x)) and differ for even n. Where the quartiles only equal observations for n %% 4 == 1 ($n \equiv 1 \mod 4$), the hinges do so *additionally* for n %% 4 == 2 ($n \equiv 2 \mod 4$), and are in the middle of two observations otherwise.

Value

List with named components as follows:

stats
: a vector of length 5, containing the extreme of the lower whisker, the lower "hinge", the median, the upper "hinge" and the extreme of the upper whisker.

n	the number of non-NA observations in the sample.
conf	the lower and upper extremes of the "notch" (if(do.conf)).
out	the values of any data points which lie beyond the extremes of the whiskers (if(do.out)).

Note that $stats and $conf are sorted in *increasing* order, unlike S, and that $n and $out include any +- Inf values.

References

Tukey, J. W. (1977) *Exploratory Data Analysis.* Section 2C.

McGill, R., Tukey, J. W. and Larsen, W. A. (1978) Variations of box plots. *The American Statistician* **32**, 12–16.

Velleman, P. F. and Hoaglin, D. C. (1981) *Applications, Basics and Computing of Exploratory Data Analysis.* Duxbury Press.

Emerson, J. D and Strenio, J. (1983). Boxplots and batch comparison. Chapter 3 of *Understanding Robust and Exploratory Data Analysis*, eds. D. C. Hoaglin, F. Mosteller and J. W. Tukey. Wiley.

See Also

fivenum, boxplot, bxp.

Examples

```
x <- c(1:100, 1000)
str(b1 <- boxplot.stats(x))
str(b2 <- boxplot.stats(x, do.conf=FALSE, do.out=FALSE))
# do.out=F is still robust
stopifnot(b1 $ stats == b2 $ stats)
str(boxplot.stats(x, coef = 3, do.conf=FALSE))
## no outlier treatment:
str(boxplot.stats(x, coef = 0))

str(boxplot.stats(c(x, NA))) # slight change : n + 1
str(r <- boxplot.stats(c(x, -1:1/0)))
stopifnot(r$out == c(1000, -Inf, Inf))
```

bxp *Box Plots from Summaries*

Description

bxp draws box plots based on the given summaries in z. It is usually called from within boxplot, but can be invoked directly.

Usage

```
bxp(z, notch = FALSE, width = NULL, varwidth = FALSE,
    outline = TRUE, notch.frac = 0.5, boxwex = 0.8,
    border = par("fg"), col = NULL, log="", pars = NULL,
    frame.plot = axes, horizontal = FALSE,
    add = FALSE, at = NULL, show.names = NULL, ...)
```

Arguments

z	a list containing data summaries to be used in constructing the plots. These are usually the result of a call to boxplot, but can be generated in any fashion.
notch	if notch is TRUE, a notch is drawn in each side of the boxes. If the notches of two plots do not overlap then the medians are significantly different at the 5 percent level.
width	a vector giving the relative widths of the boxes making up the plot.
varwidth	if varwidth is TRUE, the boxes are drawn with widths proportional to the square-roots of the number of observations in the groups.
outline	if outline is not true, the boxplot lines are not drawn.
boxwex	a scale factor to be applied to all boxes. When there are only a few groups, the appearance of the plot can be improved by making the boxes narrower.
notch.frac	numeric in (0,1). When notch=TRUE, the fraction of the box width that the notches should use.
border	character, the color of the box borders. Is recycled for multiple boxes.
col	character; the color within the box. Is recycled for multiple boxes

log	character, indicating if any axis should be drawn in logarithmic scale, as in plot.default.
frame.plot	logical, indicating if a "frame" (box) should be drawn; defaults to TRUE, unless axes = FALSE is specified.
horizontal	logical indicating if the boxplots should be horizontal; default FALSE means vertical boxes.
add	logical, if true *add* boxplot to current plot.
at	numeric vector giving the locations where the boxplots should be drawn, particularly when add = TRUE; defaults to 1:n where n is the number of boxes.
show.names	Set to TRUE or FALSE to override the defaults on whether an x-axis label is printed for each group.
pars,...	graphical parameters can be passed as arguments to this function, either as a list (pars) or normally(...). Currently, pch, cex, and bg are passed to points, ylim and axes to the main plot (plot.default), xaxt, yaxt, las to axis and the others to title.

Value

An invisible vector, actually identical to the at argument, with the coordinates ("x" if horizontal is false, "y" otherwise) of box centers, useful for adding to the plot.

References

Becker, R. A., Chambers, J. M. and Wilks, A. R. (1988) *The New S Language.* Wadsworth & Brooks/Cole.

Examples

```
set.seed(753)
str(bx.p <- boxplot(split(rt(100, 4), gl(5,20))))
op <- par(mfrow= c(2,2))
bxp(bx.p, xaxt = "n")
bxp(bx.p, notch = TRUE, axes = FALSE, pch = 4)
bxp(bx.p, notch = TRUE, col= "lightblue", frame= FALSE,
    outl= FALSE, main = "bxp(*, frame=FALSE, outl=FALSE)")
bxp(bx.p, notch = TRUE, col= "lightblue", border="red",
    ylim = c(-4,4), pch = 22, bg = "green", log = "x",
    main = "... log='x', ylim=*")
par(op)
```

```
op <- par(mfrow= c(1,2))
data(PlantGrowth)
## single group -- no label
boxplot(weight ~ group, data = PlantGrowth,
        subset = group ==" ctrl")
bx<-boxplot(weight ~ group, data = PlantGrowth,
            subset = group =="ctrl",plot = FALSE)
## with label
bxp(bx,show.names=TRUE)
par(op)
```

| chull | *Compute Convex Hull of a Set of Points* |

Description

Computes the subset of points which lie on the convex hull of the set of points specified.

Usage

```
chull(x, y=NULL)
```

Arguments

x, y   coordinate vectors of points. This can be specified as two vectors x and y, a 2-column matrix x, a list x with two components, etc, see `xy.coords`.

Details

`xy.coords` is used to interpret the specification of the points. The algorithm is that given by Eddy (1977).

'Peeling' as used in the S function `chull` can be implemented by calling `chull` recursively.

Value

An integer vector giving the indices of the points lying on the convex hull, in clockwise order.

References

Becker, R. A., Chambers, J. M. and Wilks, A. R. (1988) *The New S Language*. Wadsworth & Brooks/Cole.

Eddy, W. F. (1977) A new convex hull algorithm for planar sets. *ACM Transactions on Mathematical Software*, **3**, 398–403.

Eddy, W. F. (1977) Algorithm 523. CONVEX, A new convex hull algorithm for planar sets[Z]. *ACM Transactions on Mathematical Software*, **3**, 411–412.

See Also

`xy.coords,polygon`

Examples

```
X <- matrix(rnorm(2000), ncol=2)
plot(X, cex=0.5)
hpts <- chull(X)
hpts <- c(hpts, hpts[1])
lines(X[hpts, ])
```

col2rgb — *Color to RGB Conversion*

Description

"Any R color" to RGB (red/green/blue) conversion.

Usage

```
col2rgb(col)
```

Arguments

col
: vector of any of the three kind of R colors, i.e., either a color name (an element of colors()), a hexadecimal string of the form "#rrggbb", or an integer i meaning palette()[i].

Details

For integer colors, 0 is shorthand for the current par("bg"), and NA means "nothing" which effectively does not draw the corresponding item.

For character colors, "NA" is equivalent to NA above.

Value

an integer matrix with three rows and number of columns the length (and names if any) as col.

Author(s)

Martin Maechler

See Also

rgb, colors, palette, etc.

Examples

```
col2rgb("peachpuff")
# names kept
col2rgb(c(blu = "royalblue", reddish = "tomato"))

col2rgb(1:8) # the ones from the palette() :
col2rgb(paste("gold", 1:4, sep=""))
col2rgb("#08a0ff")
## all three kind of colors mixed :
col2rgb(c(red="red", palette= 1:3, hex="#abcdef"))

## Non-introductory examples
grC <- col2rgb(paste("gray",0:100,sep=""))
# '2' or '3': almost equidistant
table(print(diff(grC["red",])))
## The 'named' grays are in between {"slate gray" is not
## gray, strictly}
col2rgb(c(g66="gray66", darkg= "dark gray", g67="gray67",
          g74="gray74", gray =         "gray", g75="gray75",
          g82="gray82", light="light gray", g83="gray83"))

crgb <- col2rgb(cc <- colors())
colnames(crgb) <- cc
t(crgb) ## The whole table

ccodes <- c(256^(2:0) %*% crgb) ## = internal codes
## How many names are 'aliases' of each other:
table(tcc <- table(ccodes))
length(uc <- unique(sort(ccodes))) # 502
## All the multiply named colors:
mult <- uc[tcc >= 2]
cl <- lapply(mult, function(m) cc[ccodes == m])
names(cl) <- apply(col2rgb(sapply(cl, function(x)x[1])),
                   2, function(n)paste(n, collapse=","))
str(cl)

if(require(xgobi)) {
  ## Look at the color cube dynamically :
  tc <- t(crgb[, !duplicated(ccodes)])
  # (397, 105)
  table(is.gray <- tc[,1] == tc[,2] & tc[,2] == tc[,3])
  xgobi(tc, color = c("gold", "gray")[1 + is.gray])
```

}

`colors` *Color Names*

Description

Returns the built-in color names which R knows about.

Usage

```
colors()
```

Details

These color names can be used with a `col=` specification in graphics functions.

An even wider variety of colors can be created with primitives `rgb` and `hsv` or the derived `rainbow`, `heat.colors`, etc.

Value

A character vector containing all the built-in color names.

See Also

`palette` for setting the "palette" of colors for `par(col=`¡*num*¿`)`; `rgb`, `hsv`, `gray`; `rainbow` for a nice example; and `heat.colors`, `topo.colors` for images.

`col2rgb` for translating to RGB numbers and extended examples.

Examples

```
str(colors())
```

contour   *Display Contours*

Description

Create a contour plot, or add contour lines to an existing plot.

Usage

```
contour(x, ...)
## Default S3 method:
contour(x = seq(0, 1, len = nrow(z)),
  y = seq(0, 1, len = ncol(z)),
  z,
  nlevels = 10, levels = pretty(zlim, nlevels),
  labels = NULL,
  xlim = range(x, finite = TRUE),
  ylim = range(y, finite = TRUE),
  zlim = range(z, finite = TRUE),
  labcex = 0.6, drawlabels = TRUE,
  method = "flattest",
  vfont = c("sans serif", "plain"),
  axes = TRUE, frame.plot = axes,
  col = par("fg"), lty = par("lty"), lwd = par("lwd"),
  add = FALSE, ...)
```

Arguments

x,y	locations of grid lines at which the values in z are measured. These must be in ascending order. By default, equally spaced values from 0 to 1 are used. If x is a list, its components x$x and x$y are used for x and y, respectively. If the list has component z this is used for z.
z	a matrix containing the values to be plotted (NAs are allowed). Note that x can be used instead of z for convenience.
nlevels	number of contour levels desired iff levels is not supplied.
levels	numeric vector of levels at which to draw contour lines.
labels	a vector giving the labels for the contour lines. If NULL then the levels are used as labels.

labcex	cex for contour labelling.
drawlabels	logical. Contours are labelled if TRUE.
method	character string specifying where the labels will be located. Possible values are "simple", "edge" and "flattest" (the default). See the Details section.
vfont	if a character vector of length 2 is specified, then Hershey vector fonts are used for the contour labels. The first element of the vector selects a typeface and the second element selects a fontindex (see text for more information).
xlim, ylim, zlim	x-, y- and z-limits for the plot.
axes, frame.plot	logical indicating whether axes or a box should be drawn, see plot.default.
col	color for the lines drawn.
lty	line type for the lines drawn.
lwd	line width for the lines drawn.
add	logical. If TRUE, add to a current plot.
...	additional graphical parameters (see par) and the arguments to title may also be supplied.

Details

contour is a generic function with only a default method in base R.

There is currently no documentation about the algorithm. The source code is in '$R_HOME/src/main/plot3d.c'.

The methods for positioning the labels on contours are "simple" (draw at the edge of the plot, overlaying the contour line), "edge" (draw at the edge of the plot, embedded in the contour line, with no labels overlapping) and "flattest" (draw on the flattest section of the contour, embedded in the contour line, with no labels overlapping). The second and third may not draw a label on every contour line.

For information about vector fonts, see the help for text and Hershey.

References

Becker, R. A., Chambers, J. M. and Wilks, A. R. (1988) *The New S Language.* Wadsworth & Brooks/Cole.

See Also

`filled.contour` for "color-filled" contours, `image` and the graphics demo which can be invoked as `demo(graphics)`.

Examples

```
x <- -6:16
op <- par(mfrow = c(2, 2))
contour(outer(x, x), method = "edge",
        vfont = c("sans serif", "plain"))
z <- outer(x, sqrt(abs(x)), FUN = "/")
## Should not be necessary:
z[!is.finite(z)] <- NA
image(x, x, z)
contour(x, x, z, col = "pink", add = TRUE, method = "edge",
        vfont = c("sans serif", "plain"))
contour(x, x, z, ylim = c(1, 6), method = "simple",
        labcex = 1)
contour(x, x, z, ylim = c(-6, 6), nlev = 20, lty = 2,
        method = "simple")
par(op)

## Persian Rug Art:
x <- y <- seq(-4*pi, 4*pi, len = 27)
r <- sqrt(outer(x^2, y^2, "+"))
opar <- par(mfrow = c(2, 2), mar = rep(0, 4))
for(f in pi^(0:3))
  contour(cos(r^2)*exp(-r/f),
          drawlabels = FALSE, axes = FALSE, frame = TRUE)

data("volcano")
rx <- range(x <- 10*1:nrow(volcano))
ry <- range(y <- 10*1:ncol(volcano))
ry <- ry + c(-1,1) * (diff(rx) - diff(ry))/2
tcol <- terrain.colors(12)
par(opar); opar <- par(pty = "s", bg = "lightcyan")
plot(x = 0, y = 0,type = "n", xlim = rx, ylim = ry,
     xlab = "", ylab = "")
u <- par("usr")
rect(u[1], u[3], u[2], u[4], col = tcol[8], border = "red")
contour(x, y, volcano, col = tcol[2], lty = "solid",
        add = TRUE, vfont = c("sans serif", "plain"))
title("A Topographic Map of Maunga Whau", font = 4)
```

```
abline(h = 200*0:4, v = 200*0:4, col = "lightgray",
       lty = 2, lwd = 0.1)
par(opar)
```

coplot *Conditioning Plots*

Description

This function produces two variants of the conditioning plots discussed in the reference below.

Usage

```
coplot(formula, data, given.values, panel = points,
       rows, columns,
       show.given = TRUE, col = par("fg"),
       pch = par("pch"),
       bar.bg = c(num = gray(0.8), fac = gray(0.95)),
       xlab = c(x.name, paste("Given :", a.name)),
       ylab = c(y.name, paste("Given :", b.name)),
       subscripts = FALSE,
       axlabels = function(f) abbreviate(levels(f)),
       number = 6, overlap = 0.5, xlim, ylim, ...)
co.intervals(x, number = 6, overlap = 0.5)
```

Arguments

formula   a formula describing the form of conditioning plot. A formula of the form y ~ x | a indicates that plots of y versus x should be produced conditional on the variable a. A formula of the form y ~ x| a * b indicates that plots of y versus x should be produced conditional on the two variables a and b.

All three or four variables may be either numeric or factors. When x or y are factors, the result is almost as if as.numeric() was applied, whereas for factor a or b, the conditioning is adapted (and its graphics if show.given is true).

data   a data frame containing values for any variables in the formula. By default the environment where coplot was called from is used.

given.values   a value or list of two values which determine how the conditioning on a and b is to take place.

When there is no b (i.e., conditioning only on a), usually this is a matrix with two columns each row of

	which gives an interval, to be conditioned on, but is can also be a single vector of numbers or a set of factor levels (if the variable being conditioned on is a factor). In this case (no b), the result of co.intervals can be used directly as given.values argument.
panel	a function(x, y, col, pch, ...) which gives the action to be carried out in each panel of the display. The default is points.
rows	the panels of the plot are laid out in a rows by columns array. rows gives the number of rows in the array.
columns	the number of columns in the panel layout array.
show.given	logical (possibly of length 2 for 2 conditioning variables): should conditioning plots be shown for the corresponding conditioning variables (default TRUE)
col	a vector of colors to be used to plot the points. If too short, the values are recycled.
pch	a vector of plotting symbols or characters. If too short, the values are recycled.
bar.bg	a named vector with components "num" and "fac" giving the background colors for the (shingle) bars, for **num**eric and **fac**tor conditioning variables respectively.
xlab	character; labels to use for the x axis and the first conditioning variable. If only one label is given, it is used for the x axis and the default label is used for the conditioning variable.
ylab	character; labels to use for the y axis and any second conditioning variable.
subscripts	logical: if true the panel function is given an additional (third) argument subscripts giving the subscripts of the data passed to that panel.
axlabels	function for creating axis (tick) labels when x or y are factors.
number	integer; the number of conditioning intervals, for a and b, possibly of length 2. It is only used if the corresponding conditioning variable is not a factor.
overlap	numeric < 1; the fraction of overlap of the conditioning variables, possibly of length 2 for x and y direction. When overlap < 0 there will be *gaps* between the data slices.

xlim	the range for the x axis.
ylim	the range for the y axis.
...	additional arguments to the panel function.
x	a numeric vector.

Details

In the case of a single conditioning variable a, when both rows and columns are unspecified, a "close to square" layout is chosen with columns >= rows.

In the case of multiple rows, the *order* of the panel plots is from the bottom and from the left (corresponding to increasing a, typically).

A panel function should not attempt to start a new plot, but just plot within a given coordinate system: thus plot and boxplot are not panel functions.

Value

co.intervals(., number, .) returns a (number × 2) matrix, say ci, where ci[k,] is the range of x values for the k-th interval.

References

Chambers, J. M. (1992) *Data for models*. Chapter 3 of *Statistical Models in S* eds J. M. Chambers and T. J. Hastie, Wadsworth & Brooks/Cole.

Cleveland, W. S. (1993) *Visualizing Data*. New Jersey: Summit Press.

See Also

pairs, panel.smooth, points.

Examples

```
## Tonga Trench Earthquakes
data(quakes)
coplot(lat ~ long | depth, data = quakes)
given.depth <-
   co.intervals(quakes$depth, number = 4, overlap = .1)
coplot(lat ~ long | depth, data = quakes,
       given.v = given.depth, rows = 1)

## Conditioning on 2 variables:
```

```
ll.dm <- lat ~ long | depth * mag
coplot(ll.dm, data = quakes)
coplot(ll.dm, data = quakes, number=c(4,7),
       show.given=c(TRUE,FALSE))
coplot(ll.dm, data = quakes, number=c(3,7),
       overlap=c(-.5,.1)) # negative overlap DROPS values

data(warpbreaks)
## given two factors
# to get nicer default labels
Index <- seq(length=nrow(warpbreaks))
coplot(breaks ~ Index | wool * tension, data = warpbreaks,
       show.given = 0:1)
coplot(breaks ~ Index | wool * tension, data = warpbreaks,
       col = "red", bg = "pink", pch = 21,
       bar.bg = c(fac = "light blue"))

## Example with empty panels:
data(state)
attach(data.frame(state.x77)) # don't need 'data' arg below
coplot(Life.Exp ~ Income | Illiteracy * state.region,
       number = 3,  panel = function(x, y, ...)
       panel.smooth(x, y, span = .8, ...))
## y ~ factor -- not really sensical, but 'show off':
coplot(Life.Exp ~ state.region | Income * state.division,
       panel = panel.smooth)
detach() # data.frame(state.x77)
```

curve   *Draw Function Plots*

Description

Draws a curve corresponding to the given function or expression (in x) over the interval [from,to].

Usage

```
curve(expr, from, to, n = 101, add = FALSE, type = "l",
      ylab = NULL, log = NULL, xlim = NULL, ...)

## S3 method for class 'function':
plot(x, from = 0, to = 1, xlim = NULL, ...)
```

Arguments

expr	an expression written as a function of x, or alternatively the name of a function which will be plotted.
x	a 'vectorizing' numeric R function.
from,to	the range over which the function will be plotted.
n	integer; the number of x values at which to evaluate.
add	logical; if TRUE add to already existing plot.
xlim	numeric of length 2; if specified, it serves as default for c(from, to).
type, ylab, log, ...	
	graphical parameters can also be specified as arguments. plot.function passes all these to curve.

Details

The evaluation of expr is at n points equally spaced over the range [from, to], possibly adapted to log scale. The points determined in this way are then joined with straight lines. x(t) or expr (with x inside) must return a numeric of the same length as the argument t or x.

If add = TRUE, c(from,to) default to xlim which defaults to the current x-limits. Further, log is taken from the current plot when add is true.

This used to be a quick hack which now seems to serve a useful purpose, but can give bad results for functions which are not smooth.

For "expensive" expressions, you should use smarter tools.

See Also

splinefun for spline interpolation, **lines**.

Examples

```
op <- par(mfrow=c(2,2))
curve(x^3-3*x, -2, 2)
curve(x^2-2, add = TRUE, col = "violet")

plot(cos, xlim = c(-pi,3*pi), n = 1001, col = "blue")

chippy <- function(x) sin(cos(x)*exp(-x/2))
curve(chippy, -8, 7, n=2001)
curve(chippy, -8, -5)

for(ll in c("","x","y","xy"))
   curve(log(1+x), 1,100, log=ll,
         sub=paste("log= '",ll,"'",sep=""))
par(op)
```

dev.xxx   *Control Multiple Devices*

Description

These functions provide control over multiple graphics devices.

Only one device is the *active* device. This is the device in which all graphics operations occur.

Devices are associated with a name (e.g., "X11" or "postscript") and a number; the "null device" is always device 1.

dev.off shuts down the specified (by default the current) device. graphics.off() shuts down all open graphics devices.

dev.set makes the specified device the active device.

A list of the names of the open devices is stored in .Devices. The name of the active device is stored in .Device.

Usage

```
dev.cur()
dev.list()
dev.next(which = dev.cur())
dev.prev(which = dev.cur())
dev.off(which = dev.cur())
dev.set(which = dev.next())
graphics.off()
```

Arguments

which   An integer specifying a device number

Value

dev.cur returns the number and name of the active device, or 1, the null device, if none is active.

dev.list returns the numbers of all open devices, except device 1, the null device. This is a numeric vector with a names attribute giving the names, or NULL is there is no open device.

dev.next and dev.prev return the number and name of the next / previous device in the list of devices. The list is regarded as a circular list, and "null device" will be included only if there are no open devices.

dev.off returns the name and number of the new active device (after the specified device has been shut down).

dev.set returns the name and number of the new active device.

See Also

Devices, such as `postscript`, etc; `layout` and its links for setting up plotting regions on the current device.

Examples

```
## Unix-specific example
x11()
plot(1:10)
x11()
plot(rnorm(10))
dev.set(dev.prev())
abline(0,1) # through the 1:10 points
dev.set(dev.next())
abline(h=0, col="gray") # for the residual plot
dev.set(dev.prev())
dev.off(); dev.off() # close the two X devices
```

dev2   *Copy Graphics Between Multiple Devices*

Description

dev.copy copies the graphics contents of the current device to the device specified by **which** or to a new device which has been created by the function specified by **device** (it is an error to specify both **which** and **device**). (If recording is off on the current device, there are no contents to copy: this will result in no plot or an empty plot.) The device copied to becomes the current device.

dev.print copies the graphics contents of the current device to a new device which has been created by the function specified by **device** and then shuts the new device.

dev.copy2eps is similar to **dev.print** but produces an EPSF output file, in portrait orientation (**horizontal = FALSE**)

dev.control allows the user to control the recording of graphics operations in a device. If **displaylist** is **"inhibit"** (**"enable"**) then recording is turned off (on). It is only safe to change this at the beginning of a plot (just before or just after a new page). Initially recording is on for screen devices, and off for print devices.

Usage

```
dev.copy(device, ..., which = dev.next())
dev.print(device = postscript, ...)
dev.copy2eps(...)
dev.control(displaylist = c("inhibit", "enable"))
```

Arguments

device	A device function (e.g., **x11**, **postscript**, ...)
...	Arguments to the **device** function above. For dev.print, this includes **which** and by default any **postscript** arguments.
which	A device number specifying the device to copy to
displaylist	A character string: the only valid values are **"inhibit"** and **"enable"**.

Details

For `dev.copy2eps`, `width` and `height` are taken from the current device unless otherwise specified. If just one of `width` and `height` is specified, the other is adjusted to preserve the aspect ratio of the device being copied. The default file name is `Rplot.eps`.

The default for `dev.print` is to produce and print a postscript copy, if `options("printcmd")` is set suitably.

`dev.print` is most useful for producing a postscript print (its default) when the following applies. Unless `file` is specified, the plot will be printed. Unless `width`, `height` and `pointsize` are specified the plot dimensions will be taken from the current device, shrunk if necessary to fit on the paper. (`pointsize` is rescaled if the plot is shrunk.) If `horizontal` is not specified and the plot can be printed at full size by switching its value this is done instead of shrinking the plot region.

If `dev.print` is used with a specified device (even `postscript`) it sets the width and height in the same way as `dev.copy2eps`.

Value

`dev.copy` returns the name and number of the device which has been copied to.

`dev.print` and `dev.copy2eps` return the name and number of the device which has been copied from.

Note

Most devices (including all screen devices) have a display list which records all of the graphics operations that occur in the device. `dev.copy` copies graphics contents by copying the display list from one device to another device. Also, automatic redrawing of graphics contents following the resizing of a device depends on the contents of the display list.

After the command `dev.control("inhibit")`, graphics operations are not recorded in the display list so that `dev.copy` and `dev.print` will not copy anything and the contents of a device will not be redrawn automatically if the device is resized.

The recording of graphics operations is relatively expensive in terms of memory so the command `dev.control("inhibit")` can be useful if memory usage is an issue.

See Also

dev.cur and other dev.xxx functions

Examples

```
x11()
plot(rnorm(10), main="Plot 1")
dev.copy(device=x11)
mtext("Copy 1", 3)
dev.print(width=6, height=6, horizontal=FALSE) # prints it
dev.off(dev.prev())
dev.off()
```

dev2bitmap   *Graphics Device for Bitmap Files via GhostScript*

Description

`bitmap` generates a graphics file. `dev2bitmap` copies the current graphics device to a file in a graphics format.

Usage

```
bitmap(file, type = "png256", height = 6, width = 6,
       res = 72, pointsize, ...)
dev2bitmap(file, type = "png256", height = 6, width = 6,
           res = 72, pointsize, ...)
```

Arguments

`file`	The output file name, with an appropriate extension.
`type`	The type of bitmap. the default is `"png256"`.
`height`	The plot height, in inches.
`width`	The plot width, in inches.
`res`	Resolution, in dots per inch.
`pointsize`	The pointsize to be used for text: defaults to something reasonable given the width and height
`...`	Other parameters passed to `postscript`.

Details

`dev2bitmap` works by copying the current device to a `postscript` device, and post-processing the output file using `ghostscript`. `bitmap` works in the same way using a `postscript` device and postprocessing the output as "printing".

You will need a version of `ghostscript` (5.10 and later have been tested): the full path to the executable can be set by the environment variable R_GSCMD.

The types available will depend on the version of `ghostscript`, but are likely to include `"pcxmono"`, `"pcxgray"`, `"pcx16"`, `"pcx256"`, `"pcx24b"`, `"pcxcmyk"`, `"pbm"`, `"pbmraw"`, `"pgm"`, `"pgmraw"`, `"pgnm"`, `"pgnmraw"`, `"pnm"`, `"pnmraw"`, `"ppm"`, `"ppmraw"`, `"pkm"`, `"pkmraw"`, `"tiffcrle"`, `"tiffg3"`, `"tiffg32d"`, `"tiffg4"`, `"tifflzw"`, "tiffpack", "tiff12nc", "tiff24nc", "psmono", "psgray", "psrgb", "bit", "bitrgb", "bitcmyk", "pngmono", "pnggray", "png16", "png256", "png16m", "jpeg", "jpeggray", "pdfwrite".

Note: despite the name of the functions they can produce PDF *via* type = "pdfwrite", and the PDF produced is not bitmapped.

For formats which contain a single image, a file specification like Rplots%03d.png can be used: this is interpreted by GhostScript.

For dev2bitmap if just one of width and height is specified, the other is chosen to preserve aspect ratio of the device being copied.

Value

None.

See Also

postscript, png and jpeg and on Windows bmp.

pdf generate PDF directly.

To display an array of data, see image.

Devices   *List of Graphical Devices*

Description

The following graphics devices are currently available:

- `postscript` Writes PostScript graphics commands to a file
- `pdf` Write PDF graphics commands to a file
- `pictex` Writes LaTeX/PicTeX graphics commands to a file
- `xfig` Device for XFIG graphics file format
- `bitmap` bitmap pseudo-device via `GhostScript` (if available).

The following devices will be available if R was compiled to use them and started with the appropriate '`--gui`' argument:

- `X11` The graphics driver for the X11 Window system
- `png` PNG bitmap device
- `jpeg` JPEG bitmap device
- `GTK`, `GNOME` Graphics drivers for the GNOME GUI.

None of these are available under `R CMD BATCH`.

Usage

```
X11(...)
postscript(...)
pdf(...)
pictex(...)
png(...)
jpeg(...)
GTK(...)
GNOME(...)
xfig(...)
bitmap(...)

dev.interactive()
```

Details

If no device is open, using a high-level graphics function will cause a device to be opened. Which device is given by `options("device")` which is initially set as the most appropriate for each platform: a screen device in interactive use and `postscript` otherwise.

Value

`dev.interactive()` returns a logical, TRUE iff an interactive (screen) device is in use.

See Also

The individual help files for further information on any of the devices listed here;

`dev.cur, dev.print, graphics.off, image, dev2bitmap`.

`capabilities` to see if X11, jpeg and png are available.

Examples

```
## open the default screen device on this platform if no
## device is open
if(dev.cur() == 1) get(getOption("device"))()
```

dotchart   *Cleveland Dot Plots*

Description

Draw a Cleveland dot plot.

Usage

```
dotchart(x, labels = NULL, groups = NULL, gdata = NULL,
  cex = par("cex"), pch = 21, gpch = 21, bg = par("bg"),
  color = par("fg"), gcolor = par("fg"), lcolor = "gray",
  xlim = range(x[is.finite(x)]),
  main = NULL, xlab = NULL, ylab = NULL, ...)
```

Arguments

x	either a vector or matrix of numeric values (NAs are allowed). If x is a matrix the overall plot consists of juxtaposed dotplots for each row.
labels	a vector of labels for each point. For vectors the default is to use names(x) and for matrices the row labels dimnames(x)[[1]].
groups	an optional factor indicating how the elements of x are grouped. If x is a matrix, groups will default to the columns of x.
gdata	data values for the groups. This is typically a summary such as the median or mean of each group.
cex	the character size to be used. Setting cex to a value smaller than one can be a useful way of avoiding label overlap.
pch	the plotting character or symbol to be used.
gpch	the plotting character or symbol to be used for group values.
bg	the background color of plotting characters or symbols to be used; use par(bg= *) to set the background color of the whole plot.
color	the color(s) to be used for points and labels.
gcolor	the single color to be used for group labels and values.
lcolor	the color(s) to be used for the horizontal lines.

xlim	horizontal range for the plot, see plot.window, e.g.
main	overall title for the plot, see title.
xlab, ylab	axis annotations as in title.
...	graphical parameters can also be specified as arguments.

Value

This function is invoked for its side effect, which is to produce two variants of dotplots as described in Cleveland (1985).

Dot plots are a reasonable substitute for bar plots.

References

Becker, R. A., Chambers, J. M. and Wilks, A. R. (1988) *The New S Language*. Wadsworth & Brooks/Cole.

Cleveland, W. S. (1985) *The Elements of Graphing Data*. Monterey, CA: Wadsworth.

Examples

```
data(VADeaths)
dotchart(VADeaths, main = "Death Rates in Virginia - 1940")
op <- par(xaxs="i") # 0 -- 100%
dotchart(t(VADeaths), xlim = c(0,100),
         main = "Death Rates in Virginia - 1940")
par(op)
```

filled.contour *Level (Contour) Plots*

Description

This function produces a contour plot with the areas between the contours filled in solid color (Cleveland calls this a level plot). A key showing how the colors map to z values is shown to the right of the plot.

Usage

```
filled.contour(x = seq(0, 1, len = nrow(z)),
  y = seq(0, 1, len = ncol(z)),
  z,
  xlim = range(x, finite=TRUE),
  ylim = range(y, finite=TRUE),
  zlim = range(z, finite=TRUE),
  levels = pretty(zlim, nlevels), nlevels = 20,
  color.palette = cm.colors,
  col = color.palette(length(levels) - 1),
  plot.title, plot.axes, key.title, key.axes,
  asp = NA, xaxs = "i", yaxs = "i", las = 1,
  axes = TRUE, frame.plot = axes, ...)
```

Arguments

x,y	locations of grid lines at which the values in z are measured. These must be in ascending order. By default, equally spaced values from 0 to 1 are used. If x is a list, its components x$x and x$y are used for x and y, respectively. If the list has component z this is used for z.
z	a matrix containing the values to be plotted (NAs are allowed). Note that x can be used instead of z for convenience.
xlim	x limits for the plot.
ylim	y limits for the plot.
zlim	z limits for the plot.
levels	a set of levels which are used to partition the range of z. Must be **strictly** increasing (and finite). Areas with z values between consecutive levels are painted with the same color.

`nlevels`	if `levels` is not specified, the range of z, values is divided into approximately this many levels.
`color.palette`	a color palette function to be used to assign colors in the plot.
`col`	an explicit set of colors to be used in the plot. This argument overrides any palette function specification.
`plot.title`	statements which add titles to the main plot.
`plot.axes`	statements which draw axes (and a box) on the main plot. This overrides the default axes.
`key.title`	statements which add titles for the plot key.
`key.axes`	statements which draw axes on the plot key. This overrides the default axis.
`asp`	the y/x aspect ratio, see `plot.window`.
`xaxs`	the x axis style. The default is to use internal labeling.
`yaxs`	the y axis style. The default is to use internal labeling.
`las`	the style of labeling to be used. The default is to use horizontal labeling.
`axes, frame.plot`	logicals indicating if axes and a box should be drawn, as in `plot.default`.
`...`	additional graphical parameters, currently only passed to `title()`.

Note

This function currently uses the `layout` function and so is restricted to a full page display. As an alternative consider the `levelplot` function from the **lattice** package which works in multipanel displays.

The ouput produced by `filled.contour` is actually a combination of two plots; one is the filled contour and one is the legend. Two separate coordinate systems are set up for these two plots, but they are only used internally - once the function has returned these coordinate systems are lost. If you want to annotate the main contour plot, for example to add points, you can specify graphics commands in the `plot.axes` argument. An example is given below.

Author(s)

Ross Ihaka.

References

Cleveland, W. S. (1993) *Visualizing Data*. Summit, New Jersey: Hobart.

See Also

contour, image, palette; levelplot from package **lattice**.

Examples

```
data(volcano)
# simple
filled.contour(volcano, color = terrain.colors, asp = 1)

x <- 10*1:nrow(volcano)
y <- 10*1:ncol(volcano)
filled.contour(x, y, volcano, color = terrain.colors,
   plot.title = title(main = "Topography of Maunga Whau",
   xlab = "Meters North", ylab = "Meters West"),
   plot.axes = { axis(1, seq(100, 800, by = 100))
                 axis(2, seq(100, 600, by = 100)) },
   key.title = title(main="Height\n(meters)"),
   key.axes = axis(4, seq(90, 190, by = 10)))

mtext(paste("filled.contour(.) from", R.version.string),
      side = 1, line = 4, adj = 1, cex = .66)

# Annotating a filled contour plot
a <- expand.grid(1:20, 1:20)
b <- matrix(a[,1] + a[,2], 20)
filled.contour(x = 1:20, y = 1:20, z = b,
   plot.axes={ axis(1); axis(2); points(10,10) })

## Persian Rug Art:
x <- y <- seq(-4*pi, 4*pi, len = 27)
r <- sqrt(outer(x^2, y^2, "+"))
filled.contour(cos(r^2)*exp(-r/(2*pi)), axes = FALSE)
## rather, the key should be labeled:
filled.contour(cos(r^2)*exp(-r/(2*pi)), frame.plot = FALSE,
               plot.axes = {})
```

fourfoldplot *Fourfold Plots*

Description

Creates a fourfold display of a 2 by 2 by k contingency table on the current graphics device, allowing for the visual inspection of the association between two dichotomous variables in one or several populations (strata).

Usage

```
fourfoldplot(x, color = c("#99CCFF", "#6699CC"),
             conf.level = 0.95,
             std = c("margins", "ind.max", "all.max"),
             margin = c(1, 2), space = 0.2, main = NULL,
             mfrow = NULL, mfcol = NULL)
```

Arguments

x
: a 2 by 2 by k contingency table in array form, or as a 2 by 2 matrix if k is 1.

color
: a vector of length 2 specifying the colors to use for the smaller and larger diagonals of each 2 by 2 table.

conf.level
: confidence level used for the confidence rings on the odds ratios. Must be a single nonnegative number less than 1; if set to 0, confidence rings are suppressed.

std
: a character string specifying how to standardize the table. Must be one of "margins", "ind.max", or "all.max", and can be abbreviated by the initial letter. If set to "margins", each 2 by 2 table is standardized to equate the margins specified by margin while preserving the odds ratio. If "ind.max" or "all.max", the tables are either individually or simultaneously standardized to a maximal cell frequency of 1.

margin
: a numeric vector with the margins to equate. Must be one of 1, 2, or c(1, 2) (the default), which corresponds to standardizing the row, column, or both margins in each 2 by 2 table. Only used if std equals "margins".

space	the amount of space (as a fraction of the maximal radius of the quarter circles) used for the row and column labels.
main	character string for the fourfold title.
mfrow	a numeric vector of the form c(nr, nc), indicating that the displays for the 2 by 2 tables should be arranged in an nr by nc layout, filled by rows.
mfcol	a numeric vector of the form c(nr, nc), indicating that the displays for the 2 by 2 tables should be arranged in an nr by nc layout, filled by columns.

Details

The fourfold display is designed for the display of 2 by 2 by k tables.

Following suitable standardization, the cell frequencies f_{ij} of each 2 by 2 table are shown as a quarter circle whose radius is proportional to $\sqrt{f_{ij}}$ so that its area is proportional to the cell frequency. An association (odds ratio different from 1) between the binary row and column variables is indicated by the tendency of diagonally opposite cells in one direction to differ in size from those in the other direction; color is used to show this direction. Confidence rings for the odds ratio allow a visual test of the null of no association; the rings for adjacent quadrants overlap iff the observed counts are consistent with the null hypothesis.

Typically, the number k corresponds to the number of levels of a stratifying variable, and it is of interest to see whether the association is homogeneous across strata. The fourfold display visualizes the pattern of association. Note that the confidence rings for the individual odds ratios are not adjusted for multiple testing.

References

Friendly, M. (1994). A fourfold display for 2 by 2 by k tables. Technical Report 217, York University, Psychology Department. http://www.math.yorku.ca/SCS/Papers/4fold/4fold.ps.gz

See Also

mosaicplot

Examples

data(UCBAdmissions)

```
## Use the Berkeley admission data as in Friendly (1995).
x <- aperm(UCBAdmissions, c(2, 1, 3))
dimnames(x)[[2]] <- c("Yes", "No")
names(dimnames(x)) <- c("Sex", "Admit?", "Department")
ftable(x)

## Fourfold display of data aggregated over departments,
## with frequencies standardized to equate the margins for
## admission and sex. Figure 1 in Friendly (1994).
fourfoldplot(margin.table(x, c(1, 2)))

## Fourfold display of x, with frequencies in each table
## standardized to equate the margins for admission and
## sex. Figure 2 in Friendly (1994).
fourfoldplot(x)

## Fourfold display of x, with frequencies in each table
## standardized to equate the margins for admission. but
## not for sex. Figure 3 in Friendly (1994).
fourfoldplot(x, margin = 2)
```

frame   *Create / Start a New Plot Frame*

Description

This function (`frame` is an alias for `plot.new`) causes the completion of plotting in the current plot (if there is one) and an advance to a new graphics frame. This is used in all high-level plotting functions and also useful for skipping plots when a multi-figure region is in use.

Usage

```
plot.new()
frame()
```

References

Becker, R. A., Chambers, J. M. and Wilks, A. R. (1988) *The New S Language.* Wadsworth & Brooks/Cole. (`frame`.)

See Also

`plot.window`, `plot.default`.

Gnome	*GNOME Desktop Graphics Device*

Description

gnome starts a GNOME compatible device driver. GNOME is an acronym for **G**NU **N**etwork **O**bject **M**odel **E**nvironment.

Usage

```
gnome(display="", width=7, height=7, pointsize=12)
GNOME(display="", width=7, height=7, pointsize=12)
```

Arguments

`display`	the display on which the graphics window will appear. The default is to use the value in the user's environment variable `DISPLAY`.
`width`	the width of the plotting window in inches.
`height`	the height of the plotting window in inches.
`pointsize`	the default pointsize to be used.

Note

This is still in development state.

The GNOME device is only available when explicitly desired at configure/compile time, see the toplevel 'INSTALL' file.

Author(s)

Lyndon Drake

References

`http://www.gnome.org` and `http://www.gtk.org` for the associated GTK+ libraries.

See Also

`x11`, `Devices`.

Examples

    gnome(width=9)

gray — *Gray Level Specification*

Description

Create a vector of colors from a vector of gray levels.

Usage

```
gray(level)
grey(level)
```

Arguments

level   a vector of desired gray levels between 0 and 1; zero indicates "black" and one indicates "white".

Details

The values returned by gray can be used with a col= specification in graphics functions or in par.

grey is an alias for gray.

Value

A vector of "colors" of the same length as level.

See Also

rainbow, hsv, rgb.

Examples

```
gray(0:8 / 8)
```

grid   Add Grid to a Plot

Description

`grid` adds an nx by ny rectangular grid to an existing plot.

Usage

```
grid(nx = NULL, ny = nx, col = "lightgray", lty = "dotted",
    lwd = NULL, equilogs = TRUE)
```

Arguments

nx,ny	number of cells of the grid in x and y direction. When NULL, as per default, the grid aligns with the tick marks on the corresponding *default* axis (i.e., tickmarks as computed by `axTicks`). When NA, no grid lines are drawn in the corresponding direction.
col	character or (integer) numeric; color of the grid lines.
lty	character or (integer) numeric; line type of the grid lines.
lwd	non-negative numeric giving line width of the grid lines; defaults to `par("lwd")`.
equilogs	logical, only used when *log* coordinates and alignment with the axis tick marks are active. Settingequilogs = FALSE in that case gives *non equidistant* tick aligned grid lines.

Note

If more fine tuning is required, use `abline(h = ., v = .)` directly.

See Also

`plot`, `abline`, `lines`, `points`.

Examples

```
plot(1:3)
grid(NA, 5, lwd = 2) # grid only in y-direction
```

```
data(iris)
## maybe change the desired number of tick marks:
## par(lab=c(mx,my,7))
op <- par(mfcol = 1:2)
with(iris,
{
  plot(Sepal.Length, Sepal.Width,
    col = as.integer(Species),
    xlim = c(4, 8), ylim = c(2, 4.5), panel.first = grid(),
    main = "with(iris, plot(.., panel.first=grid(), ..))")
  plot(Sepal.Length, Sepal.Width,
    col = as.integer(Species),
    panel.first = grid(3, lty=1,lwd=2),
    main = "... panel.first = grid(3, lty=1,lwd=2), ..")
 }
)
par(op)
```

base — gtk

gtk   *GTK+ Graphics Device*

Description

This is a graphics device similar to the X11 device but using the Gtk widgets. This is now available via a separate package - gtkDevice - and can be used independently of the GNOME GUI for R. This package also allows a device to embedded within a Gtk-based GUI developed using the RGtk package. The gtkDevice package is available from CRAN.

Usage

```
gtk(display="", width=7, height=7, pointsize=12)
GTK(display="", width=7, height=7, pointsize=12)
```

Arguments

display	the display on which the graphics window will appear. The default is to use the value in the user's environment variable DISPLAY.
width	the width of the plotting window in inches.
height	the height of the plotting window in inches.
pointsize	the default pointsize to be used.

Author(s)

Original version by Lyndon Drake. Reorganization by Martyn Plummer and Duncan Temple Lang.

References

http://www.gtk.org for the GTK+ libraries.

See Also

x11, Devices.

Hershey   *Hershey Vector Fonts in R*

Description

If the `vfont` argument to one of the text-drawing functions (`text`, `mtext`, `title`, `axis`, and `contour`) is a character vector of length 2, Hershey vector fonts are used to render the text.

These fonts have two advantages:

1. vector fonts describe each character in terms of a set of points; R renders the character by joining up the points with straight lines. This intimate knowledge of the outline of each character means that R can arbitrarily transform the characters, which can mean that the vector fonts look better for rotated and 3d text.
2. this implementation was adapted from the GNU libplot library which provides support for non-ASCII and non-English fonts. This means that it is possible, for example, to produce weird plotting symbols and Japanese characters.

Drawback:
You cannot use mathematical expressions (`plotmath`) with Hershey fonts.

Usage

Hershey

Details

The Hershey characters are organised into a set of fonts, which are specified by a typeface (e.g., `serif` or `sans serif`) and a fontindex or "style" (e.g., plain or italic). The first element of `vfont` specifies the typeface and the second element specifies the fontindex. The first table produced by `demo(Hershey)` shows the character a produced by each of the different fonts.

The available `typeface` and `fontindex` values are available as list components of the variable `Hershey`. The allowed pairs for (`typeface`, `fontindex`) are:

serif	plain
serif	italic

serif	bold
serif	bold italic
serif	cyrillic
serif	oblique cyrillic
serif	EUC
sans serif	plain
sans serif	italic
sans serif	bold
sans serif	bold italic
script	plain
script	italic
script	bold
gothic english	plain
gothic german	plain
gothic italian	plain
serif symbol	plain
serif symbol	italic
serif symbol	bold
serif symbol	bold italic
sans serif symbol	plain
sans serif symbol	italic

and the indices of these are available as `Hershey$allowed`.

Escape sequences: The string to be drawn can include escape sequences, which all begin with a \. When R encounters a \, rather than drawing the \, it treats the subsequent character(s) as a coded description of what to draw.

One useful escape sequence (in the current context) is of the form: \123. The three digits following the \ specify an octal code for a character. For example, the octal code for p is 160 so the strings "p" and "\160" are equivalent. This is useful for producing characters when there is not an appropriate key on your keyboard.

The other useful escape sequences all begin with \\. These are described below. Remember that backslashes have to be doubled in R character strings, so they need to be entered with *four* backslashes.

Symbols: an entire string of Greek symbols can be produced by selecting the Serif Symbol or Sans Serif Symbol typeface. To allow Greek symbols to be embedded in a string which uses a non-symbol typeface, there are a set of symbol escape sequences of the form \\ab. For example, the escape sequence *a produces a Greek alpha. The second table in `demo(Hershey)` shows all of the symbol escape sequences and the symbols that they produce.

ISO Latin-1: further escape sequences of the form \\ab are provided for producing ISO Latin-1 characters (for example, if you only have a US keyboard). Another option is to use the appropriate octal code. The (non-ASCII) ISO Latin-1 characters are in the range 241...377. For example, \366 produces the character o with an umlaut. The third table in demo(Hershey) shows all of the ISO Latin-1 escape sequences.

Special Characters: a set of characters are provided which do not fall into any standard font. These can only be accessed by escape sequence. For example, \\LI produces the zodiac sign for Libra, and \\JU produces the astronomical sign for Jupiter. The fourth table in demo(Hershey) shows all of the special character escape sequences.

Cyrillic Characters: cyrillic characters are implemented according to the K018-R encoding. On a US keyboard, these can be produced using the Serif typeface and Cyrillic (or Oblique Cyrillic) fontindex and specifying an octal code in the range 300 to 337 for lower case characters or 340 to 377 for upper case characters. The fifth table in demo(Hershey) shows the octal codes for the available cyrillic characters.

Japanese Characters: 83 Hiragana, 86 Katakana, and 603 Kanji characters are implemented according to the EUC (Extended Unix Code) encoding. Each character is identified by a unique hexadecimal code. The Hiragana characters are in the range 0x2421 to 0x2473, Katakana are in the range 0x2521 to 0x2576, and Kanji are (scattered about) in the range 0x3021 to 0x6d55.

When using the Serif typeface and EUC fontindex, these characters can be produced by a *pair* of octal codes. Given the hexadecimal code (e.g., 0x2421), take the first two digits and add 0x80 and do the same to the second two digits (e.g., 0x21 and 0x24 become 0xa4 and 0xa1), then convert both to octal (e.g., 0xa4 and 0xa1 become 244 and 241). For example, the first Hiragana character is produced by \244\241.

It is also possible to use the hexadecimal code directly. This works for all non-EUC fonts by specifying an escape sequence of the form \\#J1234. For example, the first Hiragana character is produced by \\#J2421.

The Kanji characters may be specified in a third way, using the so-called "Nelson Index", by specifying an escape sequence of the form \\#N1234. For example, the Kanji for "one" is produced by \\#N0001.

demo(Japanese) shows the available Japanese characters.

Raw Hershey Glyphs: all of the characters in the Hershey fonts are stored in a large array. Some characters are not accessible in any of the Hershey fonts. These characters can only be accessed via an escape sequence of the form \\#H1234. For example, the fleur-de-lys is produced by \\#H0746. The sixth and seventh tables of demo(Hershey) shows all of the available raw glyphs.

References

http://www.gnu.org/software/plotutils/

See Also

demo(Hershey), text, contour.

Japanese for the Japanese characters in the Hershey fonts.

Examples

```
str(Hershey)

## for tables of examples, see demo(Hershey)
```

hist *Histograms*

Description

The generic function `hist` computes a histogram of the given data values. If `plot=TRUE`, the resulting object of class `"histogram"` is plotted by `plot.histogram`, before it is returned.

Usage

```
hist(x, ...)

## Default S3 method:
hist(x, breaks = "Sturges", freq = NULL,
  probability = !freq,
  include.lowest = TRUE, right = TRUE,
  density = NULL, angle = 45, col = NULL, border = NULL,
  main = paste("Histogram of" , xname),
  xlim = range(breaks), ylim = NULL,
  xlab = xname, ylab,
  axes = TRUE, plot = TRUE, labels = FALSE,
  nclass = NULL, ...)
```

Arguments

x
: a vector of values for which the histogram is desired.

breaks
: one of:
  - a vector giving the breakpoints between histogram cells,
  - a single number giving the number of cells for the histogram,
  - a character string naming an algorithm to compute the number of cells (see Details),
  - a function to compute the number of cells.

  In the last three cases the number is a suggestion only.

freq
: logical; if `TRUE`, the histogram graphic is a representation of frequencies, the `counts` component of the result; if `FALSE`, *relative* frequencies ("probabilities"), component `density`, are plotted. Defaults to `TRUE` *iff* breaks are equidistant (and `probability` is not specified).

probability	an *alias* for !freq, for S compatibility.
include.lowest	logical; if TRUE, an x[i] equal to the breaks value will be included in the first (or last, for right = FALSE) bar. This will be ignored (with a warning) unless breaks is a vector.
right	logical; if TRUE, the histograms cells are right-closed (left open) intervals.
density	the density of shading lines, in lines per inch. The default value of NULL means that no shading lines are drawn. Non-positive values of density also inhibit the drawing of shading lines.
angle	the slope of shading lines, given as an angle in degrees (counter-clockwise).
col	a colour to be used to fill the bars. The default of NULL yields unfilled bars.
border	the color of the border around the bars. The default is to use the standard foreground color.
main, xlab, ylab	these arguments to title have useful defaults here.
xlim, ylim	the range of x and y values with sensible defaults. Note that xlim is *not* used to define the histogram (breaks), but only for plotting (when plot = TRUE).
axes	logical. If TRUE (default), axes are draw if the plot is drawn.
plot	logical. If TRUE (default), a histogram is plotted, otherwise a list of breaks and counts is returned.
labels	logical or character. Additionally draw labels on top of bars, if not FALSE; see plot.histogram.
nclass	numeric (integer). For S(-PLUS) compatibility only, nclass is equivalent to breaks for a scalar or character argument.
...	further graphical parameters to title and axis.

Details

The definition of "histogram" differs by source (with country-specific biases). R's default with equi-spaced breaks (also the default) is to plot the counts in the cells defined by breaks. Thus the height of a rectangle is proportional to the number of points falling into the cell, as is the area *provided* the breaks are equally-spaced.

The default with non-equi-spaced breaks is to give a plot of area one, in which the *area* of the rectangles is the fraction of the data points falling in the cells.

If right = TRUE (default), the histogram cells are intervals of the form (a, b], i.e., they include their right-hand endpoint, but not their left one, with the exception of the first cell when include.lowest is TRUE.

For right = FALSE, the intervals are of the form [a, b), and include.lowest really has the meaning of *"include highest"*.

A numerical tolerance of 10^{-7} times the range of the breaks is applied when counting entries on the edges of bins.

The default for breaks is "Sturges": see nclass.Sturges. Other names for which algorithms are supplied are "Scott" and "FD" / "Friedman-Diaconis" (with corresponding functions nclass.scott and nclass.FD). Case is ignored and partial matching is used. Alternatively, a function can be supplied which will compute the intended number of breaks as a function of x.

Value

an object of class "histogram" which is a list with components:

breaks	the $n + 1$ cell boundaries (= breaks if that was a vector).
counts	n integers; for each cell, the number of x[] inside.
density	values $\hat{f}(x_i)$, as estimated density values. If all(diff(breaks) == 1), they are the relative frequencies counts/n and in general satisfy $\sum_i \hat{f}(x_i)(b_{i+1} - b_i) = 1$, where $b_i = $ breaks[i].
intensities	same as density. Deprecated, but retained for compatibility.
mids	the n cell midpoints.
xname	a character string with the actual x argument name.
equidist	logical, indicating if the distances between breaks are all the same.

Note

The resulting value does *not* depend on the values of the arguments freq (or probability) or plot. This is intentionally different from S.

Prior to R 1.7.0, the element breaks of the result was adjusted for numerical tolerances. The nominal values are now returned even though tolerances are still used when counting.

References

Becker, R. A., Chambers, J. M. and Wilks, A. R. (1988) *The New S Language.* Wadsworth & Brooks/Cole.

Venables, W. N. and Ripley. B. D. (2002) *Modern Applied Statistics with S.* Springer.

See Also

nclass.Sturges, stem, density, truehist.

Examples

```
data(islands)
op <- par(mfrow=c(2, 2))
hist(islands)
str(hist(islands, col="gray", labels = TRUE))
hist(sqrt(islands), br = 12, col="lightblue", border="pink")
# For non-equidistant breaks, counts should NOT be
# graphed unscaled:
r <- hist(sqrt(islands),
          br = c(4*0:5, 10*3:5, 70, 100, 140),
          col='blue1')
text(r$mids, r$density, r$counts, adj=c(.5, -.5),
     col='blue3')
sapply(r[2:3], sum)
sum(r$density * diff(r$breaks)) # == 1
lines(r, lty = 3, border = "purple") # lines.histogram(*)
par(op)

str(hist(islands, plot= FALSE))        # 5 breaks
str(hist(islands, br=12, plot= FALSE)) # 10 (~= 12) breaks
str(hist(islands, br=c(12,20,36,80,200,1000,17000),
    plot = FALSE))
hist(islands, br=c(12,20,36,80,200,1000,17000),
     freq = TRUE, main = "WRONG histogram") # and warning
```

hsv   *HSV Color Specification*

Description

Create a vector of colors from vectors specifying hue, saturation and value.

Usage

```
hsv(h=1, s=1, v=1, gamma=1)
```

Arguments

h,s,v   numeric vectors of values in the range [0,1] for "hue", "saturation" and "value" to be combined to form a vector of colors. Values in shorter arguments are recycled.

gamma   a "gamma correction", γ

Value

This function creates a vector of "colors" corresponding to the given values in HSV space. The values returned by hsv can be used with a col= specification in graphics functions or in par.

Gamma correction

For each color, (r, g, b) in RGB space (with all values in $[0, 1]$), the final color corresponds to $(r^\gamma, g^\gamma, b^\gamma)$.

See Also

rainbow, rgb, gray.

Examples

```
hsv(.5,.5,.5)

## Look at gamma effect:
n <- 20;  y <- -sin(3*pi*((1:n)-1/2)/n)
op <- par(mfrow=c(3,2),mar=rep(1.5,4))
for(gamma in c(.4, .6, .8, 1, 1.2, 1.5))
```

```
    plot(y, axes = FALSE, frame.plot = TRUE,
         xlab = "", ylab = "", pch = 21, cex = 30,
         bg = rainbow(n, start=.85, end=.1, gamma = gamma),
         main = paste("Red tones;   gamma=",format(gamma)))
  par(op)
```

`identify` *Identify Points in a Scatter Plot*

Description

`identify` reads the position of the graphics pointer when the (first) mouse button is pressed. It then searches the coordinates given in `x` and `y` for the point closest to the pointer. If this point is close to the pointer, its index will be returned as part of the value of the call.

Usage

```
identify(x, ...)

## Default S3 method:
identify(x, y = NULL, labels = seq(along = x), pos = FALSE,
        n = length(x), plot = TRUE, offset = 0.5, ...)
```

Arguments

`x,y`	coordinates of points in a scatter plot. Alternatively, any object which defines coordinates (a plotting structure, time series etc.) can be given as `x` and `y` left undefined.
`labels`	an optional vector, the same length as `x` and `y`, giving labels for the points.
`pos`	if `pos` is TRUE, a component is added to the return value which indicates where text was plotted relative to each identified point (1=below, 2=left, 3=above and 4=right).
`n`	the maximum number of points to be identified.
`plot`	if `plot` is TRUE, the labels are printed at the points and if FALSE they are omitted.
`offset`	the distance (in character widths) which separates the label from identified points.
`...`	further arguments to `par(.)`.

Details

If in addition, `plot` is TRUE, the point is labelled with the corresponding element of `text`.

The labels are placed either below, to the left, above or to the right of the identified point, depending on where the cursor was.

The identification process is terminated by pressing any mouse button other than the first.

On most devices which support `locator`, successful selection of a point is indicated by a bell sound unless `options(locatorBell=FALSE`

Value

If `pos` is `FALSE`, an integer vector containing the indexes of the identified points.

If `pos` is `TRUE`, a list containing a component `ind`, indicating which points were identified and a component `pos`, indicating where the labels were placed relative to the identified points.

References

Becker, R. A., Chambers, J. M. and Wilks, A. R. (1988) *The New S Language.* Wadsworth & Brooks/Cole.

See Also

`locator`

image *Display a Color Image*

Description

Creates a grid of colored or gray-scale rectangles with colors corresponding to the values in z. This can be used to display three-dimensional or spatial data aka "images". This is a generic function.

The functions heat.colors, terrain.colors and topo.colors create heat-spectrum (red to white) and topographical color schemes suitable for displaying ordered data, with n giving the number of colors desired.

Usage

```
image(x, ...)

## Default S3 method:
image(x, y, z, zlim, xlim, ylim, col = heat.colors(12),
      add = FALSE, xaxs = "i", yaxs = "i", xlab, ylab,
      breaks, oldstyle = FALSE, ...)
```

Arguments

x,y
: locations of grid lines at which the values in z are measured. These must be in (strictly) ascending order. By default, equally spaced values from 0 to 1 are used. If x is a list, its components x$x and x$y are used for x and y, respectively. If the list has component z this is used for z.

z
: a matrix containing the values to be plotted (NAs are allowed). Note that x can be used instead of z for convenience.

zlim
: the minimum and maximum z values for which colors should be plotted. Each of the given colors will be used to color an equispaced interval of this range. The *midpoints* of the intervals cover the range, so that values just outside the range will be plotted.

xlim, ylim
: ranges for the plotted x and y values, defaulting to the range of the finite values of x and y.

col
: a list of colors such as that generated by rainbow, heat.colors, topo.colors, terrain.colors or similar functions.

add	logical; if TRUE, add to current plot (and disregard the following arguments). This is rarely useful because image "paints" over existing graphics.
xaxs, yaxs	style of x and y axis. The default "i" is appropriate for images. See par.
xlab, ylab	each a character string giving the labels for the x and y axis. Default to the 'call names' of x or y, or to "" if these where unspecified.
breaks	a set of breakpoints for the colours: must give one more breakpoint than colour.
oldstyle	logical. If true the midpoints of the colour intervals are equally spaced, and zlim[1] and zlim[2] were taken to be midpoints. (This was the default prior to R 1.1.0.) The current default is to have colour intervals of equal lengths between the limits.
...	graphical parameters for plot may also be passed as arguments to this function.

Details

The length of x should be equal to the nrow(z)+1 or nrow(z). In the first case x specifies the boundaries between the cells: in the second case x specifies the midpoints of the cells. Similar reasoning applies to y. It probably only makes sense to specify the midpoints of an equally-spaced grid. If you specify just one row or column and a length-one x or y, the whole user area in the corresponding direction is filled.

If breaks is specified then zlim is unused and the algorithm used follows cut, so intervals are closed on the right and open on the left except for the lowest interval.

Note

Based on a function by Thomas Lumley.

See Also

filled.contour or heatmap which can look nicer (but are less modular), contour;

heat.colors, topo.colors, terrain.colors, rainbow, hsv, par.

Examples

```
x <- y <- seq(-4*pi, 4*pi, len=27)
r <- sqrt(outer(x^2, y^2, "+"))
image(z = z <- cos(r^2)*exp(-r/6), col=gray((0:32)/32))
image(z, axes = FALSE, main = "Math can be beautiful ...",
      xlab = expression(cos(r^2) * e^{-r/6}))
contour(z, add = TRUE, drawlabels = FALSE)

data(volcano)
x <- 10*(1:nrow(volcano))
y <- 10*(1:ncol(volcano))
image(x, y, volcano, col = terrain.colors(100),
      axes = FALSE)
contour(x, y, volcano, levels = seq(90, 200, by=5),
        add = TRUE, col = "peru")
axis(1, at = seq(100, 800, by = 100))
axis(2, at = seq(100, 600, by = 100))
box()
title(main = "Maunga Whau Volcano", font.main = 4)
```

interaction.plot   *Two-way Interaction Plot*

Description

Plots the mean (or other summary) of the response for two-way combinations of factors, thereby illustrating possible interactions.

Usage

```
interaction.plot(x.factor, trace.factor, response,
    fun = mean, type = c("l", "p"), legend = TRUE,
    trace.label=deparse(substitute(trace.factor)),
    fixed=FALSE, xlab = deparse(substitute(x.factor)),
    ylab = ylabel, ylim = range(cells, na.rm=TRUE),
    lty = nc:1, col = 1, pch = c(1:9, 0, letters),
    xpd = NULL, leg.bg = par("bg"), leg.bty = "n",
    xtick = FALSE, xaxt = par("xaxt"), axes = TRUE, ...)
```

Arguments

x.factor	a factor whose levels will form the x axis.
trace.factor	another factor whose levels will form the traces.
response	a numeric variable giving the response
fun	the function to compute the summary. Should return a single real value.
type	the type of plot: lines or points.
legend	logical. Should a legend be included?
trace.label	overall label for the legend.
fixed	logical. Should the legend be in the order of the levels of trace.factor or in the order of the traces at their right-hand ends?
xlab,ylab	the x and y label of the plot each with a sensible default.
ylim	numeric of length 2 giving the y limits for the plot.
lty	line type for the lines drawn, with sensible default.
col	the color to be used for plotting.
pch	a vector of plotting symbols or characters, with sensible default.

xpd   determines clipping behaviour for the `legend` used,
                see `par(xpd)`. Per default, the legend is *not* clipped
                at the figure border.

`leg.bg, leg.bty`
                arguments passed to `legend()`.

xtick   logical. Should tick marks be used on the x axis?

`xaxt, axes, ...`
                graphics parameters to be passed to the plotting routines.

Details

By default the levels of `x.factor` are plotted on the x axis in their given order, with extra space left at the right for the legend (if specified). If `x.factor` is an ordered factor and the levels are numeric, these numeric values are used for the x axis.

The response and hence its summary can contain missing values. If so, the missing values and the line segments joining them are omitted from the plot (and this can be somewhat disconcerting).

The graphics parameters `xlab`, `ylab`, `ylim`, `lty`, `col` and `pch` are given suitable defaults (and `xlim` and `xaxs` are set and cannot be overriden). The defaults are to cycle through the line types, use the foreground colour, and to use the symbols 1:9, 0, and the capital letters to plot the traces.

Note

Some of the argument names and the precise behaviour are chosen for S-compatibility.

References

Chambers, J. M., Freeny, A and Heiberger, R. M. (1992) *Analysis of variance; designed experiments.* Chapter 5 of *Statistical Models in S* eds J. M. Chambers and T. J. Hastie, Wadsworth & Brooks/Cole.

Examples

```
data(ToothGrowth)
attach(ToothGrowth)
interaction.plot(dose, supp, len, fixed=TRUE)
dose <- ordered(dose)
interaction.plot(dose, supp, len, fixed=TRUE, col = 2:3,
```

```
                    leg.bty = "o")
detach()

data(OrchardSprays)
with(OrchardSprays, {
  interaction.plot(treatment, rowpos, decrease)
  interaction.plot(rowpos, treatment, decrease,
                   cex.axis=0.8)
  ## order the rows by their mean effect
  rowpos <- factor(rowpos, levels = sort.list(tapply(
                   decrease, rowpos, mean)))
  interaction.plot(rowpos, treatment, decrease, col = 2:9,
                   lty = 1)
})

data(esoph)
with(esoph, {
  interaction.plot(agegp, alcgp, ncases/ncontrols)
  interaction.plot(agegp, tobgp, ncases/ncontrols,
                   trace.label="tobacco",
                   fixed=TRUE, xaxt = "n")
})
```

Japanese   *Japanese characters in R*

Description

The implementation of Hershey vector fonts provides a large number of Japanese characters (Hiragana, Katakana, and Kanji).

Details

Without keyboard support for typing Japanese characters, the only way to produce these characters is to use special escape sequences: see Hershey.

For example, the Hiragana character for the sound "ka" is produced by \\#J242b and the Katakana character for this sound is produced by \\#J252b. The Kanji ideograph for "one" is produced by \\#J306c or \\#N0001.

The output from demo(Japanese) shows tables of the escape sequences for the available Japanese characters.

References

http://www.gnu.org/software/plotutils/

See Also

demo(Japanese), Hershey, text, contour

Examples

```
plot(1:9, type="n", axes=FALSE, frame=TRUE, ylab="",
  main= "example(Japanese)", xlab= "using Hershey fonts")
par(cex=3)
Vf <- c("serif", "plain")
text(4, 2, "\\#J2438\\#J2421\\#J2451\\#J2473", vfont = Vf)
text(4, 4, "\\#J2538\\#J2521\\#J2551\\#J2573", vfont = Vf)
text(4, 6, "\\#J467c\\#J4b5c", vfont = Vf)
text(4, 8, "Japan", vfont = Vf)
par(cex=1)
text(8, 2, "Hiragana")
text(8, 4, "Katakana")
text(8, 6, "Kanji")
text(8, 8, "English")
```

jitter   *Add 'Jitter' (Noise) to Numbers*

Description

Add a small amount of noise to a numeric vector.

Usage

```
jitter(x, factor=1, amount = NULL)
```

Arguments

x	numeric to which *jitter* should be added.
factor	numeric
amount	numeric; if positive, used as *amount* (see below), otherwise, if = 0 the default is factor * z/50.
	Default (NULL): factor * d/5 where d is about the smallest difference between x values.

Details

The result, say r, is r <- x + runif(n, -a, a) where n <- length(x) and a is the amount argument (if specified).

Let z <- max(x) - min(x) (assuming the usual case). The amount a to be added is either provided as *positive* argument amount or otherwise computed from z, as follows:

If amount == 0, we set a <- factor * z/50 (same as S).

If amount is NULL (*default*), we set a <- factor * d/5 where d is the smallest difference between adjacent unique (apart from fuzz) x values.

Value

jitter(x,...) returns a numeric of the same length as x, but with an amount of noise added in order to break ties.

Author(s)

Werner Stahel and Martin Maechler, ETH Zurich

References

Chambers, J. M., Cleveland, W. S., Kleiner, B. and Tukey, P.A. (1983) *Graphical Methods for Data Analysis.* Wadsworth; figures 2.8, 4.22, 5.4.

Chambers, J. M. and Hastie, T. J. (1992) *Statistical Models in S.* Wadsworth & Brooks/Cole.

See Also

rug which you may want to combine with jitter.

Examples

```
round(jitter(c(rep(1,3), rep(1.2, 4), rep(3,3))), 3)
## These two 'fail' with S-plus 3.x:
jitter(rep(0, 7))
jitter(rep(10000,5))
```

base — layout

layout — Specifying Complex Plot Arrangements

Description

layout divides the device up into as many rows and columns as there are in matrix mat, with the column-widths and the row-heights specified in the respective arguments.

Usage

```
layout(mat,
       widths = rep(1, dim(mat)[2]),
       heights= rep(1, dim(mat)[1]),
       respect= FALSE)

layout.show(n = 1)
lcm(x)
```

Arguments

mat
: a matrix object specifying the location of the next N figures on the output device. Each value in the matrix must be 0 or a positive integer. If N is the largest positive integer in the matrix, then the integers $\{1, \ldots, N-1\}$ must also appear at least once in the matrix.

widths
: a vector of values for the widths of columns on the device. Relative widths are specified with numeric values. Absolute widths (in centimetres) are specified with the lcm() function (see examples).

heights
: a vector of values for the heights of rows on the device. Relative and absolute heights can be specified, see widths above.

respect
: either a logical value or a matrix object. If the latter, then it must have the same dimensions as mat and each value in the matrix must be either 0 or 1.

n
: number of figures to plot.

x
: a dimension to be interpreted as a number of centimetres.

Details

Figure *i* is allocated a region composed from a subset of these rows and columns, based on the rows and columns in which *i* occurs in mat.

The respect argument controls whether a unit column-width is the same physical measurement on the device as a unit row-height.

layout.show(n) plots (part of) the current layout, namely the outlines of the next n figures.

lcm is a trivial function, to be used as *the* interface for specifying absolute dimensions for the widths and heights arguments of layout().

Value

layout returns the number of figures, N, see above.

Author(s)

Paul R. Murrell

References

Murrell, P. R. (1999) Layouts: A mechanism for arranging plots on a page. *Journal of Computational and Graphical Statistics*, 8, 121-134. Chapter 5 of Paul Murrell's Ph.D. thesis.

See Also

par with arguments mfrow, mfcol, or mfg.

Examples

```
# save default, for resetting...
def.par <- par(no.readonly = TRUE)

## divide the device into two rows and two columns allocate
## figure 1 all of row 1 allocate figure 2 the intersection
## of column 2 and row 2
layout(matrix(c(1,1,0,2), 2, 2, byrow = TRUE))
## show the regions that have been allocated to each plot
layout.show(2)

## divide device into two rows and two columns allocate
## figure 1 and figure 2 as above respect relations between
## widths and heights
```

```
nf <- layout(matrix(c(1,1,0,2), 2, 2, byrow=TRUE),
             respect=TRUE)
layout.show(nf)

## create single figure which is 5cm square
nf <- layout(matrix(1), widths=lcm(5), heights=lcm(5))
layout.show(nf)

## Create a scatterplot with marginal histograms
x <- pmin(3, pmax(-3, rnorm(50)))
y <- pmin(3, pmax(-3, rnorm(50)))
xhist <- hist(x, breaks=seq(-3,3,0.5), plot=FALSE)
yhist <- hist(y, breaks=seq(-3,3,0.5), plot=FALSE)
top <- max(c(xhist$counts, yhist$counts))
xrange <- c(-3,3)
yrange <- c(-3,3)
nf <- layout(matrix(c(2,0,1,3),2,2,byrow=TRUE), c(3,1),
             c(1,3), TRUE)
layout.show(nf)

par(mar=c(3,3,1,1))
plot(x, y, xlim=xrange, ylim=yrange, xlab="", ylab="")
par(mar=c(0,3,1,1))
barplot(xhist$counts, axes=FALSE, ylim=c(0, top), space=0)
par(mar=c(3,0,1,1))
barplot(yhist$counts, axes=FALSE, xlim=c(0, top), space=0,
        horiz=TRUE)

par(def.par) # reset to default
```

legend   *Add Legends to Plots*

Description

This function can be used to add legends to plots. Note that a call to the function `locator` can be used in place of the x and y arguments.

Usage

```
legend(x, y = NULL, legend, fill = NULL, col = "black",
    lty, lwd, pch, angle = NULL, density = NULL, bty = "o",
    bg = par("bg"), pt.bg = NA, cex = 1, xjust = 0, yjust = 1,
    x.intersp = 1, y.intersp = 1, adj = c(0, 0.5),
    text.width = NULL, merge = do.lines && has.pch,
    trace = FALSE, plot = TRUE, ncol = 1, horiz = FALSE)
```

Arguments

x, y
: the x and y co-ordinates to be used to position the legend. They can be specified in any way which is accepted by xy.coords: See Details.

legend
: a vector of text values or an expression of length ≥ 1, or a call (as resulting from substitute) to appear in the legend.

fill
: if specified, this argument will cause boxes filled with the specified colors (or shaded in the specified colors) to appear beside the legend text.

col
: the color of points or lines appearing in the legend.

lty, lwd
: the line types and widths for lines appearing in the legend. One of these two *must* be specified for line drawing.

pch
: the plotting symbols appearing in the legend, either as vector of 1-character strings, or one (multi character) string. *Must* be specified for symbol drawing.

angle
: angle of shading lines.

density
: the density of shading lines, if numeric and positive. If NULL or negative or NA color filling is assumed.

bty
: the type of box to be drawn around the legend. The allowed values are "o" (the default) and "n".

bg	the background color for the legend box. (Note that this is only used if bty = "n".)
pt.bg	the background color for the points.
cex	character expansion factor **relative** to current par("cex").
xjust	how the legend is to be justified relative to the legend x location. A value of 0 means left justified, 0.5 means centered and 1 means right justified.
yjust	the same as xjust for the legend y location.
x.intersp	character interspacing factor for horizontal (x) spacing.
y.intersp	the same for vertical (y) line distances.
adj	numeric of length 1 or 2; the string adjustment for legend text. Useful for y-adjustment when labels are plotmath expressions.
text.width	the width of the legend text in x ("user") coordinates. Defaults to the proper value computed by strwidth(legend).
merge	logical; if TRUE, "merge" points and lines but not filled boxes. Defaults to TRUE if there are points and lines.
trace	logical; if TRUE, shows how legend does all its magical computations.
plot	logical. If FALSE, nothing is plotted but the sizes are returned.
ncol	the number of columns in which to set the legend items (default is 1, a vertical legend).
horiz	logical; if TRUE, set the legend horizontally rather than vertically (specifying horiz overrides the ncol specification).

Details

Arguments x, y, legend are interpreted in a non-standard way to allow the coordinates to be specified *via* one or two arguments. If legend is missing and y is not numeric, it is assumed that the second argument is intended to be legend and that the first argument specifies the coordinates.

The coordinates can be specified in any way which is accepted by xy.coords. If this gives the coordinates of one point, it is used as the top-left coordinate of the rectangle containing the legend. If it gives the coordinates of two points, these specify opposite corners of the rectangle (either pair of corners, in any order).

"Attribute" arguments such as `col`, `pch`, `lty`, etc, are recycled if necessary. `merge` is not.

Points are drawn *after* lines in order that they can cover the line with their background color `pt.bg`, if applicable.

See the examples for how to right-justify labels.

Value

A list with list components

`rect`   a list with components

> `w, h` positive numbers giving width and height of the legend's box.
>
> `left, top` x and y coordinates of upper left corner of the box.

`text`   a list with components

> `x, y` numeric vectors of length `length(legend)`, giving the x and y coordinates of the legend's text(s).

returned invisibly.

References

Becker, R. A., Chambers, J. M. and Wilks, A. R. (1988) *The New S Language*. Wadsworth & Brooks/Cole.

See Also

`plot`, `barplot` which uses `legend()`, and `text` for more examples of math expressions.

Examples

```
## Run the example in '?matplot' or the following:
leg.txt <- c("Setosa     Petals", "Setosa     Sepals",
             "Versicolor Petals", "Versicolor Sepals")
y.leg <- c(4.5, 3, 2.1, 1.4, .7)
cexv  <- c(1.2, 1, 4/5, 2/3, 1/2)
matplot(c(1,8), c(0,4.5), type = "n",
  xlab = "Length", ylab = "Width",
```

```
      main = "Petal and Sepal Dimensions in Iris Blossoms")
for (i in seq(cexv)) {
  text(1, y.leg[i]-.1, paste("cex=",formatC(cexv[i])),
    cex=.8, adj = 0)
  legend(3, y.leg[i], leg.txt, pch = "sSvV", col = c(1, 3),
    cex = cexv[i])
}

## 'merge = TRUE' for merging lines & points:
x <- seq(-pi, pi, len = 65)
plot(x, sin(x), type = "l", ylim = c(-1.2, 1.8), col = 3,
    lty = 2)
points(x, cos(x), pch = 3, col = 4)
lines(x, tan(x), type = "b", lty = 1, pch = 4, col = 6)
title("legend(..,lty=c(2,-1,1),pch=c(-1,3,4),merge=TRUE)",
    cex.main = 1.1)
legend(-1, 1.9, c("sin", "cos", "tan"), col = c(3,4,6),
      lty = c(2, -1, 1), pch = c(-1, 3, 4), merge = TRUE,
      bg='gray90')

## right-justifying a set of labels: thanks to Uwe Ligges
x <- 1:5; y1 <- 1/x; y2 <- 2/x
plot(rep(x, 2), c(y1, y2), type="n", xlab="x", ylab="y")
lines(x, y1); lines(x, y2, lty=2)
temp <- legend(5, 2, legend = c(" ", " "),
                text.width = strwidth("1,000,000"),
                lty = 1:2, xjust = 1, yjust = 1)
text(temp$rect$left + temp$rect$w, temp$text$y,
    c("1,000", "1,000,000"), pos=2)

## log scaled Examples
leg.txt <- c("a one", "a two")

par(mfrow = c(2,2))
for(ll in c("","x","y","xy")) {
  plot(2:10, log=ll, main=paste("log = '",ll,"'", sep=""))
  abline(1,1)
  lines(2:3,3:4, col=2) #
  points(2,2, col=3)    #
  rect(2,3,3,2, col=4)
  text(c(3,3),2:3, c("rect(2,3,3,2, col=4)",
    "text(c(3,3),2:3,\"c(rect(...)\")"), adj = c(0,.3))
  legend(list(x=2,y=8), legend = leg.txt, col=2:3, pch=1:2,
```

```
          lty=1, merge=TRUE) #, trace=TRUE)
}
par(mfrow=c(1,1))

## Math expressions:
x <- seq(-pi, pi, len = 65)
plot(x, sin(x), type="l", col = 2,
     xlab = expression(phi), ylab = expression(f(phi)))
abline(h=-1:1, v=pi/2*(-6:6), col="gray90")
lines(x, cos(x), col = 3, lty = 2)
# 2 ways
ex.cs1 <- expression(plain(sin) * phi,  paste("cos", phi))
# adj y !
str(legend(-3,.9,ex.cs1,lty=1:2,plot=FALSE,adj=c(0,.6)))
legend(-3, .9, ex.cs1, lty=1:2, col=2:3, adj = c(0, .6))

x <- rexp(100, rate = .5)
hist(x, main = "Mean and Median of a Skewed Distribution")
abline(v = mean(x),   col=2, lty=2, lwd=2)
abline(v = median(x), col=3, lty=3, lwd=2)
ex12 <- expression(bar(x) == sum(over(x[i], n), i==1, n),
                   hat(x) == median(x[i], i==1,n))
str(legend(4.1, 30, ex12, col = 2:3, lty=2:3, lwd=2))

## 'Filled' boxes -- for more, see example(plotfactor)
op <- par(bg="white") # to get an opaque box for the legend
data(PlantGrowth)
plot(cut(weight, 3) ~ group, data = PlantGrowth,
     col = NULL, density = 16*(1:3))
par(op)

## Using 'ncol' :
x <- 0:64/64
matplot(x, outer(x, 1:7, function(x, k) sin(k * pi * x)),
  type = "o", col = 1:7, ylim = c(-1, 1.5), pch = "*")
op <- par(bg="antiquewhite1")
legend(0, 1.5, paste("sin(",1:7,"pi * x)"), col=1:7,
       lty=1:7, pch = "*", ncol = 4, cex=.8)
legend(.8,1.2, paste("sin(",1:7,"pi * x)"), col=1:7,
       lty=1:7, pch = "*",cex=.8)
legend(0, -.1, paste("sin(",1:4,"pi * x)"), col=1:4,
       lty=1:4, ncol=2, cex=.8)
legend(0, -.4, paste("sin(",5:7,"pi * x)"), col=5:7,
```

```
            pch=24, ncol=2, cex=1.5, pt.bg="pink")
par(op)

## point covering line :
y <- sin(3*pi*x)
plot(x, y, type="l", col="blue",
     main = "points with bg & legend(*, pt.bg)")
points(x, y, pch=21, bg="white")
legend(.4,1, "sin(c x)", pch=21, pt.bg="white", lty=1,
       col = "blue")
```

lines —— Add Connected Line Segments to a Plot

Description

A generic function taking coordinates given in various ways and joining the corresponding points with line segments.

Usage

```
lines(x, ...)
```

```
## Default S3 method:
lines(x, y = NULL, type = "l", col = par("col"),
      lty = par("lty"), ...)
```

Arguments

x, y	coordinate vectors of points to join.
type	character indicating the type of plotting; actually any of the types as in plot.
col	color to use. This can be vector of length greater than one, but only the first value will be used.
lty	line type to use.
...	Further graphical parameters (see par) may also be supplied as arguments, particularly, line type, lty and line width, lwd.

Details

The coordinates can be passed to lines in a plotting structure (a list with x and y components), a time series, etc. See xy.coords.

The coordinates can contain NA values. If a point contains NA in either its x or y value, it is omitted from the plot, and lines are not drawn to or from such points. Thus missing values can be used to achieve breaks in lines.

References

Becker, R. A., Chambers, J. M. and Wilks, A. R. (1988) *The New S Language.* Wadsworth & Brooks/Cole.

See Also

points, plot, and the underlying "primitive" plot.xy.

par for how to specify colors.

Examples

```
data(cars)
# draw a smooth line through a scatter plot
plot(cars, main="Stopping Distance versus Speed")
lines(lowess(cars))
```

locator   *Graphical Input*

Description

Reads the position of the graphics cursor when the (first) mouse button is pressed.

Usage

```
locator(n = 512, type = "n", ...)
```

Arguments

n
: the maximum number of points to locate.

type
: One of "n", "p", "l" or "o". If "p" or "o" the points are plotted; if "l" or "o" they are joined by lines.

...
: additional graphics parameters used if `type != "n"` for plotting the locations.

Details

Unless the process is terminated prematurely by the user (see below) at most n positions are determined.

The identification process can be terminated by pressing any mouse button other than the first.

The current graphics parameters apply just as if `plot.default` has been called with the same value of `type`. The plotting of the points and lines is subject to clipping, but locations outside the current clipping rectangle will be returned.

On most devices which support `locator`, successful selection of a point is indicated by a bell sound unless `options(locatorBell=FALSE)` has been set.

If the window is resized or hidden and then exposed before the input process has terminated, any lines or points drawn by `locator` will disappear. These will reappear once the input process has terminated and the window is resized or hidden and exposed again. This is because the points and lines drawn by `locator` are not recorded in the device's display list until the input process has terminated.

Value

A list containing x and y components which are the coordinates of the identified points in the user coordinate system, i.e., the one specified by par("usr").

References

Becker, R. A., Chambers, J. M. and Wilks, A. R. (1988) *The New S Language.* Wadsworth & Brooks/Cole.

See Also

identify

matplot	*Plot Columns of Matrices*

Description

Plot the columns of one matrix against the columns of another.

Usage

```
matplot(x, y, type = "p", lty = 1:5, lwd = 1, pch = NULL,
   col = 1:6, cex = NULL, xlab = NULL, ylab = NULL,
   xlim = NULL, ylim = NULL, ...,
   add = FALSE, verbose = getOption("verbose"))
matpoints(x, y, type = "p", lty = 1:5, lwd = 1, pch = NULL,
         col = 1:6, ...)
matlines (x, y, type = "l", lty = 1:5, lwd = 1, pch = NULL,
         col = 1:6, ...)
```

Arguments

x,y
: vectors or matrices of data for plotting. The number of rows should match. If one of them are missing, the other is taken as y and an x vector of 1:n is used. Missing values (NAs) are allowed.

type
: character string (length 1 vector) or vector of 1-character strings indicating the type of plot for each column of y, see plot for all possible types. The first character of type defines the first plot, the second character the second, etc. Characters in type are cycled through; e.g., "pl" alternately plots points and lines.

lty,lwd
: vector of line types and widths. The first element is for the first column, the second element for the second column, etc., even if lines are not plotted for all columns. Line types will be used cyclically until all plots are drawn.

pch
: character string or vector of 1-characters or integers for plotting characters, see points. The first character is the plotting-character for the first plot, the second for the second, etc. The default is the digits (1 through 9, 0) then the letters.

col	vector of colors. Colors are used cyclically.
cex	vector of character expansion sizes, used cyclically.
xlab, ylab	titles for x and y axes, as in `plot`.
xlim, ylim	ranges of x and y axes, as in `plot`.
...	Graphical parameters (see `par`) and any further arguments of `plot`, typically `plot.default`, may also be supplied as arguments to this function. Hence, the high-level graphics control arguments described under `par` and the arguments to `title` may be supplied to this function.
add	logical. If TRUE, plots are added to current one, using `points` and `lines`.
verbose	logical. If TRUE, write one line of what is done.

Details

Points involving missing values are not plotted.

The first column of x is plotted against the first column of y, the second column of x against the second column of y, etc. If one matrix has fewer columns, plotting will cycle back through the columns again. (In particular, either x or y may be a vector, against which all columns of the other argument will be plotted.)

The first element of `col, cex, lty, lwd` is used to plot the axes as well as the first line.

Because plotting symbols are drawn with lines and because these functions may be changing the line style, you should probably specify `lty=1` when using plotting symbols.

Side Effects

Function `matplot` generates a new plot; `matpoints` and `matlines` add to the current one.

References

Becker, R. A., Chambers, J. M. and Wilks, A. R. (1988) *The New S Language.* Wadsworth & Brooks/Cole.

See Also

`plot, points, lines, matrix, par`.

Examples

```
# almost identical to plot(*)
matplot((-4:5)^2, main = "Quadratic")
sines <- outer(1:20, 1:4, function(x,y) sin(x/20 * pi * y))
matplot(sines, pch = 1:4, type = "o",
        col = rainbow(ncol(sines)))

x <- 0:50/50
matplot(x, outer(x, 1:8, function(x, k) sin(k*pi * x)),
        ylim = c(-2,2), type = "plobcsSh",
        main= "matplot(,type = \"plobcsSh\" )")
## pch & type = vector of 1-chars :
matplot(x, outer(x, 1:4, function(x, k) sin(k*pi * x)),
        pch = letters[1:4], type = c("b","p","o"))

data(iris) # is data.frame with 'Species' factor
table(iris$Species)
iS <- iris$Species == "setosa"
iV <- iris$Species == "versicolor"
op <- par(bg = "bisque")
matplot(c(1, 8), c(0, 4.5), type= "n",
        xlab = "Length", ylab = "Width",
        main = "Petal and Sepal Dimension in Iris Blossoms")
matpoints(iris[iS,c(1,3)], iris[iS,c(2,4)], pch = "sS",
          col = c(2,4))
matpoints(iris[iV,c(1,3)], iris[iV,c(2,4)], pch = "vV",
          col = c(2,4))
legend(1, 4, c("    Setosa Petals", "    Setosa Sepals",
               "Versicolor Petals", "Versicolor Sepals"),
       pch = "sSvV", col = rep(c(2,4), 2))

nam.var <- colnames(iris)[-5]
nam.spec <- as.character(iris[1+50*0:2, "Species"])
iris.S <- array(NA, dim = c(50,4,3),
                dimnames = list(NULL, nam.var, nam.spec))
for(i in 1:3)
  iris.S[,,i] <- data.matrix(iris[1:50+50*(i-1), -5])

matplot(iris.S[,"Petal.Length",], iris.S[,"Petal.Width",],
  pch="SCV",
  col = rainbow(3, start = .8, end = .1),
  sub = paste(c("S", "C", "V"), dimnames(iris.S)[[3]],
```

```
                    sep = "=", collapse= ",  "),
  main = "Fisher's Iris Data")
```

mosaicplot *Mosaic Plots*

Description

Plots a mosaic on the current graphics device.

Usage

```
mosaicplot(x, ...)

## Default S3 method:
mosaicplot(x, main = deparse(substitute(x)),
   sub = NULL, xlab = NULL, ylab = NULL,
   sort = NULL, off = NULL, dir = NULL,
   color = FALSE, shade = FALSE, margin = NULL,
   cex.axis = 0.66, las = par("las"),
   type = c("pearson", "deviance", "FT"), ...)

## S3 method for class 'formula':
mosaicplot(formula, data = NULL, ...,
          main = deparse(substitute(data)), subset)
```

Arguments

x	a contingency table in array form, with optional category labels specified in the dimnames(x) attribute. The table is best created by the table() command.
main	character string for the mosaic title.
sub	character string for the mosaic sub-title (at bottom).
xlab,ylab	x- and y-axis labels used for the plot; by default, the first and second element of names(dimnames(X)) (i.e., the name of the first and second variable in X).
sort	vector ordering of the variables, containing a permutation of the integers 1:length(dim(x)) (the default).
off	vector of offsets to determine percentage spacing at each level of the mosaic (appropriate values are between 0 and 20, and the default is 10 at each level). There should be one offset for each dimension of the contingency table.

dir	vector of split directions ("v" for vertical and "h" for horizontal) for each level of the mosaic, one direction for each dimension of the contingency table. The default consists of alternating directions, beginning with a vertical split.
color	logical or (recycling) vector of colors for color shading, used only when shade is FALSE. The default color=FALSE gives empty boxes with no shading.
shade	a logical indicating whether to produce extended mosaic plots, or a numeric vector of at most 5 distinct positive numbers giving the absolute values of the cut points for the residuals. By default, shade is FALSE, and simple mosaics are created. Using shade = TRUE cuts absolute values at 2 and 4.
margin	a list of vectors with the marginal totals to be fit in the log-linear model. By default, an independence model is fitted. See loglin for further information.
cex.axis	The magnification to be used for axis annotation, as a multiple of par("cex").
las	numeric; the style of axis labels, see par.
type	a character string indicating the type of residual to be represented. Must be one of "pearson" (giving components of Pearson's χ^2), "deviance" (giving components of the likelihood ratio χ^2), or "FT" for the Freeman-Tukey residuals. The value of this argument can be abbreviated.
formula	a formula, such as y ~ x.
data	a data frame (or list), or a contingency table from which the variables in formula should be taken.
...	further arguments to be passed to or from methods.
subset	an optional vector specifying a subset of observations in the data frame to be used for plotting.

Details

This is a generic function. It currently has a default method (mosaicplot.default) and a formula interface (mosaicplot.formula).

Extended mosaic displays show the standardized residuals of a loglinear model of the counts from by the color and outline of the mosaic's tiles. (Standardized residuals are often referred to a standard normal distribution.) Negative residuals are drawn in shaded of red and with broken outlines; positive ones are drawn in blue with solid outlines.

For the formula method, if `data` is an object inheriting from classes `"table"` or `"ftable"`, or an array with more than 2 dimensions, it is taken as a contingency table, and hence all entries should be nonnegative. In this case, the left-hand side of `formula` should be empty, and the variables on the right-hand side should be taken from the names of the dimnames attribute of the contingency table. A marginal table of these variables is computed, and a mosaic of this table is produced.

Otherwise, `data` should be a data frame or matrix, list or environment containing the variables to be cross-tabulated. In this case, after possibly selecting a subset of the data as specified by the `subset` argument, a contingency table is computed from the variables given in `formula`, and a mosaic is produced from this.

See Emerson (1998) for more information and a case study with television viewer data from Nielsen Media Research.

Author(s)

S-PLUS original by John Emerson. Originally modified and enhanced for R by KH.

References

Hartigan, J.A., and Kleiner, B. (1984) A mosaic of television ratings. *The American Statistician*, **38**, 32–35.

Emerson, J. W. (1998) Mosaic displays in S-PLUS: a general implementation and a case study. *Statistical Computing and Graphics Newsletter (ASA)*, **9**, 1, 17–23.

Friendly, M. (1994) Mosaic displays for multi-way contingency tables. *Journal of the American Statistical Association*, **89**, 190–200.

The home page of Michael Friendly (`http://www.math.yorku.ca/SCS/friendly.html`) provides information on various aspects of graphical methods for analyzing categorical data, including mosaic plots.

See Also

`assocplot`, `loglin`.

Examples

```
data(Titanic)
mosaicplot(Titanic, main = "Survival on the Titanic",
           color = TRUE)
## Formula interface for tabulated data:
mosaicplot(~ Sex + Age + Survived, data = Titanic,
           color = TRUE)

data(HairEyeColor)
mosaicplot(HairEyeColor, shade = TRUE)
## Independence model of hair and eye color and sex.
## Indicates that there are significantly more blue eyed
## blonde females than expected in the case of independence
## (and too few brown eyed blonde females).

mosaicplot(HairEyeColor, shade = TRUE,
           margin = list(c(1,2), 3))
## Model of joint independence of sex from hair and eye
## color.  Males are underrepresented among people with
## brown hair and eyes, and are overrepresented among
## people with brown hair and blue eyes, but not
## "significantly".

## Formula interface for raw data: visualize
## crosstabulation of numbers of gears and carburettors in
## Motor Trend car data.
data(mtcars)
mosaicplot(~ gear + carb, data = mtcars, color = TRUE,
           las = 1)
# color recycling
mosaicplot(~ gear + carb, data = mtcars, color = 2:3,
           las = 1)
```

mtext — Write Text into the Margins of a Plot

Description

Text is written in one of the four margins of the current figure region or one of the outer margins of the device region.

Usage

```
mtext(text, side = 3, line = 0, outer = FALSE, at = NA,
   adj = NA, cex = NA, col = NA, font = NA, vfont = NULL,
   ...)
```

Arguments

text	one or more character strings or expressions.
side	on which side of the plot (1=bottom, 2=left, 3=top, 4=right).
line	on which MARgin line, starting at 0 counting outwards.
outer	use outer margins if available.
at	give location in user-coordinates. If length(at)==0 (the default), the location will be determined by adj.
adj	adjustment for each string. For strings parallel to the axes, adj=0 means left or bottom alignment, and adj=1 means right or top alignment. If adj is not a finite value (the default), the value par("las") determines the adjustment. For strings plotted parallel to the axis the default is to centre the string.
...	Further graphical parameters (see text and par) ; currently supported are:
cex	character expansion factor (default = 1).
col	color to use.
font	font for text.
vfont	vector font for text.

Details

The "user coordinates" in the outer margins always range from zero to one, and are not affected by the user coordinates in the figure region(s) — R is differing here from other implementations of S.

The arguments `side`, `line`, `at`, `at`, `adj`, the further graphical parameters and even `outer` can be vectors, and recycling will take place to plot as many strings as the longest of the vector arguments. Note that a vector `adj` has a different meaning from `text`.

`adj = 0.5` will centre the string, but for `outer=TRUE` on the device region rather than the plot region.

Parameter `las` will determine the orientation of the string(s). For strings plotted perpendicular to the axis the default justification is to place the end of the string nearest the axis on the specified line.

Note that if the text is to be plotted perpendicular to the axis, `adj` determines the justification of the string *and* the position along the axis unless `at` is specified.

Side Effects

The given text is written onto the current plot.

References

Becker, R. A., Chambers, J. M. and Wilks, A. R. (1988) *The New S Language*. Wadsworth & Brooks/Cole.

See Also

`title`, `text`, `plot`, `par`; `plotmath` for details on mathematical annotation.

Examples

```
plot(1:10, (-4:5)^2, main="Parabola Points", xlab="xlab")
mtext("10 of them")
for(s in 1:4)
   mtext(paste("mtext(..., line= -1, {side, col, font} = ",
        s, ", cex = ", (1+s)/2, ")"), line = -1,
        side=s, col=s, font=s, cex= (1+s)/2)
mtext("mtext(..., line= -2)", line = -2)
mtext("mtext(..., line= -2, adj = 0)", line = -2, adj =0)
## log axis :
```

```
plot(1:10, exp(1:10), log='y', main="log='y'", xlab="xlab")
for(s in 1:4) mtext(paste("mtext(...,side=",s,")"), side=s)
```

n2mfrow *Compute Default mfrow From Number of Plots*

Description

Easy setup for plotting multiple figures (in a rectangular layout) on one page. This computes a sensible default for **par(mfrow)**.

Usage

```
n2mfrow(nr.plots)
```

Arguments

nr.plots   integer; the number of plot figures you'll want to draw.

Value

A length two integer vector **nr**, **nc** giving the number of rows and columns, fulfilling **nr >= nc >= 1** and **nr * nc >= nr.plots**.

Author(s)

Martin Maechler

See Also

par, layout.

Examples

```
n2mfrow(8) # 3 x 3

n <- 5 ; x <- seq(-2,2, len=51)
## suppose now that 'n' is not known {inside function}
op <- par(mfrow = n2mfrow(n))
for (j in 1:n)
  plot(x, x^j, main = substitute(x^ exp, list(exp = j)),
       type='l', col="blue")

sapply(1:10, n2mfrow)
```

pairs   *Scatterplot Matrices*

Description

A matrix of scatterplots is produced.

Usage

```
pairs(x, ...)

## S3 method for class 'formula':
pairs(formula, data = NULL, ..., subset)

## Default S3 method:
pairs(x, labels, panel = points, ...,
      lower.panel = panel, upper.panel = panel,
      diag.panel = NULL, text.panel = textPanel,
      label.pos = 0.5 + has.diag/3,
      cex.labels = NULL, font.labels = 1,
      row1attop = TRUE, gap = 1)
```

Arguments

x	the coordinates of points given as columns of a matrix.
formula	a formula, such as y ~ x.
data	a data.frame (or list) from which the variables in formula should be taken.
subset	an optional vector specifying a subset of observations to be used for plotting.
labels	the names of the variables.
panel	function(x,y,...) which is used to plot the contents of each panel of the display.
...	graphical parameters can be given as arguments to plot.
lower.panel, upper.panel	
	separate panel functions to be used below and above the diagonal respectively.
diag.panel	optional function(x, ...) to be applied on the diagonals.

text.panel	optional function(x, y, labels, cex, font, ...) to be applied on the diagonals.
label.pos	y position of labels in the text panel.
cex.labels, font.labels	
	graphics parameters for the text panel.
row1attop	logical. Should the layout be matrix-like with row 1 at the top, or graph-like with row 1 at the bottom?
gap	Distance between subplots, in margin lines.

Details

The ijth scatterplot contains x[,i] plotted against x[,j]. The "scatterplot" can be customised by setting panel functions to appear as something completely different. The off-diagonal panel functions are passed the appropriate columns of x as x and y: the diagonal panel function (if any) is passed a single column, and the text.panel function is passed a single (x, y) location and the column name.

The graphical parameters pch and col can be used to specify a vector of plotting symbols and colors to be used in the plots.

The graphical parameter oma will be set by pairs.default unless supplied as an argument.

A panel function should not attempt to start a new plot, but just plot within a given coordinate system: thus plot and boxplot are not panel functions.

Author(s)

Enhancements for R 1.0.0 contributed by Dr. Jens Oehlschlaegel-Akiyoshi and R-core members.

References

Becker, R. A., Chambers, J. M. and Wilks, A. R. (1988) *The New S Language*. Wadsworth & Brooks/Cole.

Examples

```
data(iris)
pairs(iris[1:4], main = "Anderson's Iris Data, 3 species",
  pch = 21,
  bg = c("red", "green3", "blue")[unclass(iris$Species)])
```

```
## formula method
data(swiss)
pairs(~ Fertility + Education + Catholic, data = swiss,
   subset = Education < 20, main = "Swiss data,
   Education < 20")

data(USJudgeRatings)
pairs(USJudgeRatings)

## put histograms on the diagonal
panel.hist <- function(x, ...)
{
    usr <- par("usr"); on.exit(par(usr))
    par(usr = c(usr[1:2], 0, 1.5) )
    h <- hist(x, plot = FALSE)
    breaks <- h$breaks; nB <- length(breaks)
    y <- h$counts; y <- y/max(y)
    rect(breaks[-nB], 0, breaks[-1], y, col="cyan", ...)
}
pairs(USJudgeRatings[1:5], panel=panel.smooth,
      cex = 1.5, pch = 24, bg="light blue",
      diag.panel=panel.hist, cex.labels = 2, font.labels=2)

## put (absolute) correlations on the upper panels, with
## size proportional to the correlations.
panel.cor <- function(x, y, digits=2, prefix="", cex.cor)
{
    usr <- par("usr"); on.exit(par(usr))
    par(usr = c(0, 1, 0, 1))
    r <- abs(cor(x, y))
    txt <- format(c(r, 0.123456789), digits=digits)[1]
    txt <- paste(prefix, txt, sep="")
    if(missing(cex.cor)) cex <- 0.8/strwidth(txt)
    text(0.5, 0.5, txt, cex = cex * r)
}
pairs(USJudgeRatings,
      lower.panel=panel.smooth, upper.panel=panel.cor)
```

base — palette   123

palette   *Set or View the Graphics Palette*

Description

View or manipulate the color palette which is used when a col= has a numeric index.

Usage

```
palette(value)
```

Arguments

value   an optional character vector.

Details

If value has length 1, it is taken to be the name of a built in color palette. If value has length greater than 1 it is assumed to contain a description of the colors which are to make up the new palette (either by name or by RGB levels).

If value is omitted or has length 0, no change is made the current palette.

Currently, the only built-in palette is "default".

Value

The palette which *was* in effect. This is invisible unless the argument is omitted.

See Also

colors for the vector of built-in "named" colors; hsv, gray, rainbow, terrain.colors,... to construct colors;

col2rgb for translating colors to RGB 3-vectors.

Examples

```
palette()                # obtain the current palette
palette(rainbow(6))      # six color rainbow

# gray scales; print old palette
```

```
 (palette(gray(seq(0,.9,len=25))))
 matplot(outer(1:100,1:30), type='l', lty=1,lwd=2, col=1:30,
         main = "Gray Scales Palette",
         sub = "palette(gray(seq(0,.9,len=25)))")
 palette("default")      # reset back to the default
```

Palettes *Color Palettes*

Description

Create a vector of n "contiguous" colors.

Usage

```
rainbow(n, s = 1, v = 1, start = 0, end = max(1,n - 1)/n,
        gamma = 1)
heat.colors(n)
terrain.colors(n)
topo.colors(n)
cm.colors(n)
```

Arguments

n	the number of colors (≥ 1) to be in the palette.
s,v	the "saturation" and "value" to be used to complete the HSV color descriptions.
start	the (corrected) hue in [0,1] at which the rainbow begins.
end	the (corrected) hue in [0,1] at which the rainbow ends.
gamma	the gamma correction, see argument `gamma` in `hsv`.

Details

Conceptually, all of these functions actually use (parts of) a line cut out of the 3-dimensional color space, parametrized by `hsv(h,s,v, gamma)`, where gamma= 1 for the *foo*`.colors` function, and hence, equispaced hues in RGB space tend to cluster at the red, green and blue primaries.

Some applications such as contouring require a palette of colors which do not "wrap around" to give a final color close to the starting one.

With `rainbow`, the parameters `start` and `end` can be used to specify particular subranges of hues. The following values can be used when generating such a subrange: red=0, yellow=$\frac{1}{6}$, green=$\frac{2}{6}$, cyan=$\frac{3}{6}$, blue=$\frac{4}{6}$ and magenta=$\frac{5}{6}$.

Value

A character vector, cv, of color names. This can be used either to create a user–defined color palette for subsequent graphics by palette(cv), a col= specification in graphics functions or in par.

See Also

colors, palette, hsv, rgb, gray and col2rgb for translating to RGB numbers.

Examples

```
# A Color Wheel
pie(rep(1,12), col=rainbow(12))

# Some palettes
demo.pal <-
  function(n, border = if (n<32) "light gray" else NA,
           main = paste("color palettes;  n=", n),
           ch.col = c("rainbow(n, start=.7, end=.1)",
                      "heat.colors(n)", "terrain.colors(n)",
                      "topo.colors(n)", "cm.colors(n)"))
{
  nt <- length(ch.col)
  i <- 1:n; j <- n / nt; d <- j/6; dy <- 2*d
  plot(i,i+d, type="n", yaxt="n", ylab="", main=main)
  for (k in 1:nt) {
    rect(i-.5, (k-1)*j+ dy, i+.4, k*j,
    col = eval(parse(text=ch.col[k])), border = border)
    text(2*j,  k * j +dy/4, ch.col[k])
  }
}
n <- if(.Device == "postscript") 64 else 16
# For screen, larger n may give color allocation problem
demo.pal(n)
```

panel.smooth   *Simple Panel Plot*

Description

An example of a simple useful `panel` function to be used as argument in e.g., `coplot` or `pairs`.

Usage

```
panel.smooth(x, y, col = par("col"), bg = NA,
  pch = par("pch"), cex = 1, col.smooth = "red",
  span = 2/3, iter=3, ...)
```

Arguments

`x,y`	numeric vectors of the same length
`col,bg,pch,cex`	numeric or character codes for the color(s), point type and size of `points`; see also `par`.
`col.smooth`	color to be used by `lines` for drawing the smooths.
`span`	smoothing parameter `f` for `lowess`, see there.
`iter`	number of robustness iterations for `lowess`.
`...`	further arguments to `lines`.

See Also

`coplot` and `pairs` where `panel.smooth` is typically used; `lowess`.

Examples

```
data(swiss)
# emphasize the smooths
pairs(swiss, panel = panel.smooth, pch = ".")
pairs(swiss, panel = panel.smooth, lwd = 2, cex= 1.5,
      col="blue")
```

par	Set or Query Graphical Parameters

Description

par can be used to set or query graphical parameters. Parameters can be set by specifying them as arguments to par in tag = value form, or by passing them as a list of tagged values.

Usage

```
par(..., no.readonly = FALSE)

<highlevel plot> (..., <tag> = <value>)
```

Arguments

... arguments in tag = value form, or a list of tagged values. The tags must come from the graphical parameters described below.

no.readonly logical; if TRUE and there are no other arguments, only parameters are returned which can be set by a subsequent par() call.

Details

Parameters are queried by giving one or more character vectors to par.

par() (no arguments) or par(no.readonly=TRUE) is used to get *all* the graphical parameters (as a named list). Their names are currently taken from the variable .Pars. .Pars.readonly contains the names of the par arguments which are *readonly*.

R.O. indicates *read-only arguments*: These may only be used in queries, i.e., they do *not* set anything.

All but these *R.O.* and the following *low-level arguments* can be set as well in high-level and mid-level plot functions, such as plot, points, lines, axis, title, text, mtext:

- "ask"
- "fig", "fin"
- "mai", "mar", "mex"
- "mfrow", "mfcol", "mfg"

- "new"
- "oma", "omd", "omi"
- "pin", "plt", "ps", "pty"
- "usr"
- "xlog", "ylog"

Value

When parameters are set, their former values are returned in an invisible named list. Such a list can be passed as an argument to `par` to restore the parameter values. Use `par(no.readonly = TRUE)` for the full list of parameters that can be restored.

When just one parameter is queried, the value is a character string. When two or more parameters are queried, the result is a list of character strings, with the list names giving the parameters.

Note the inconsistency: setting one parameter returns a list, but querying one parameter returns a vector.

Graphical Parameters

- **adj** The value of `adj` determines the way in which text strings are justified. A value of 0 produces left-justified text, 0.5 centered text and 1 right-justified text. (Any value in $[0, 1]$ is allowed, and on most devices values outside that interval will also work.) Note that the `adj` argument of `text` also allows `adj = c(x, y)` for different adjustment in x- and y- direction.
- **ann** If set to `FALSE`, high-level plotting functions do not annotate the plots they produce with axis and overall titles. The default is to do annotation.
- **ask** logical. If `TRUE`, the user is asked for input, before a new figure is drawn.
- **bg** The color to be used for the background of plots. A description of how colors are specified is given below.
- **bty** A character string which determined the type of box which is drawn about plots. If `bty` is one of `"o"`, `"l"`, `"7"`, `"c"`, `"u"`, or `"]"` the resulting box resembles the corresponding upper case letter. A value of `"n"` suppresses the box.
- **cex** A numerical value giving the amount by which plotting text and symbols should be scaled relative to the default.

cex.axis The magnification to be used for axis annotation relative to the current.

cex.lab The magnification to be used for x and y labels relative to the current.

cex.main The magnification to be used for main titles relative to the current.

cex.sub The magnification to be used for sub-titles relative to the current.

cin *R.O.*; character size (width,height) in inches.

col A specification for the default plotting color. A description of how colors are specified is given below.

col.axis The color to be used for axis annotation.

col.lab The color to be used for x and y labels.

col.main The color to be used for plot main titles.

col.sub The color to be used for plot sub-titles.

cra *R.O.*; size of default character (width,height) in "rasters" (pixels).

crt A numerical value specifying (in degrees) how single characters should be rotated. It is unwise to expect values other than multiples of 90 to work. Compare with srt which does string rotation.

csi *R.O.*; height of (default sized) characters in inches.

cxy *R.O.*; size of default character (width,height) in user coordinate units. par("cxy") is par("cin")/par("pin") scaled to user coordinates. Note that c(strwidth(ch), strwidth(ch)) for a given string ch is usually much more precise.

din *R.O.*; the device dimensions in inches.

err (*Unimplemented*; R is silent when points outside the plot region are *not* plotted.) The degree of error reporting desired.

fg The color to be used for the foreground of plots. This is the default color is used for things like axes and boxes around plots. A description of how colors are specified is given below.

fig A numerical vector of the form c(x1, x2, y1, y2) which gives the (NDC) coordinates of the figure region in the display region of the device.

fin A numerical vector of the form c(x, y) which gives the size of the figure region in inches.

font An integer which specifies which font to use for text. If possible, device drivers arrange so that 1 corresponds to plain text, 2 to bold face, 3 to italic and 4 to bold italic.

font.axis The font to be used for axis annotation.

font.lab The font to be used for x and y labels.

font.main The font to be used for plot main titles.

font.sub The font to be used for plot sub-titles.

gamma the gamma correction, see argument `gamma` to `hsv`.

lab A numerical vector of the form `c(x, y, len)` which modifies the way that axes are annotated. The values of `x` and `y` give the (approximate) number of tickmarks on the x and y axes and `len` specifies the label size. The default is `c(5, 5, 7)`. *Currently*, `len` *is unimplemented.*

las numeric in {0,1,2,3}; the style of axis labels.

    0: always parallel to the axis [*default*],

    1: always horizontal,

    2: always perpendicular to the axis,

    3: always vertical.

Note that other string/character rotation (via argument `srt` to `par`) does *not* affect the axis labels.

lty The line type. Line types can either be specified as an integer (0=blank, 1=solid, 2=dashed, 3=dotted, 4=dotdash, 5=longdash, 6=twodash) or as one of the character strings "blank", "solid", "dashed", "dotted", "dotdash", "longdash", or "twodash", where "blank" uses 'invisible lines' (i.e., doesn't draw them).

Alternatively, a string of up to 8 characters (from `c(1:9, "A":"F")`) may be given, giving the length of line segments which are alternatively drawn and skipped. See section 'Line Type Specification' below.

lwd The line width, a *positive* number, defaulting to 1.

mai A numerical vector of the form `c(bottom, left, top, right)` which gives the margin size specified in inches.

mar A numerical vector of the form `c(bottom, left, top, right)` which gives the lines of margin to be specified on the four sides of the plot. The default is `c(5, 4, 4, 2) + 0.1`.

mex mex is a character size expansion factor which is used to describe coordinates in the margins of plots.

mfcol, mfrow A vector of the form c(nr, nc). Subsequent figures will be drawn in an nr-by-nc array on the device by *columns* (mfcol), or *rows* (mfrow), respectively.

Consider the alternatives, layout and split.screen.

mfg A numerical vector of the form c(i, j) where i and j indicate which figure in an array of figures is to be drawn next (if setting) or is being drawn (if enquiring). The array must already have been set by mfcol or mfrow.

For compatibility with S, the form c(i, j, nr, nc) is also accepted, when nr and nc should be the current number of rows and number of columns. Mismatches will be ignored, with a warning.

mgp The margin line (in mex units) for the axis title, axis labels and axis line. The default is c(3, 1, 0).

mkh The height in inches of symbols to be drawn when the value of pch is an integer. *Completely ignored currently.*

new logical, defaulting to FALSE. If set to TRUE, the next high-level plotting command (actually plot.new) should *not clean* the frame before drawing "as if it was on a *new* device".

oma A vector of the form c(bottom, left, top, right) giving the size of the outer margins in lines of text.

omd A vector of the form c(x1, x2, y1, y2) giving the outer margin region in NDC (= normalized device coordinates), i.e., as fraction (in $[0, 1]$) of the device region.

omi A vector of the form c(bottom, left, top, right) giving the size of the outer margins in inches.

pch Either an integer specifying a symbol or a single character to be used as the default in plotting points.

pin The width and height of the current plot in inches.

plt A vector of the form c(x1, x2, y1, y2) giving the coordinates of the plot region as fractions of the current figure region.

ps integer; the pointsize of text and symbols.

pty A character specifying the type of plot region to be used; "s" generates a square plotting region and "m" generates the maximal plotting region.

smo (*Unimplemented*) a value which indicates how smooth circles and circular arcs should be.

srt The string rotation in degrees. See the comment about crt.

tck The length of tick marks as a fraction of the smaller of the width or height of the plotting region. If tck >= 0.5 it is interpreted as a fraction of the relevant side, so if tck=1 grid lines are drawn. The default setting (tck = NA) is to use tcl = -0.5 (see below).

tcl The length of tick marks as a fraction of the height of a line of text. The default value is -0.5; setting tcl = NA sets tck = -0.01 which is S' default.

tmag A number specifying the enlargement of text of the main title relative to the other annotating text of the plot.

type character; the default plot type desired, see plot.default(type=...), defaulting to "p".

usr A vector of the form c(x1, x2, y1, y2) giving the extremes of the user coordinates of the plotting region. When a logarithmic scale is in use (i.e., par("xlog") is true, see below), then the x-limits will be 10 ^ par("usr")[1:2]. Similarly for the y-axis.

xaxp A vector of the form c(x1, x2, n) giving the coordinates of the extreme tick marks and the number of intervals between tick-marks when par("xlog") is false. Otherwise, when *log* coordinates are active, the three values have a different meaning: For a small range, n is *negative*, and the ticks are as in the linear case, otherwise, n is in 1:3, specifying a case number, and x1 and x2 are the lowest and highest power of 10 inside the user coordinates, par("usr")[1:2]. See axTicks() for more details.

xaxs The style of axis interval calculation to be used for the x-axis. Possible values are "r", "i", "e", "s", "d". The styles are generally controlled by the range of data or xlim, if given. Style "r" (regular) first extends the data range by 4 percent and then finds an axis with pretty labels that fits within the range. Style "i" (internal) just finds an axis with pretty labels that fits within the original data range. Style "s" (standard) finds an axis with pretty labels within which the original data range fits. Style "e" (extended) is like style "s", except that it is also ensured that there is room for plotting symbols within the bounding box. Style "d" (direct) specifies that the current axis should be used on subsequent plots. (*Only "r" and "i" styles are currently implemented*)

xaxt A character which specifies the axis type. Specifying "n" causes an axis to be set up, but not plotted. The standard value is "s": for compatibility with S values "l" and "e" are accepted but are equivalent to "s".

xlog logical value (see log in plot.default). If TRUE, a logarithmic scale is in use (e.g., after plot(*, log = "x")). For a new device, it defaults to FALSE, i.e., linear scale.

xpd A logical value or NA. If FALSE, all plotting is clipped to the plot region, if TRUE, all plotting is clipped to the figure region, and if NA, all plotting is clipped to the device region.

yaxp A vector of the form c(y1, y2, n) giving the coordinates of the extreme tick marks and the number of intervals between tick-marks unless for log coordinates, see xaxp above.

yaxs The style of axis interval calculation to be used for the y-axis. See xaxs above.

yaxt A character which specifies the axis type. Specifying "n" causes an axis to be set up, but not plotted.

ylog a logical value; see xlog above.

Color Specification

Colors can be specified in several different ways. The simplest way is with a character string giving the color name (e.g., "red"). A list of the possible colors can be obtained with the function colors. Alternatively, colors can be specified directly in terms of their RGB components with a string of the form "#RRGGBB" where each of the pairs RR, GG, BB consist of two hexadecimal digits giving a value in the range 00 to FF. Colors can also be specified by giving an index into a small table of colors, the palette. This provides compatibility with S. Index 0 corresponds to the background color.

Additionally, "transparent" or (integer) NA is *transparent*, useful for filled areas (such as the background!), and just invisible for things like lines or text.

The functions rgb, hsv, gray and rainbow provide additional ways of generating colors.

Line Type Specification

Line types can either be specified by giving an index into a small built in table of line types (1 = solid, 2 = dashed, etc, see lty above) or directly as the lengths of on/off stretches of line. This is done with a string of an even number (up to eight) of characters, namely non-zero (hexadecimal) digits which give the lengths in consecutive positions in the string. For example, the string "33" specifies three units on followed by three off and "3313" specifies three units on followed by three off followed by one on and finally three off. The 'units' here are (on most devices) proportional to lwd, and with lwd = 1 are in pixels or points.

The five standard dash-dot line types (`lty = 2:6`) correspond to
`c("44", "13", "1343", "73", "2262")`.

Note that `NA` is not a valid value for `lty`.

Note

The effect of restoring all the (settable) graphics parameters as in the examples is hard to predict if the device has been resized. Several of them are attempting to set the same things in different ways, and those last in the alphabet will win. In particular, the settings of `mai`, `mar`, `pin`, `plt` and `pty` interact, as do the outer margin settings, the figure layout and figure region size.

References

Becker, R. A., Chambers, J. M. and Wilks, A. R. (1988) *The New S Language.* Wadsworth & Brooks/Cole.

See Also

`plot.default` for some high-level plotting parameters; `colors`, `gray`, `rainbow`, `rgb`; `options` for other setup parameters; graphic devices `x11`, `postscript` and setting up device regions by `layout` and `split.screen`.

Examples

```
op <- par(mfrow = c(2, 2), # 2 x 2 pictures on one plot
          pty = "s")       # square plotting region,
                           # independent of device size

## At end of plotting, reset to previous settings:
par(op)

## Alternatively,
# the whole list of settable par's.
op <- par(no.readonly = TRUE)
## do lots of plotting and par(.) calls, then reset:
par(op)

par("ylog") # FALSE
plot(1 : 12, log = "y")
par("ylog") # TRUE
```

```
plot(1:2, xaxs = "i") # 'inner axis' w/o extra space
stopifnot(par("xaxp")[1:2] == 1:2 &&
          par("usr") [1:2] == 1:2)

( nr.prof <-
  c(prof.pilots=16,lawyers=11,farmers=10,salesmen=9,
    physicians=9,mechanics=6,policemen=6,managers=6,
    engineers=5,teachers=4,housewives=3,students=3,
    armed.forces=1))
par(las = 3)
barplot(rbind(nr.prof)) # R 0.63.2: shows alignment problem
par(las = 0) # reset to default

## 'fg' use:
plot(1:12, type = "b",
     main="'fg' : axes, ticks and box in gray",
     fg = gray(0.7), bty="7" , sub=R.version.string)

ex <- function() {
   # all par settings which could be changed.
   old.par <- par(no.readonly = TRUE)

   on.exit(par(old.par))
   ## ...
   ## ... do lots of par() settings and plots
   ## ...
   invisible() # now, par(old.par) will be executed
}
ex()
```

pdf *PDF Graphics Device*

Description

pdf starts the graphics device driver for producing PDF graphics.

Usage

```
pdf(file = ifelse(onefile, "Rplots.pdf", "Rplot%03d.pdf"),
    width=6, height=6, onefile = TRUE, family = "Helvetica",
    title = "R Graphics Output", encoding, bg, fg, pointsize)
```

Arguments

file	a character string giving the name of the file.
width, height	the width and height of the graphics region in inches.
onefile	logical: if true (the default) allow multiple figures in one file. If false, generate a file name containing the page number.
family	the font family to be used, one of "AvantGarde", "Bookman", "Courier", "Helvetica", "Helvetica-Narrow", "NewCenturySchoolbook", "Palatino" or "Times".
title	title string to embed in the file.
encoding	the name of an encoding file. Defaults to "ISOLatin1.enc" in the 'R_HOME/afm' directory, which is used if the path does not contain a path separator. An extension ".enc" can be omitted.
pointsize	the default point size to be used.
bg	the default background color to be used.
fg	the default foreground color to be used.

Details

pdf() opens the file file and the PDF commands needed to plot any graphics requested are sent to that file.

See postscript for details of encodings, as the internal code is shared between the drivers. The native PDF encoding is given in file 'PDFDoc.enc'.

pdf writes uncompressed PDF. It is primarily intended for producing PDF graphics for inclusion in other documents, and PDF-includers such as pdftex are usually able to handle compression.

At present the PDF is fairly simple, with each page being represented as a single stream. The R graphics model does not distinguish graphics objects at the level of the driver interface.

Note

Acrobat Reader does not use the fonts specified but rather emulates them from multiple-master fonts. This can be seen in imprecise centering of characters, for example the multiply and divide signs in Helvetica.

See Also

`Devices, postscript`

Examples

```
## Test function for encodings
TestChars <- function(encoding="ISOLatin1")
{
    pdf(encoding=encoding)
    par(pty="s")
    plot(c(0,15), c(0,15), type="n", xlab="", ylab="")
    title(paste("Centred chars in encoding", encoding))
    grid(15, 15, lty=1)
    for(i in c(32:255)) {
        x <- i
        y <- i
        points(x, y, pch=i)
    }
    dev.off()
}
## there will be many warnings.
TestChars("ISOLatin2")
## doesn't view properly in US-spec Acrobat 5.05, but
## gs7.04 works. Lots of characters are not centred.
```

persp *Perspective Plots*

Description

This function draws perspective plots of surfaces over the x–y plane.
persp is a generic function.

Usage

```
persp(x, ...)

## Default S3 method:
persp(x = seq(0, 1, len = nrow(z)),
   y = seq(0, 1, len = ncol(z)), z, xlim = range(x),
   ylim = range(y), zlim = range(z, na.rm = TRUE),
   xlab = NULL, ylab = NULL, zlab = NULL, main = NULL,
   sub = NULL, theta = 0, phi = 15, r = sqrt(3), d = 1,
   scale = TRUE, expand = 1, col = "white", border = NULL,
   ltheta = -135, lphi = 0, shade = NA, box = TRUE,
   axes = TRUE, nticks = 5, ticktype = "simple", ...)
```

Arguments

x, y	locations of grid lines at which the values in z are measured. These must be in ascending order. By default, equally spaced values from 0 to 1 are used. If x is a list, its components x$x and x$y are used for x and y, respectively.
z	a matrix containing the values to be plotted (NAs are allowed). Note that x can be used instead of z for convenience.
xlim, ylim, zlim	x-, y- and z-limits. The plot is produced so that the rectangular volume defined by these limits is visible.
xlab, ylab, zlab	titles for the axes. N.B. These must be character strings; expressions are not accepted. Numbers will be coerced to character strings.
main, sub	main and sub title, as for title.
theta, phi	angles defining the viewing direction. theta gives the azimuthal direction and phi the colatitude.

r	the distance of the eyepoint from the centre of the plotting box.
d	a value which can be used to vary the strength of the perspective transformation. Values of d greater than 1 will lessen the perspective effect and values less and 1 will exaggerate it.
scale	before viewing the x, y and z coordinates of the points defining the surface are transformed to the interval [0,1]. If scale is TRUE the x, y and z coordinates are transformed separately. If scale is FALSE the coordinates are scaled so that aspect ratios are retained. This is useful for rendering things like DEM information.
expand	an expansion factor applied to the z coordinates. Often used with 0 < expand < 1 to shrink the plotting box in the z direction.
col	the color(s) of the surface facets. Transparent colours are ignored. This is recycled to the $(nx-1)(ny-1)$ facets.
border	the color of the line drawn around the surface facets. A value of NA will disable the drawing of borders. This is sometimes useful when the surface is shaded.
ltheta, lphi	if finite values are specified for ltheta and lphi, the surface is shaded as though it was being illuminated from the direction specified by azimuth ltheta and colatitude lphi.
shade	the shade at a surface facet is computed as ((1+d)/2)^shade, where d is the dot product of a unit vector normal to the facet and a unit vector in the direction of a light source. Values of shade close to one yield shading similar to a point light source model and values close to zero produce no shading. Values in the range 0.5 to 0.75 provide an approximation to daylight illumination.
box	should the bounding box for the surface be displayed. The default is TRUE.
axes	should ticks and labels be added to the box. The default is TRUE. If box is FALSE then no ticks or labels are drawn.

ticktype   character: "simple" draws just an arrow parallel to
                 the axis to indicate direction of increase; "detailed"
                 draws normal ticks as per 2D plots.

nticks   the (approximate) number of tick marks to draw on
                 the axes. Has no effect if ticktype is "simple".

...   additional graphical parameters (see par).

Details

The plots are produced by first transforming the coordinates to the
interval [0,1]. The surface is then viewed by looking at the origin from
a direction defined by theta and phi. If theta and phi are both zero
the viewing direction is directly down the negative y axis. Changing
theta will vary the azimuth and changing phi the colatitude.

Value

The *viewing transformation matrix*, say VT, a 4×4 matrix suitable for
projecting 3D coordinates (x, y, z) into the 2D plane using homogenous
4D coordinates (x, y, z, t). It can be used to superimpose additional
graphical elements on the 3D plot, by lines() or points(), e.g. using
the function trans3d given in the last examples section below.

References

Becker, R. A., Chambers, J. M. and Wilks, A. R. (1988) *The New S
Language.* Wadsworth & Brooks/Cole.

See Also

contour and image.

Examples

```
## More examples in demo(persp)
# (1) The Obligatory Mathematical surface.
#     Rotated sinc function.
x <- seq(-10, 10, length= 30)
y <- x
f <- function(x,y) { r <- sqrt(x^2+y^2); 10 * sin(r)/r }
z <- outer(x, y, f)
z[is.na(z)] <- 1
op <- par(bg = "white")
persp(x, y, z, theta = 30, phi = 30, expand = 0.5,
```

```
        col = "lightblue")
persp(x, y, z, theta = 30, phi = 30, expand = 0.5,
      col = "lightblue", ltheta = 120, shade = 0.75,
      ticktype = "detailed", xlab = "X", ylab = "Y",
      zlab = "Sinc( r )") -> res
round(res, 3)

# (2) Add to existing persp plot :

trans3d <- function(x,y,z, pmat) {
  tr <- cbind(x,y,z,1) %*% pmat
  list(x = tr[,1]/tr[,4], y= tr[,2]/tr[,4])
}
xE <- c(-10,10); xy <- expand.grid(xE, xE)
points(trans3d(xy[,1], xy[,2], 6, pm=res), col=2, pch=16)
lines (trans3d(x, y=10, z= 6 + sin(x), pm = res), col = 3)

phi <- seq(0, 2*pi, len = 201)
r1 <- 7.725 # radius of 2nd maximum
xr <- r1 * cos(phi)
yr <- r1 * sin(phi)
## (no hidden lines)
lines(trans3d(xr,yr, f(xr,yr), res), col = "pink", lwd=2)

# (3) Visualizing a simple DEM model

data(volcano)
z <- 2 * volcano        # Exaggerate the relief
x <- 10 * (1:nrow(z))   # 10 meter spacing (S to N)
y <- 10 * (1:ncol(z))   # 10 meter spacing (E to W)
## Don't draw the grid lines : border = NA
par(bg = "slategray")
persp(x, y, z, theta = 135, phi = 30, col = "green3",
   scale = FALSE, ltheta = -120, shade = 0.75, border = NA,
   box = FALSE)
par(op)
```

pictex *A PicTeX Graphics Driver*

Description

This function produces graphics suitable for inclusion in TeX and LaTeX documents.

Usage

```
pictex(file = "Rplots.tex", width = 5, height = 4,
  debug = FALSE, bg = "white", fg = "black")
```

Arguments

file	the file where output will appear.
width	The width of the plot in inches.
height	the height of the plot in inches.
debug	should debugging information be printed.
bg	the background color for the plot.
fg	the foreground color for the plot.

Details

This driver does not have any font metric information, so the use of `plotmath` is not supported.

Multiple plots will be placed as separate environments in the output file.

Author(s)

This driver was provided by Valerio Aimale of the Department of Internal Medicine, University of Genoa, Italy.

References

Knuth, D. E. (1984) *The TeXbook.* Reading, MA: Addison-Wesley.

Lamport, L. (1994) *LATEX: A Document Preparation System.* Reading, MA: Addison-Wesley.

Goossens, M., Mittelbach, F. and Samarin, A. (1994) *The LATEX Companion.* Reading, MA: Addison-Wesley.

See Also

postscript, Devices.

Examples

```
pictex()
plot(1:11,(-5:5)^2, type='b', main="Simple Example Plot")
dev.off()
##

%% LaTeX Example
\documentclass{article}
\usepackage{pictex}
\begin{document}
%...
\begin{figure}[h]
  \centerline{\input{Rplots.tex}}
  \caption{}
\end{figure}
%...
\end{document}

%%-- TeX Example --
\input pictex
$$ \input Rplots.tex $$

##
unlink("Rplots.tex")
```

pie *Pie Charts*

Description

Draw a pie chart.

Usage

```
pie(x, labels = names(x), edges = 200, radius = 0.8,
    density = NULL, angle = 45, col = NULL, border = NULL,
    lty = NULL, main = NULL, ...)
```

Arguments

x	a vector of positive quantities. The values in x are displayed as the areas of pie slices.
labels	a vector of character strings giving names for the slices. For empty or NA labels, no pointing line is drawn either.
edges	the circular outline of the pie is approximated by a polygon with this many edges.
radius	the pie is drawn centered in a square box whose sides range from −1 to 1. If the character strings labeling the slices are long it may be necessary to use a smaller radius.
density	the density of shading lines, in lines per inch. The default value of NULL means that no shading lines are drawn. Non-positive values of density also inhibit the drawing of shading lines.
angle	the slope of shading lines, given as an angle in degrees (counter-clockwise).
col	a vector of colors to be used in filling or shading the slices. If missing a set of 6 pastel colours is used, unless density is specified when par("fg") is used.
border, lty	(possibly vectors) arguments passed to polygon which draws each slice.
main	an overall title for the plot.
...	graphical parameters can be given as arguments to pie. They will affect the main title and labels only.

Note

Pie charts are a very bad way of displaying information. The eye is good at judging linear measures and bad at judging relative areas. A bar chart or dot chart is a preferable way of displaying this type of data.

Cleveland (1985), page 264: "Data that can be shown by pie charts always can be shown by a dot chart. This means that judgements of position along a common scale can be made instead of the less accurate angle judgements." This statement is based on the empirical investigations of Cleveland and McGill as well as investigations by perceptual psychologists.

Prior to R 1.5.0 this was known as `piechart`, which is the name of a Trellis function, so the name was changed to be compatible with S.

References

Becker, R. A., Chambers, J. M. and Wilks, A. R. (1988) *The New S Language.* Wadsworth & Brooks/Cole.

Cleveland, W. S. (1985) *The elements of graphing data.* Wadsworth: Monterey, CA, USA.

See Also

dotchart.

Examples

```
pie(rep(1, 24), col = rainbow(24), radius = 0.9)

pie.sales <- c(0.12, 0.3, 0.26, 0.16, 0.04, 0.12)
names(pie.sales) <- c("Blueberry", "Cherry",
    "Apple", "Boston Cream", "Other", "Vanilla Cream")
pie(pie.sales) # default colours
pie(pie.sales,
   col = c("purple", "violetred1", "green3", "cornsilk",
           "cyan", "white"))
pie(pie.sales, col = gray(seq(0.4,1.0,length=6)))
pie(pie.sales, density = 10, angle = 15 + 10 * 1:6)

n <- 200
pie(rep(1,n), labels="", col=rainbow(n), border=NA,
    main = "pie(*, labels=\"\", col=rainbow(n), border=NA,..")
```

plot   *Generic X-Y Plotting*

Description

Generic function for plotting of R objects. For more details about the graphical parameter arguments, see `par`.

Usage

```
plot(x, y, ...)
```

Arguments

x
: the coordinates of points in the plot. Alternatively, a single plotting structure, function or *any R object with a plot method* can be provided.

y
: the y coordinates of points in the plot, *optional* if x is an appropriate structure.

...
: graphical parameters can be given as arguments to `plot`. Many methods will also accept the following arguments:

type
: what type of plot should be drawn. Possible types are
  - "p" for points,
  - "l" for lines,
  - "b" for both,
  - "c" for the lines part alone of "b",
  - "o" for both "overplotted",
  - "h" for "histogram" like (or "high-density") vertical lines,
  - "s" for stair steps,
  - "S" for other steps, see *Details* below,
  - "n" for no plotting.

  All other `types` give a warning or an error; using, e.g., `type = "punkte"` being equivalent to `type = "p"` for S compatibility.

main
: an overall title for the plot: see `title`.

sub
: a sub title for the plot: see `title`.

xlab
: a title for the x axis: see `title`.

ylab
: a title for the y axis: see `title`.

Details

For simple scatter plots, plot.default will be used. However, there are plot methods for many R objects, including functions, data.frames, density objects, etc. Use methods(plot) and the documentation for these.

The two step types differ in their x-y preference: Going from $(x1, y1)$ to $(x2, y2)$ with $x1 < x2$, type = "s" moves first horizontal, then vertical, whereas type = "S" moves the other way around.

See Also

plot.default, plot.formula and other methods; points, lines, par.

Examples

```
data(cars)
plot(cars)
lines(lowess(cars))

plot(sin, -pi, 2*pi)

## Discrete Distribution Plot:
plot(table(rpois(100,5)), type = "h", col = "red", lwd=10,
     main="rpois(100,lambda=5)")

## Simple quantiles/ECDF, see ecdf() {library(stepfun)} for
## a better one:
plot(x <- sort(rnorm(47)), type = "s", main =
   "plot(x, type = \"s\")")
points(x, cex = .5, col = "dark red")
```

plot.data.frame *Plot Method for Data Frames*

Description

`plot.data.frame`, a method of the `plot` generic, uses `stripchart` for *one* variable, `plot.default` (scatterplot) for *two* variables, and `pairs` (scatterplot matrix) otherwise.

Usage

```
## S3 method for class 'data.frame':
plot(x, ...)
```

Arguments

x	object of class `data.frame`.
...	further arguments to `stripchart`, `plot.default` or `pairs`.

See Also

`data.frame`

Examples

```
data(OrchardSprays)
plot(OrchardSprays[1], method="jitter")
plot(OrchardSprays[c(4,1)])
plot(OrchardSprays)
```

plot.default   *The Default Scatterplot Function*

Description

Draw a scatter plot with "decorations" such as axes and titles in the active graphics window.

Usage

```
## Default S3 method:
plot(x, y = NULL, type = "p",   xlim = NULL, ylim = NULL,
   log="", main = NULL, sub = NULL, xlab = NULL, ylab = NULL,
   ann = par("ann"), axes = TRUE, frame.plot = axes,
   panel.first = NULL, panel.last = NULL,
   col = par("col"), bg = NA, pch = par("pch"),
   cex = 1, lty = par("lty"), lab = par("lab"),
   lwd = par("lwd"), asp = NA, ...)
```

Arguments

x,y	the x and y arguments provide the x and y coordinates for the plot. Any reasonable way of defining the coordinates is acceptable. See the function xy.coords for details.
type	1-character string giving the type of plot desired. The following values are possible, for details, see plot: "p" for points, "l" for lines, "o" for overplotted points and lines, "b", "c") for (empty if "c") points joined by lines, "s" and "S" for stair steps and "h" for histogram-like vertical lines. Finally, "n" does not produce any points or lines.
xlim	the x limits (min,max) of the plot.
ylim	the y limits of the plot.
log	a character string which contains "x" if the x axis is to be logarithmic, "y" if the y axis is to be logarithmic and "xy" or "yx" if both axes are to be logarithmic.
main	a main title for the plot.
sub	a sub title for the plot.
xlab	a label for the x axis.

`ylab`	a label for the y axis.
`ann`	a logical value indicating whether the default annotation (title and x and y axis labels) should appear on the plot.
`axes`	a logical value indicating whether axes should be drawn on the plot.
`frame.plot`	a logical indicating whether a box should be drawn around the plot.
`panel.first`	an expression to be evaluated after the plot axes are set up but before any plotting takes place. This can be useful for drawing background grids or scatterplot smooths.
`panel.last`	an expression to be evaluated after plotting has taken place.
`col`	The colors for lines and points. Multiple colors can be specified so that each point can be given its own color. If there are fewer colors than points they are recycled in the standard fashion. Lines will all be plotted in the first colour specified.
`bg`	background color for open plot symbols, see `points`.
`pch`	a vector of plotting characters or symbols: see `points`.
`cex`	a numerical vector giving the amount by which plotting text and symbols should be scaled relative to the default.
`lty`	the line type, see `par`.
`lab`	the specification for the (approximate) numbers of tick marks on the x and y axes.
`lwd`	the line width **not yet supported for postscript**.
`asp`	the y/x aspect ratio, see `plot.window`.
`...`	graphical parameters as in `par` may also be passed as arguments.

References

Becker, R. A., Chambers, J. M. and Wilks, A. R. (1988) *The New S Language.* Wadsworth & Brooks/Cole.

Cleveland, W. S. (1985) *The Elements of Graphing Data.* Monterey, CA: Wadsworth.

See Also

`plot`, `plot.window`, `xy.coords`.

Examples

```
data(cars)
Speed <- cars$speed
Distance <- cars$dist
plot(Speed, Distance, panel.first = grid(8,8),
     pch = 0, cex = 1.2, col = "blue")
plot(Speed, Distance, panel.first =
  lines(lowess(Speed, Distance), lty = "dashed"),
  pch = 0, cex = 1.2, col = "blue")

## Show the different plot types
x <- 0:12
y <- sin(pi/5 * x)
op <- par(mfrow = c(3,3), mar = .1+ c(2,2,3,1))
for (tp in c("p","l","b", "c","o","h", "s","S","n")) {
  plot(y ~ x, type = tp,
       main = paste("plot(*, type = \"",tp,"\")",sep=""))
  if(tp == "S") {
    lines(x,y, type = "s", col = "red", lty = 2)
    mtext("lines(*, type = \"s\", ...)", col="red", cex=.8)
  }
}
par(op)

## Log-Log Plot with custom axes
lx <- seq(1,5, length=41)
yl <- expression(e^{-frac(1,2) * {log[10](x)}^2})
y <- exp(-.5*lx^2)
op <- par(mfrow=c(2,1), mar=par("mar")+c(0,1,0,0))
plot(10^lx, y, log="xy", type="l", col="purple",
     main="Log-Log plot", ylab=yl, xlab="x")
plot(10^lx, y, log="xy", type="o", pch='.',
     col = "forestgreen",
     main = "Log-Log plot with custom axes",
     ylab=yl, xlab="x",
     axes = FALSE, frame.plot = TRUE)
axis(1, at = my.at <- 10^(1:5),
     labels = formatC(my.at, format="fg"))
at.y <- 10^(-5:-1)
```

```
axis(2, at = at.y,
     labels = formatC(at.y, format="fg"), col.axis="red")
par(op)
```

plot.density   *Plot Method for Kernel Density Estimation*

Description

The plot method for density objects.

Usage

```
## S3 method for class 'density':
plot(x, main = NULL, xlab = NULL, ylab = "Density",
     type = "l", zero.line = TRUE, ...)
```

Arguments

x	a "density" object.
main, xlab, ylab, type	
	plotting parameters with useful defaults.
...	further plotting parameters.
zero.line	logical; if TRUE, add a base line at $y = 0$

Value

None.

References

See Also

density.

plot.design   *Plot Univariate Effects of a 'Design' or Model*

Description

Plot univariate effects of one ore more `factors`, typically for a designed experiment as analyzed by `aov()`. Further, in S this a method of the `plot` generic function for `design` objects.

Usage

```
plot.design(x, y = NULL, fun = mean, data = NULL, ...,
   ylim = NULL, xlab = "Factors", ylab = NULL, main = NULL,
   ask = NULL, xaxt = par("xaxt"), axes = TRUE,
   xtick = FALSE)
```

Arguments

x	either a data frame containing the design factors and optionally the response, or a `formula` or `terms` object.
y	the response, if not given in x.
fun	a function (or name of one) to be applied to each subset. It must return one number for a numeric (vector) input.
data	data frame containing the variables referenced by x when that is formula like.
...	graphical arguments such as `col`, see `par`.
ylim	range of y values, as in `plot.default`.
xlab	x axis label, see `title`.
ylab	y axis label with a "smart" default.
main	main title, see `title`.
ask	logical indicating if the user should be asked before a new page is started – in the case of multiple y's.
xaxt	character giving the type of x axis.
axes	logical indicating if axes should be drawn.
xtick	logical indicating if "ticks" (one per factor) should be drawn on the x axis.

Details

The supplied function will be called once for each level of each factor in the design and the plot will show these summary values. The levels of a particular factor are shown along a vertical line, and the overall value of `fun()` for the response is drawn as a horizontal line.

This is a new R implementation which will not be completely compatible to the earlier S implementations. This is not a bug but might still change.

Note

A big effort was taken to make this closely compatible to the S version. However, `col` (and `fg`) specification has different effects.

Author(s)

Roberto Frisullo and Martin Maechler

References

Chambers, J. M. and Hastie, T. J. eds (1992) *Statistical Models in S*. Chapman & Hall, London, the *white book*, pp. 546–7 (and 163–4).

Freeny, A. E. and Landwehr, J. M. (1990) Displays for data from large designed experiments; Computer Science and Statistics: Proc. 22nd SympInterface, 117–126, Springer Verlag.

See Also

`interaction.plot` for a "standard graphic" of designed experiments.

Examples

```
data(warpbreaks)
# automatic for data frame with one numeric var.
plot.design(warpbreaks)

Form <- breaks ~ wool + tension
summary(fm1 <- aov(Form, data = warpbreaks))
# same as above
plot.design(     Form, data = warpbreaks, col = 2)

## More than one y :
data(esoph)
```

```
str(esoph)
## two plots; if interactive you are "ask"ed
plot.design(esoph)

## or rather, compare mean and median:
op <- par(mfcol = 1:2)
plot.design(ncases/ncontrols ~ ., data = esoph,
            ylim = c(0,0.8))
plot.design(ncases/ncontrols ~ ., data = esoph,
            ylim = c(0,0.8), fun = median)
par(op)
```

plot.factor   *Plotting Factor Variables*

Description

This function implements a "scatterplot" method for `factor` arguments of the *generic* `plot` function. Actually, `boxplot` or `barplot` are used when appropriate.

Usage

```
## S3 method for class 'factor':
plot(x, y, legend.text = levels(y), ...)
```

Arguments

`x,y`	numeric or factor. `y` may be missing.
`legend.text`	a vector of text used to construct a legend for the plot. Only used if `y` is present and a factor.
`...`	Further arguments to `plot`, see also `par`.

See Also

`plot.default`, `plot.formula`, `barplot`, `boxplot`.

Examples

```
data(PlantGrowth)
# plot.data.frame
plot(PlantGrowth)
# numeric vector ~ factor
plot(weight ~ group, data = PlantGrowth)
# factor ~ factor
plot(cut(weight, 2) ~ group, data = PlantGrowth)
## passing "..." to barplot() eventually:
plot(cut(weight, 3) ~ group, data = PlantGrowth,
     density = 16*(1:3),col=NULL)

# extremely silly
plot(PlantGrowth$group, axes=FALSE, main="no axes")
```

plot.formula *Formula Notation for Scatterplots*

Description

Specify a scatterplot or add points or lines via a formula.

Usage

```
## S3 method for class 'formula':
plot(formula, data = parent.frame(), ..., subset,
            ylab = varnames[response], ask = TRUE)

## S3 method for class 'formula':
points(formula, data = parent.frame(), ..., subset)

## S3 method for class 'formula':
lines(formula, data = parent.frame(), ..., subset)
```

Arguments

formula	a formula, such as y ~ x.
data	a data.frame (or list) from which the variables in formula should be taken.
...	Further graphical parameters may also be passed as arguments, see par. horizontal = TRUE is also accepted.
subset	an optional vector specifying a subset of observations to be used in the fitting process.
ylab	the y label of the plot(s).
ask	logical, see par.

Details

Both the terms in the formula and the ... arguments are evaluated in data enclosed in parent.frame() if data is a list or a data frame. The terms of the formula and those arguments in ... that are of the same length as data are subjected to the subsetting specified in subset. If the formula in plot.formula contains more than one non-response term, a series of plots of y against each term is given. A plot against the running index can be specified as plot(y~1).

If y is an object (ie. has a class attribute) then plot.formula looks for a plot method for that class first. Otherwise, the class of x will determine the type of the plot. For factors this will be a parallel boxplot, and argument horizontal = TRUE can be used (see boxplot).

Value

These functions are invoked for their side effect of drawing in the active graphics device.

See Also

plot.default, plot.factor.

Examples

```
data(airquality)
op <- par(mfrow=c(2,1))
plot(Ozone ~ Wind, data = airquality,
     pch=as.character(Month))
plot(Ozone ~ Wind, data = airquality,
     pch=as.character(Month),
     subset = Month != 7)
par(op)
```

plot.histogram   *Plot Histograms*

Description

These are methods for objects of class "histogram", typically produced by hist.

Usage

```
## S3 method for class 'histogram':
plot(x, freq = equidist, density = NULL, angle = 45,
  col = NULL, border = par("fg"), lty = NULL,
  main = paste("Histogram of", x$xname), sub = NULL,
  xlab = x$xname, ylab, xlim = range(x$breaks), ylim = NULL,
  axes = TRUE, labels = FALSE, add = FALSE, ...)

## S3 method for class 'histogram':
lines(x, ...)
```

Arguments

x	a histogram object, or a list with components density, mid, etc, see hist for information about the components of x.
freq	logical; if TRUE, the histogram graphic is to present a representation of frequencies, i.e, x$counts; if FALSE, *relative* frequencies ("probabilities"), i.e., x$density, are plotted. The default is true for equidistant breaks and false otherwise.
col	a colour to be used to fill the bars. The default of NULL yields unfilled bars.
border	the color of the border around the bars.
angle, density	select shading of bars by lines: see rect.
lty	the line type used for the bars, see also lines.
main, sub, xlab, ylab	these arguments to title have useful defaults here.
xlim, ylim	the range of x and y values with sensible defaults.
axes	logical, indicating if axes should be drawn.

labels	logical or character. Additionally draw labels on top of bars, if not FALSE; if TRUE, draw the counts or rounded densities; if labels is a character, draw itself.
add	logical. If TRUE, only the bars are added to the current plot. This is what lines.histogram(*) does.
...	further graphical parameters to title and axis.

Details

lines.histogram(*) is the same as plot.histogram(*, add = TRUE).

See Also

hist, stem, density.

Examples

```
data(women)
str(wwt <- hist(women$weight, nc= 7, plot = FALSE))
# default main & xlab using wwt$xname
plot(wwt, labels = TRUE)
plot(wwt, border = "dark blue", col = "light blue",
  main = "Histogram of 15 women's weights",
  xlab = "weight [pounds]")

## Fake "lines" example, using non-default labels:
w2 <- wwt; w2$counts <- w2$counts - 1
lines(w2, col = "Midnight Blue",
      labels = ifelse(w2$counts, "> 1", "1"))
```

plot.lm *Plot Diagnostics for an lm Object*

Description

Four plots (selectable by which) are currently provided: a plot of residuals against fitted values, a Scale-Location plot of $\sqrt{|residuals|}$ against fitted values, a Normal Q-Q plot, and a plot of Cook's distances versus row labels.

Usage

```
## S3 method for class 'lm':
plot(x, which = 1:4,
   caption = c("Residuals vs Fitted",
               "Normal Q-Q plot",
               "Scale-Location plot",
               "Cook's distance plot"),
   panel = points,
   sub.caption = deparse(x$call), main = "",
   ask =
      prod(par("mfcol")) < length(which) && dev.interactive(),
   ...,
   id.n = 3, labels.id = names(residuals(x)), cex.id = 0.75)
```

Arguments

x	lm object, typically result of lm or glm.
which	If a subset of the plots is required, specify a subset of the numbers 1:4.
caption	Captions to appear above the plots
panel	Panel function. A useful alternative to points is panel.smooth.
sub.caption	common title—above figures if there are multiple; used as sub (s.title) otherwise.
main	title to each plot—in addition to the above caption.
ask	logical; if TRUE, the user is *ask*ed before each plot, see par(ask=.).
...	other parameters to be passed through to plotting functions.

`id.n`	number of points to be labelled in each plot, starting with the most extreme.
`labels.id`	vector of labels, from which the labels for extreme points will be chosen. `NULL` uses observation numbers.
`cex.id`	magnification of point labels.

Details

`sub.caption`—by default the function call—is shown as a subtitle (under the x-axis title) on each plot when plots are on separate pages, or as a subtitle in the outer margin (if any) when there are multiple plots per page.

The "Scale-Location" plot, also called "Spread-Location" or "S-L" plot, takes the square root of the absolute residuals in order to diminish skewness ($\sqrt{|E|}$ is much less skewed than $|E|$ for Gaussian zero-mean E).

This 'S-L' and the Q-Q plot use *standardized* residuals which have identical variance (under the hypothesis). They are given as $R_i/(s \times \sqrt{1-h_{ii}})$ where h_{ii} are the diagonal entries of the hat matrix, `influence()$hat`, see also `hat`.

Author(s)

John Maindonald and Martin Maechler.

References

Belsley, D. A., Kuh, E. and Welsch, R. E. (1980) *Regression Diagnostics*. New York: Wiley.

Cook, R. D. and Weisberg, S. (1982) *Residuals and Influence in Regression*. London: Chapman and Hall.

Hinkley, D. V. (1975) On power transformations to symmetry. *Biometrika* **62**, 101–111.

McCullagh, P. and Nelder, J. A. (1989) *Generalized Linear Models*. London: Chapman and Hall.

See Also

`termplot`, `lm.influence`, `cooks.distance`.

Examples

```
## Analysis of the life-cycle savings data given in
## Belsley, Kuh and Welsch.
data(LifeCycleSavings)
plot(lm.SR <- lm(sr ~ pop15 + pop75 + dpi + ddpi,
                 data = LifeCycleSavings))

## 4 plots on 1 page; allow room for printing model formula
## in outer margin:
par(mfrow = c(2, 2), oma = c(0, 0, 2, 0))
plot(lm.SR)
plot(lm.SR, id.n = NULL)               # no id's
plot(lm.SR, id.n = 5, labels.id = NULL) # 5 id numbers

## Fit a smooth curve, where applicable:
plot(lm.SR, panel = panel.smooth)
## Gives a smoother curve
plot(lm.SR, panel = function(x,y) panel.smooth(x, y,
  span = 1))

par(mfrow=c(2,1)) # same oma as above
plot(lm.SR, which = 1:2,
     sub.caption = "Saving Rates, n=50, p=5")
```

plot.table *Plot Methods for 'table' Objects*

Description

This is a method of the generic `plot` function for (contingency) `table` objects. Whereas for two- and more dimensional tables, a `mosaicplot` is drawn, one-dimensional ones are plotted "bar like".

Usage

```
## S3 method for class 'table':
plot(x, type = "h", ylim = c(0, max(x)), lwd = 2,
     xlab = NULL, ylab = NULL, frame.plot = is.num, ...)
```

Arguments

x	a `table` (like) object.
type	plotting type.
ylim	range of y-axis.
lwd	line width for bars when `type = "h"` is used in the 1D case.
xlab, ylab	x- and y-axis labels.
frame.plot	logical indicating if a frame (`box`) should be drawn in the 1D case. Defaults to true when `x` has `dimnames` coerceable to numbers.
...	further graphical arguments, see `plot.default`.

Details

The current implementation (R 1.2) is somewhat experimental and will be improved and extended.

See Also

`plot.factor`, the `plot` method for factors.

Examples

```
## 1-d tables
(Poiss.tab <- table(N = rpois(200, lam= 5)))
plot(Poiss.tab, main = "plot(table(rpois(200, lam=5)))")

data(state)
plot(table(state.division))

## 4-D :
data(Titanic)
plot(Titanic, main ="plot(Titanic, main= *)")
```

`plot.ts`   *Plotting Time-Series Objects*

Description

Plotting method for objects inheriting from class `"ts"`.

Usage

```
## S3 method for class 'ts':
plot(x, y = NULL, plot.type = c("multiple", "single"),
     xy.labels, xy.lines, panel = lines, nc, ...)

## S3 method for class 'ts':
lines(x, ...)
```

Arguments

`x, y`	time series objects, usually inheriting from class `"ts"`.
`plot.type`	for multivariate time series, should the series by plotted separately (with a common time axis) or on a single plot?
`xy.labels`	logical, indicating if `text()` labels should be used for an x-y plot, *or* character, supplying a vector of labels to be used. The default is to label for up to 150 points, and not for more.
`xy.lines`	logical, indicating if `lines` should be drawn for an x-y plot. Defaults to the value of `xy.labels` if that is logical, otherwise to TRUE.
`panel`	a `function(x, col, bg, pch, type, ...)` which gives the action to be carried out in each panel of the display for `plot.type="multiple"`. The default is `lines`.
`nc`	the number of columns to use when `type="multiple"`. Defaults to 1 for up to 4 series, otherwise to 2.
`...`	additional graphical arguments, see `plot`, `plot.default` and `par`.

Details

If y is missing, this function creates a time series plot, for multivariate series of one of two kinds depending on plot.type.

If y is present, both x and y must be univariate, and a "scatter" plot y ~ x will be drawn, enhanced by using text if xy.labels is TRUE or character, and lines if xy.lines is TRUE.

See Also

ts for basic time series construction and access functionality.

Examples

```
## Multivariate
z <- ts(matrix(rt(300, df = 3), 100, 3), start=c(1961, 1),
        frequency=12)
plot(z, type = "b")    # multiple
plot(z, plot.type="single", lty=1:3, col=4:2)

## A phase plot:
data(nhtemp)
plot(nhtemp, c(nhtemp[-1], NA), cex = .8, col="blue",
    main = "Lag plot of New Haven temperatures")
## a clearer way to do this would be
library(ts)
plot(nhtemp, lag(nhtemp, 1), cex = .8, col="blue",
    main = "Lag plot of New Haven temperatures")


library(ts)  # normally loaded

data(sunspots)
## xy.lines and xy.labels are FALSE for large series:
plot(lag(sunspots, 1), sunspots, pch = ".")

data(EuStockMarkets)
SMI <- EuStockMarkets[, "SMI"]
plot(lag(SMI,  1), SMI, pch = ".")
plot(lag(SMI, 20), SMI, pch = ".", log = "xy",
  main = "4 weeks lagged SMI stocks -- log scale",
  xy.lines= TRUE)
```

plot.window   *Set up World Coordinates for Graphics Window*

Description

This function sets up the world coordinate system for a graphics window. It is called by higher level functions such as `plot.default` (*after* `plot.new`).

Usage

```
plot.window(xlim, ylim, log = "", asp = NA, ...)
```

Arguments

xlim, ylim
: numeric of length 2, giving the x and y coordinates ranges.

log
: character; indicating which axes should be in log scale.

asp
: numeric, giving the aspect ratio y/x.

...
: further graphical parameters as in `par`.

Details

Note that if `asp` is a finite positive value then the window is set up so that one data unit in the x direction is equal in length to `asp` × one data unit in the y direction.

The special case `asp == 1` produces plots where distances between points are represented accurately on screen. Values with `asp > 1` can be used to produce more accurate maps when using latitude and longitude.

Usually, one should rather use the higher level functions such as `plot`, `hist`, `image`, ..., instead and refer to their help pages for explanation of the arguments.

See Also

`xy.coords`, `plot.xy`, `plot.default`.

Examples

```
## An example for the use of 'asp' :
library(mva)   # normally loaded
data(eurodist)
loc <- cmdscale(eurodist)
rx <- range(x <- loc[,1])
ry <- range(y <- -loc[,2])
plot(x, y, type="n", asp=1, xlab="", ylab="")
abline(h = pretty(rx, 10), v = pretty(ry, 10),
       col = "lightgray")
text(x, y, names(eurodist), cex=0.8)
```

plot.xy   *Basic Internal Plot Function*

Description

This is *the* internal function that does the basic plotting of points and lines. Usually, one should rather use the higher level functions instead and refer to their help pages for explanation of the arguments.

Usage

```
plot.xy(xy, type, pch=1, lty="solid", col=par("fg"), bg=NA,
        cex=1, ...)
```

Arguments

xy	A four-element list as results from `xy.coords`.
type	1 character code.
pch	character or integer code for kind of points/lines, see `points.default`.
lty	line type code, see `lines`.
col	color code or name, see `colors`, `palette`.
bg	background ("fill") color for open plot symbols.
cex	character expansion.
...	further graphical parameters.

See Also

`plot`, `plot.default`, `points`, `lines`.

Examples

```
# to see how it calls "plot.xy(xy.coords(x, y), ...)"
points.default
```

plotmath *Mathematical Annotation in R*

Description

If the `text` argument to one of the text-drawing functions (`text`, `mtext`, `axis`) in R is an expression, the argument is interpreted as a mathematical expression and the output will be formatted according to TeX-like rules. Expressions can also be used for titles, subtitles and x- and y-axis labels (but not for axis labels on `persp` plots).

Details

A mathematical expression must obey the normal rules of syntax for any R expression, but it is interpreted according to very different rules than for normal R expressions.

It is possible to produce many different mathematical symbols, generate sub- or superscripts, produce fractions, etc.

The output from `demo(plotmath)` includes several tables which show the available features. In these tables, the columns of grey text show sample R expressions, and the columns of black text show the resulting output.

The available features are also described in the tables below:

Syntax	Meaning
x + y	x plus y
x - y	x minus y
x*y	juxtapose x and y
x/y	x forwardslash y
x %+-% y	x plus or minus y
x %/% y	x divided by y
x %*% y	x times y
x[i]	x subscript i
x^2	x superscript 2
paste(x, y, z)	juxtapose x, y, and z
sqrt(x)	square root of x
sqrt(x, y)	yth root of x
x == y	x equals y
x != y	x is not equal to y
x < y	x is less than y
x <= y	x is less than or equal to y
x > y	x is greater than y

x >= y	x is greater than or equal to y
x %~~% y	x is approximately equal to y
x %=~% y	x and y are congruent
x %==% y	x is defined as y
x %prop% y	x is proportional to y
plain(x)	draw x in normal font
bold(x)	draw x in bold font
italic(x)	draw x in italic font
bolditalic(x)	draw x in bolditalic font
list(x, y, z)	comma-separated list
...	ellipsis (height varies)
cdots	ellipsis (vertically centred)
ldots	ellipsis (at baseline)
x %subset% y	x is a proper subset of y
x %subseteq% y	x is a subset of y
x %notsubset% y	x is not a subset of y
x %supset% y	x is a proper superset of y
x %supseteq% y	x is a superset of y
x %in% y	x is an element of y
x %notin% y	x is not an element of y
hat(x)	x with a circumflex
tilde(x)	x with a tilde
dot(x)	x with a dot
ring(x)	x with a ring
bar(xy)	xy with bar
widehat(xy)	xy with a wide circumflex
widetilde(xy)	xy with a wide tilde
x %<->% y	x double-arrow y
x %->% y	x right-arrow y
x %<-% y	x left-arrow y
x %up% y	x up-arrow y
x %down% y	x down-arrow y
x %<=>% y	x is equivalent to y
x %=>% y	x implies y
x %<=% y	y implies x
x %dblup% y	x double-up-arrow y
x %dbldown% y	x double-down-arrow y
alpha – omega	Greek symbols
Alpha – Omega	uppercase Greek symbols
infinity	infinity symbol
partialdiff	partial differential symbol
32*degree	32 degrees
60*minute	60 minutes of angle

`30*second`	30 seconds of angle
`displaystyle(x)`	draw x in normal size (extra spacing)
`textstyle(x)`	draw x in normal size
`scriptstyle(x)`	draw x in small size
`scriptscriptstyle(x)`	draw x in very small size
`x ~~ y`	put extra space between x and y
`x + phantom(0) + y`	leave gap for "0", but don't draw it
`x + over(1, phantom(0))`	leave vertical gap for "0" (don't draw)
`frac(x, y)`	x over y
`over(x, y)`	x over y
`atop(x, y)`	x over y (no horizontal bar)
`sum(x[i], i==1, n)`	sum x[i] for i equals 1 to n
`prod(plain(P)(X==x), x)`	product of P(X=x) for all values of x
`integral(f(x)*dx, a, b)`	definite integral of f(x) wrt x
`union(A[i], i==1, n)`	union of A[i] for i equals 1 to n
`intersect(A[i], i==1, n)`	intersection of A[i]
`lim(f(x), x %->% 0)`	limit of f(x) as x tends to 0
`min(g(x), x > 0)`	minimum of g(x) for x greater than 0
`inf(S)`	infimum of S
`sup(S)`	supremum of S
`x^y + z`	normal operator precedence
`x^(y + z)`	visible grouping of operands
`x^{y + z}`	invisible grouping of operands
`group("(",list(a, b),"]")`	specify left and right delimiters
`bgroup("(",atop(x,y),")")`	use scalable delimiters
`group(lceil, x, rceil)`	special delimiters

References

Murrell, P. and Ihaka, R. (2000) An approach to providing mathematical annotation in plots. *Journal of Computational and Graphical Statistics*, **9**, 582–599.

See Also

`demo(plotmath)`, `axis`, `mtext`, `text`, `title`, `substitute` `quote`, `bquote`

Examples

```
x <- seq(-4, 4, len = 101)
y <- cbind(sin(x), cos(x))
matplot(x, y, type = "l", xaxt = "n",
```

```
        main = expression(paste(plain(sin) * phi, " and ",
                                plain(cos) * phi)),
      # only 1st is taken
      ylab = expression("sin" * phi, "cos" * phi),
      xlab = expression(paste("Phase Angle ", phi)),
      col.main = "blue")
axis(1, at = c(-pi, -pi/2, 0, pi/2, pi),
     lab = expression(-pi, -pi/2, 0, pi/2, pi))

## How to combine "math" and numeric variables :
plot(1:10, type="n", xlab="", ylab="",
     main = "plot math & numbers")
theta <- 1.23 ; mtext(bquote(hat(theta) == .(theta)))
for(i in 2:9)
  text(i,i+1,
       substitute(list(xi,eta) == group("(",list(x,y),")"),
       list(x=i, y=i+1)))

plot(1:10, 1:10)
text(4, 9,
  expression(hat(beta) == (X^t * X)^{-1} * X^t * y))
text(4, 8.4,
  "expression(hat(beta) == (X^t * X)^{-1} * X^t * y)",
  cex = .8)
text(4, 7,
  expression(bar(x) == sum(frac(x[i], n), i==1, n)))
text(4, 6.4,
  "expression(bar(x) == sum(frac(x[i], n), i==1, n))",
  cex = .8)
text(8, 5,
  expression(paste(frac(1, sigma*sqrt(2*pi)), " ",
  plain(e)^{frac(-(x-mu)^2, 2*sigma^2)})),
  cex = 1.2)
```

png   *JPEG and PNG graphics devices*

Description

A graphics device for JPEG or PNG format bitmap files.

Usage

```
jpeg(filename = "Rplot%03d.jpeg", width = 480, height = 480,
    pointsize = 12, quality = 75, bg = "white", ...)
png(filename ="Rplot%03d.png", width = 480, height = 480,
    pointsize = 12, bg = "white", ...)
```

Arguments

filename   the name of the output file. The page number is substituted if an integer format is included in the character string. (The result must be less than PATH_MAX characters long, and may be truncated if not.) Tilde expansion is performed where supported by the platform.

width   the width of the device in pixels.

height   the height of the device in pixels.

pointsize the default pointsize of plotted text.

quality   the 'quality' of the JPEG image, as a percentage. Smaller values will give more compression but also more degradation of the image.

bg   default background colour.

...   additional arguments to the X11 device.

Details

Plots in PNG and JPEG format can easily be converted to many other bitmap formats, and both can be displayed in most modern web browsers. The PNG format is lossless and is best for line diagrams and blocks of solid colour. The JPEG format is lossy, but may be useful for image plots, for example.

png supports transparent backgrounds: use bg = "transparent". Not all PNG viewers render files with transparency correctly. When transparency is in use a very light grey is used as the background and so will appear as transparent if used in the plot. This allows opaque white to be used, as on the example.

R can be compiled without support for either or both of these devices: this will be reported if you attempt to use them on a system where they are not supported. They will not be available if R has been started with '--gui=none' (and will give a different error message), and they may not be usable unless the X11 display is available to the owner of the R process.

Value

A plot device is opened: nothing is returned to the R interpreter.

Warning

If you plot more than one page on one of these devices and do not include something like %d for the sequence number in `file`, the file will contain the last page plotted.

Note

These are based on the X11 device, so the additional arguments to that device work, but are rarely appropriate. The colour handling will be that of the X11 device in use.

Author(s)

Guido Masarotto and Brian Ripley

See Also

`Devices`, `dev.print`

`capabilities` to see if these devices are supported by this build of R.

`bitmap` provides an alternative way to generate PNG and JPEG plots that does not depend on accessing the X11 display but does depend on having GhostScript installed.

Examples

```
## these examples will work only if the devices are
## available and the X11 display is available.

## copy current plot to a PNG file
dev.print(png, file="myplot.png", width=480, height=480)
```

```
png(file="myplot.png", bg="transparent")
plot(1:10)
rect(1, 5, 3, 7, col="white")
dev.off()

jpeg(file="myplot.jpeg")
example(rect)
dev.off()
```

points   *Add Points to a Plot*

Description

points is a generic function to draw a sequence of points at the specified coordinates. The specified character(s) are plotted, centered at the coordinates.

Usage

```
points(x, ...)

## Default S3 method:
points(x, y = NULL, type = "p", pch = par("pch"),
      col = par("col"), bg = NA, cex = 1, ...)
```

Arguments

x, y   coordinate vectors of points to plot.

type   character indicating the type of plotting; actually any of the types as in plot.

pch   plotting "character", i.e., symbol to use. pch can either be a character or an integer code for a set of graphics symbols. The full set of S symbols is available with pch=0:18, see the last picture from example(points), i.e., the examples below.

In addition, there is a special set of R plotting symbols which can be obtained with pch=19:25 and 21:25 can be colored and filled with different colors:

- pch=19: solid circle,
- pch=20: bullet (smaller circle),
- pch=21: circle,
- pch=22: square,
- pch=23: diamond,
- pch=24: triangle point-up,
- pch=25: triangle point down.

Values pch=26:32 are currently unused, and pch=32:255 give the text symbol in the encoding in use (see postscript).

col	color code or name, see **par**.
bg	background ("fill") color for open plot symbols
cex	character expansion: a numerical vector.
...	Further graphical parameters (see plot.xy and par) may also be supplied as arguments.

Details

The coordinates can be passed in a plotting structure (a list with x and y components), a two-column matrix, a time series, ... See xy.coords.

Arguments pch, col, bg and cex can be vectors (which will be recycled as needed) giving a value for each point plotted. Points whose x, y, pch, col or cex value is NA are omitted from the plot.

Graphical parameters are permitted as arguments to this function.

References

Becker, R. A., Chambers, J. M. and Wilks, A. R. (1988) *The New S Language.* Wadsworth & Brooks/Cole.

See Also

plot, lines, and the underlying "primitive" plot.xy.

Examples

```
plot(-4:4, -4:4, type = "n") # setting up coord. system
points(rnorm(200), rnorm(200), col = "red")
points(rnorm(100)/2, rnorm(100)/2, col = "blue", cex = 1.5)

op <- par(bg = "light blue")
x <- seq(0,2*pi, len=51)
## something "between type='b' and type='o'":
plot(x, sin(x), type="o", pch=21, bg=par("bg"),
 col = "blue", cex=.6,
 main='plot(..., type="o", pch=21, bg=par("bg"))')
par(op)

## Showing all the extra & some char graphics symbols
Pex <- 3 ## good for both .Device=="postscript" and "x11"
ipch <- 1:(np <- 25+11); k <- floor(sqrt(np));
dd <- c(-1,1)/2
```

```
rx <- dd + range(ix <- (ipch-1) %/% k)
ry <- dd + range(iy <- 3 + (k-1)-(ipch-1) %% k)
pch <- as.list(ipch)
pch[25+ 1:11] <-
   as.list(c("*",".", "o","O","0","+","-",":","|","%","#"))
plot(rx, ry, type="n", axes = FALSE, xlab = "", ylab = "",
  main = paste("symbols: points(... pch=*, cex=", Pex,")"))
abline(v = ix, h = iy, col = "lightgray", lty = "dotted")
for(i in 1:np) {
  pc <- pch[[i]]
  points(ix[i], iy[i], pch = pc, col = "red",
         bg = "yellow", cex = Pex)
  ## red symbols with a yellow interior (where available)
  text(ix[i] - .3, iy[i], pc, col = "brown", cex = 1.2)
}
```

polygon   *Polygon Drawing*

Description

polygon draws the polygons whose vertices are given in x and y.

Usage

```
polygon(x, y = NULL, density = NULL, angle = 45,
  border = NULL, col = NA, lty = NULL, xpd = NULL, ...)
```

Arguments

x,y	vectors containing the coordinates of the vertices of the polygon.
density	the density of shading lines, in lines per inch. The default value of NULL means that no shading lines are drawn. A zero value of density means no shading lines whereas negative values (and NA) suppress shading (and so allow color filling).
angle	the slope of shading lines, given as an angle in degrees (counter-clockwise).
col	the color for filling the polygon. The default, NA, is to leave polygons unfilled.
border	the color to draw the border. The default, NULL, uses par("fg"). Use border = NA to omit borders.
	For compatibility with S, border can also be logical, in which case FALSE is equivalent to NA (borders omitted) and TRUE is equivalent to NULL (use the foreground colour),
lty	the line type to be used, as in par.
xpd	(where) should clipping take place? Defaults to par("xpd").
...	graphical parameters can be given as arguments to polygon.

Details

The coordinates can be passed in a plotting structure (a list with x and y components), a two-column matrix, ... See xy.coords.

It is assumed that the polygon is closed by joining the last point to the first point.

The coordinates can contain missing values. The behaviour is similar to that of lines, except that instead of breaking a line into several lines, NA values break the polygon into several complete polygons (including closing the last point to the first point). See the examples below.

When multiple polygons are produced, the values of density, angle, col, border, and lty are recycled in the usual manner.

Bugs

The present shading algorithm can produce incorrect results for self-intersecting polygons.

Author(s)

The code implementing polygon shading was donated by Kevin Buhr.

References

Becker, R. A., Chambers, J. M. and Wilks, A. R. (1988) *The New S Language.* Wadsworth & Brooks/Cole.

See Also

segments for even more flexibility, lines, rect, box, abline.

par for how to specify colors.

Examples

```
x <- c(1:9,8:1)
y <- c(1,2*(5:3),2,-1,17,9,8,2:9)
op <- par(mfcol=c(3,1))
for(xpd in c(FALSE,TRUE,NA)) {
  plot(1:10, main=paste("xpd =", xpd)) ;
  box("figure", col = "pink", lwd=3)
  polygon(x,y, xpd=xpd, col = "orange", lty=2, lwd=2,
          border = "red")
}
```

```
par(op)

n <- 100
xx <- c(0:n, n:0)
yy <- c(c(0,cumsum(rnorm(n))), rev(c(0,cumsum(rnorm(n)))))
plot    (xx, yy, type="n", xlab="Time", ylab="Distance")
polygon(xx, yy, col="gray", border = "red")
title("Distance Between Brownian Motions")

# Multiple polygons from NA values
# and recycling of col, border, and lty
op <- par(mfrow=c(2,1))
plot(c(1,9), 1:2, type="n")
polygon(1:9, c(2,1,2,1,1,2,1,2,1),
        col=c("red", "blue"),
        border=c("green", "yellow"),
        lwd=3, lty=c("dashed", "solid"))
plot(c(1,9), 1:2, type="n")
polygon(1:9, c(2,1,2,1,NA,2,1,2,1),
        col=c("red", "blue"),
        border=c("green", "yellow"),
        lwd=3, lty=c("dashed", "solid"))
par(op)

# Line-shaded polygons
plot(c(1,9), 1:2, type="n")
polygon(1:9, c(2,1,2,1,NA,2,1,2,1),
        density=c(10, 20), angle=c(-45, 45))
```

postscript *PostScript Graphics*

Description

`postscript` starts the graphics device driver for producing PostScript graphics.

The auxiliary function `ps.options` can be used to set and view (if called without arguments) default values for the arguments to `postscript`.

Usage

```
postscript(file = ifelse(onefile, "Rplots.ps",
   "Rplot%03d.ps"), onefile = TRUE, paper, family, encoding,
   bg, fg, width, height, horizontal, pointsize, pagecentre,
   print.it, command, title = "R Graphics Output")

ps.options(paper, horizontal, width, height, family,
   encoding, pointsize, bg, fg, onefile = TRUE,
   print.it = FALSE, append = FALSE, reset = FALSE,
   override.check = FALSE)
.PostScript.Options
```

Arguments

file
: a character string giving the name of the file. If it is "", the output is piped to the command given by the argument command. If it is "|cmd", the output is piped to the command given by 'cmd'.

    For use with onefile=FALSE give a printf format such as "Rplot%03d.ps" (the default in that case).

paper
: the size of paper in the printer. The choices are "a4", "letter", "legal" and "executive" (and these can be capitalized). Also, "special" can be used, when the width and height specify the paper size. A further choice is "default", which is the default. If this is selected, the papersize is taken from the option "papersize" if that is set and to "a4" if it is unset or empty.

horizontal
: the orientation of the printed image, a logical. Defaults to true, that is landscape orientation.

`width, height`	the width and height of the graphics region in inches. The default is to use the entire page less a 0.25 inch border on each side.
`family`	the font family to be used. EITHER a single character string OR a character vector of length four or five. See the section 'Families'.
`encoding`	the name of an encoding file. Defaults to "ISO-Latin1.enc" in the 'R_HOME/afm' directory, which is used if the path does not contain a path separator. An extension ".enc" can be omitted.
`pointsize`	the default point size to be used.
`bg`	the default background color to be used. If `"transparent"` (or an equivalent specification), no background is painted.
`fg`	the default foreground color to be used.
`onefile`	logical: if true (the default) allow multiple figures in one file. If false, generate a file name containing the page number and use an EPSF header and no `DocumentMedia` comment.
`pagecentre`	logical: should the device region be centred on the page: defaults to true.
`print.it`	logical: should the file be printed when the device is closed? (This only applies if `file` is a real file name.)
`command`	the command to be used for "printing". Defaults to option `"printcmd"`; this can also be selected as `"default"`.
`append`	logical; currently **disregarded**; just there for compatibility reasons.
`reset, override.check`	logical arguments passed to `check.options`. See the Examples.
`title`	title string to embed in the file.

Details

`postscript` opens the file `file` and the PostScript commands needed to plot any graphics requested are stored in that file. This file can then be printed on a suitable device to obtain hard copy.

A postscript plot can be printed via `postscript` in two ways.

1. Setting `print.it = TRUE` causes the command given in argument `command` to be called with argument `"file"` when the device is closed. Note that the plot file is not deleted unless `command` arranges to delete it.

2. `file=""` or `file="|cmd"` can be used to print using a pipe on systems that support 'popen'. Failure to open the command will probably be reported to the terminal but not to 'popen', in which case close the device by `dev.off` immediately.

The postscript produced by R is EPS (*Encapsulated PostScript*) compatible, and can be included into other documents, e.g., into LaTeX, using
`includegraphics{<filename>}`. For use in this way you will probably want to set `horizontal = FALSE`, `onefile = FALSE`, `paper = "special"`.

Most of the PostScript prologue used is taken from the R character vector `.ps.prolog`. This is marked in the output, and can be changed by changing that vector. (This is only advisable for PostScript experts.)

`ps.options` needs to be called before calling `postscript`, and the default values it sets can be overridden by supplying arguments to `postscript`.

Families

The argument `family` specifies the font family to be used. In normal use it is one of `"AvantGarde"`, `"Bookman"`, `"Courier"`, `"Helvetica"`, `"Helvetica-Narrow"`, `"NewCenturySchoolbook"`, `"Palatino"` or `"Times"`, and refers to the standard Adobe PostScript fonts of those names which are included (or cloned) in all common PostScript devices.

Many PostScript emulators (including those based on `ghostscript`) use the URW equivalents of these fonts, which are `"URWGothic"`, `"URWBookman"`, `"NimbusMon"`, `"NimbusSan"`, `"NimbusSanCond"`, `"CenturySch"`, `"URWPalladio"` and `"NimbusRom"` respectively. If your PostScript device is using URW fonts, you will obtain access to more characters and more appropriate metrics by using these names. To make these easier to remember, `"URWHelvetica"` == `"NimbusSan"` and `"URWTimes"` == `"NimbusRom"` are also supported.

It is also possible to specify `family="ComputerModern"`. This is intended to use with the Type 1 versions of the TeX CM fonts. It will normally be possible to include such output in TeX or LaTeX provided it is processed with `dvips -Ppfb -j0` or the equivalent on your system. (`-j0` turns off font subsetting.)

If the second form of argument `"family"` is used, it should be a character vector of four or five paths to Adobe Font Metric files for the regular, bold, italic, bold italic and (optionally) symbol fonts to be used. If these paths do not contain the file separator, they are taken to refer to files in the R directory 'R_HOME/afm'. Thus the default Helvetica family can be specified by `family = c("hv_____.afm", "hvb_____.afm", "hvo_____.afm", "hvbo____.afm", "sy_____.afm")`. It is the user's responsibility to check that suitable fonts are made available, and that they contain the needed characters when re-encoded. The fontnames used are taken from the `FontName` fields of the afm files. The software including the PostScript plot file should either embed the font outlines (usually from '.pfb' or '.pfa' files) or use DSC comments to instruct the print spooler to do so.

Encodings

Encodings describe which glyphs are used to display the character codes (in the range 0–255). By default R uses ISOLatin1 encoding, and the examples for `text` are in that encoding. However, the encoding used on machines running R may well be different, and by using the `encoding` argument the glyphs can be matched to encoding in use.

None of this will matter if only ASCII characters (codes 32–126) are used as all the encodings agree over that range. Some encodings are supersets of ISOLatin1, too. However, if accented and special characters do not come out as you expect, you may need to change the encoding. Three other encodings are supplied with R: `"WinAnsi.enc"` and `"MacRoman.enc"` correspond to the encodings normally used on Windows and MacOS (at least by Adobe), and `"PDFDoc.enc"` is the first 256 characters of the Unicode encoding, the standard for PDF.

If you change the encoding, it is your responsibility to ensure that the PostScript font contains the glyphs used. One issue here is the Euro symbol which is in the WinAnsi and MacRoman encodings but may well not be in the PostScript fonts. (It is in the URW variants; it is not in the supplied Adobe Font Metric files.)

There is one exception. Character 45 (`"-"`) is always set as minus (its value in Adobe ISOLatin1) even though it is hyphen in the other encodings. Hyphen is available as character 173 (octal 0255) in ISOLatin1.

Author(s)

Support for Computer Modern fonts is based on a contribution by Brian D'Urso.

References

Becker, R. A., Chambers, J. M. and Wilks, A. R. (1988) *The New S Language.* Wadsworth & Brooks/Cole.

See Also

Devices, check.options which is called from both ps.options and postscript.

Examples

```
# open the file "foo.ps" for graphics output
postscript("foo.ps")
# produce the desired graph(s)
dev.off()              # turn off the postscript device
postscript("|lp -dlw")
# produce the desired graph(s)
dev.off()              # plot will appear on printer

# for URW PostScript devices
postscript("foo.ps", family = "NimbusSan")

# for inclusion in Computer Modern TeX documents, perhaps
postscript("cm_test.eps", width = 4.0, height = 3.0,
  horizontal = FALSE, onefile = FALSE, paper = "special",
  family = "ComputerModern")
# The resultant postscript file can be used by dvips -Ppfb
# -jO.

# To test out encodings, you can use
TestChars <-
  function (encoding="ISOLatin1", family="URWHelvetica")
{
    postscript(encoding=encoding, family=family)
    par(pty="s")
    plot(c(0,15), c(0,15), type="n", xlab="", ylab="")
    title(paste("Centred chars in encoding", encoding))
    grid(15, 15, lty=1)
    for(i in c(32:255)) {
        x <- i
        y <- i
        points(x, y, pch=i)
    }
```

```
        dev.off()
}
## there will be many warnings. We use URW to get a
## complete enough set of font metrics.
TestChars()
TestChars("ISOLatin2")
TestChars("WinAnsi")

stopifnot(
  unlist(ps.options()) == unlist(.PostScript.Options)
)
ps.options(bg = "pink")
str(ps.options(reset = TRUE))

### ---- error checking of arguments: ----
ps.options(width=0:12, onefile=0, bg=pi)
# override the check for 'onefile', but not the others:
str(ps.options(width=0:12, onefile=1, bg=pi,
               override.check = c(FALSE,TRUE,FALSE)))
```

ppoints — *Ordinates for Probability Plotting*

Description

Generates the sequence of "probability" points (1:m - a)/(m + (1-a)-a) where m is either n, if length(n)==1, or length(n).

Usage

```
ppoints(n, a = ifelse(n <= 10, 3/8, 1/2))
```

Arguments

n
: either the number of points generate or a vector of observations.

a
: the offset fraction to be used; typically in (0,1).

Details

If $0 < a < 1$, the resulting values are within $(0,1)$ (excluding boundaries). In any case, the resulting sequence is symmetric in $[0,1]$, i.e., p + rev(p) == 1.

ppoints() is used in qqplot and qqnorm to generate the set of probabilities at which to evaluate the inverse distribution.

References

Becker, R. A., Chambers, J. M. and Wilks, A. R. (1988) *The New S Language*. Wadsworth & Brooks/Cole.

See Also

qqplot, qqnorm.

Examples

```
ppoints(4) # the same as ppoints(1:4)
ppoints(10)
ppoints(10, a=1/2)
```

preplot   *Pre-computations for a Plotting Object*

Description

Compute an object to be used for plots relating to the given model object.

Usage

```
preplot(object, ...)
```

Arguments

object   a fitted model object.

...   additional arguments for specific methods.

Details

Only the generic function is currently provided in base R, but some add-on packages have methods. Principally here for S compatibility.

Value

An object set up to make a plot that describes `object`.

| pretty | *Pretty Breakpoints* |

Description

Compute a sequence of about n+1 equally spaced nice values which cover the range of the values in x. The values are chosen so that they are 1, 2 or 5 times a power of 10.

Usage

```
pretty(x, n = 5, min.n = n %/% 3,   shrink.sml = 0.75,
       high.u.bias = 1.5, u5.bias = .5 + 1.5*high.u.bias,
       eps.correct = 0)
```

Arguments

x	numeric vector
n	integer giving the *desired* number of intervals. Non-integer values are rounded down.
min.n	nonnegative integer giving the *minimal* number of intervals. If min.n == 0, pretty(.) may return a single value.
shrink.sml	positive numeric by a which a default scale is shrunk in the case when range(x) is "very small" (usually 0).
high.u.bias	non-negative numeric, typically > 1. The interval unit is determined as $\{1,2,5,10\}$ times b, a power of 10. Larger high.u.bias values favor larger units.
u5.bias	non-negative numeric multiplier favoring factor 5 over 2. Default and "optimal": u5.bias = .5 + 1.5*high.u.bias.
eps.correct	integer code, one of $\{0,1,2\}$. If non-0, an "*epsilon correction*" is made at the boundaries such that the result boundaries will be outside range(x); in the *small* case, the correction is only done if eps.correct ≥ 2.

Details

Let d <- max(x) - min(x) ≥ 0. If d is not (very close) to 0, we let c <- d/n, otherwise more or less c <- max(abs(range(x)))*shrink.sml / min.n. Then, the *10 base* b is $10^{\lfloor \log_{10}(c) \rfloor}$ such that $b \leq c < 10b$.

Now determine the basic *unit* u as one of $\{1, 2, 5, 10\}b$, depending on $c/b \in [1, 10)$ and the two "*bias*" coefficients, $h =$ high.u.bias and $f =$ u5.bias.

. . . .

References

Becker, R. A., Chambers, J. M. and Wilks, A. R. (1988) *The New S Language*. Wadsworth & Brooks/Cole.

Examples

```
pretty(1:15)          # 0  2  4  6  8 10 12 14 16
pretty(1:15, h=2)     # 0  5 10 15
pretty(1:15, n=4)     # 0  5 10 15
pretty(1:15 * 2)      # 0  5 10 15 20 25 30
pretty(1:20)          # 0  5 10 15 20
pretty(1:20, n=2)     # 0 10 20
pretty(1:20, n=10)    # 0  2  4 ... 20

for(k in 5:11) {
  cat("k=",k,": ");
  print(diff(range(pretty(100 + c(0, pi*10^-k)))))
}

## more bizarre, when min(x) == max(x):
pretty(pi)

add.names <- function(v) { names(v) <- paste(v); v}
str(lapply(add.names(-10:20), pretty))
str(lapply(add.names(0:20),   pretty, min = 0))
sapply(    add.names(0:20),   pretty, min = 4)

pretty(1.234e100)
pretty(1001.1001)
pretty(1001.1001, shrink = .2)
for(k in -7:3)
   cat("shrink=",formatC(2^k,wid=9),":",
       formatC(pretty(1001.1001, shrink = 2^k), wid=6),"\n")
```

qqnorm *Quantile-Quantile Plots*

Description

qqnorm is a generic functions the default method of which produces a normal QQ plot of the values in y. qqline adds a line to a normal quantile-quantile plot which passes through the first and third quartiles.

qqplot produces a QQ plot of two datasets.

Graphical parameters may be given as arguments to qqnorm, qqplot and qqline.

Usage

```
qqnorm(y, ...)
## Default S3 method:
qqnorm(y, ylim, main = "Normal Q-Q Plot",
  xlab = "Theoretical Quantiles",
  ylab = "Sample Quantiles", plot.it = TRUE, datax = FALSE,
  ...)
qqline(y, datax = FALSE, ...)
qqplot(x, y, plot.it = TRUE, xlab = deparse(substitute(x)),
    ylab = deparse(substitute(y)), ...)
```

Arguments

x	The first sample for qqplot.
y	The second or only data sample.
xlab, ylab, main	plot labels.
plot.it	logical. Should the result be plotted?
datax	logical. Should data values be on the x-axis?
ylim, ...	graphical parameters.

Value

For qqnorm and qqplot, a list with components

x	The x coordinates of the points that were/would be plotted
y	The original y vector, i.e., the corresponding y coordinates *including NAs*.

References

Becker, R. A., Chambers, J. M. and Wilks, A. R. (1988) *The New S Language.* Wadsworth & Brooks/Cole.

See Also

ppoints.

Examples

```
y <- rt(200, df = 5)
qqnorm(y); qqline(y, col = 2)
qqplot(y, rt(300, df = 5))
data(precip)
qqnorm(precip, ylab = "Precipitation [in/yr] - US cities")
```

quartz   *MacOS X Quartz device*

Description

`quartz` starts a graphics device driver for the MacOS X System. This can only be done on machines that run MacOS X.

Usage

```
quartz(display = "", width = 5, height = 5, pointsize = 12,
   family = "Helvetica", antialias=TRUE, autorefresh=TRUE)
```

Arguments

`display`	the display on which the graphics window will appear. The default is to use the value in the user's environment variable `DISPLAY`.
`width`	the width of the plotting window in inches.
`height`	the height of the plotting window in inches.
`pointsize`	the default pointsize to be used.
`family`	this is the family name of the Postscript font that will be used by the device.
`antialias`	whether to use antialiasing. It is never the case to set it `FALSE`
`autorefresh`	logical specifying if realtime refreshing should be done. If `FALSE`, the system is charged to refresh the context of the device window.

Details

Quartz is the graphic engine based on the PDF format. It is used by the graphic interface of MacOS X to render high quality graphics. As PDF it is device independent and can be rescaled without loss of definition.

Calling `quartz()` sets `.Device` to `"quartz"`.

See Also

`Devices`.

recordPlot — *Record and Replay Plots*

Description

Functions to save the current plot in an R variable, and to replay it.

Usage

```
recordPlot()
replayPlot(x)
```

Arguments

x   A saved plot.

Details

These functions record and replay the displaylist of the current graphics device. The returned object is of class `"recordedplot"`, and `replayPlot` acts as a `print` method for that class.

The format of recorded plots was changed in R 1.4.0: plots saved in earlier versions can still be replayed.

Value

`recordPlot` returns an object of class `"recordedplot"`, a list with components:

displaylist   The saved display list, as a pairlist.
gpar   The graphics state, as an integer vector.

`replayPlot` has no return value.

rect   *Draw a Rectangle*

Description

rect draws a rectangle (or sequence of rectangles) with the given coordinates, fill and border colors.

Usage

```
rect(xleft, ybottom, xright, ytop, density = NULL,
  angle = 45, col = NULL, border = NULL, lty = NULL,
  lwd = par("lwd"), xpd = NULL, ...)
```

Arguments

xleft	a vector (or scalar) of left x positions.
ybottom	a vector (or scalar) of bottom y positions.
xright	a vector (or scalar) of right x positions.
ytop	a vector (or scalar) of top y positions.
density	the density of shading lines, in lines per inch. The default value of NULL means that no shading lines are drawn. A zero value of density means no shading lines whereas negative values (and NA) suppress shading (and so allow color filling).
angle	angle (in degrees) of the shading lines.
col	color(s) to fill or shade the rectangle(s) with. The default NULL, or also NA do not fill, i.e., draw transparent rectangles, unless density is specified.
border	color for rectangle border(s).
lty	line type for borders; defaults to "solid".
lwd	width for borders.
xpd	logical ("expand"); defaults to par("xpd"). See par(xpd=).
...	other graphical parameters can be given as arguments.

Details

The positions supplied, i.e., xleft, ..., are relative to the current plotting region. If the x-axis goes from 100 to 200 then xleft must be larger than 100 and xright must be less than 200.

It is a primitive function used in hist, barplot, legend, etc.

See Also

box for the "standard" box around the plot; polygon and segments for flexible line drawing.

par for how to specify colors.

Examples

```
## set up the plot region:
op <- par(bg = "thistle")
plot(c(100, 250), c(300, 450), type = "n", xlab="", ylab="",
  main = "2x11 rectangles; 'rect(100+i,300+i,150+i,380+i)'")
i <- 4*(0:10)
## draw rectangles with bottom left (100, 300)+i and top
## right (150, 380)+i
rect(100+i, 300+i, 150+i, 380+i,
     col=rainbow(11, start=.7,end=.1))
rect(240-i, 320+i, 250-i, 410+i,
     col=heat.colors(11), lwd=i/5)
## Background alternating (transparent / "bg"):
j <- 10*(0:5)
rect(125+j, 360+j,   141+j, 405+j/2, col = c(NA,0),
     border = "gold", lwd = 2)
rect(125+j, 296+j/2, 141+j, 331+j/5,
     col = c(NA,"midnightblue"))
mtext("+2x6 rect(*,col=c(NA,0)) and col=c(NA,\"m..blue\"))")

## an example showing colouring and shading
plot(c(100, 200), c(300, 450), type= "n", xlab="", ylab="")
# transparent
rect(100, 300, 125, 350)
# coloured
rect(100, 400, 125, 450, col="green", border="blue")
rect(115, 375, 150, 425, col=par("bg"),
     border="transparent")
rect(150, 300, 175, 350, density=10, border="red")
```

```
rect(150, 400, 175, 450, density=30, col="blue",
     angle=-30, border="transparent")

legend(180, 450, legend=1:4,
       fill=c(NA, "green", par("fg"), "blue"),
       density=c(NA, NA, 10, 30), angle=c(NA, NA, 30, -30))

par(op)
```

rgb   *RGB Color Specification*

Description

This function creates "colors" corresponding to the given intensities (between 0 and max) of the red, green and blue primaries. The names argument may be used to provide names for the colors.

The values returned by rgb can be used with a col= specification in graphics functions or in par.

Usage

```
rgb(red, green, blue, names=NULL, maxColorValue = 1)
```

Arguments

red, blue, green
: vectors of same length with values in $[0, M]$ where M is maxColorValue. When this is 255, the red, blue and green values are coerced to integers in 0:255 and the result is computed most efficiently.

names
: character. The names for the resulting vector.

maxColorValue
: number giving the maximum of the color values range, see above.

See Also

col2rgb the "inverse" for translating R colors to RGB vectors; rainbow, hsv, gray.

Examples

```
rgb(0,1,0)
(u01 <- seq(0,1, length=11))
stopifnot(rgb(u01,u01,u01) == gray(u01))
reds <- rgb((0:15)/15, g=0,b=0,
            names=paste("red",0:15,sep="."))
reds

rgb(0, 0:12, 0, max = 255) # integer input
```

rug   *Add a Rug to a Plot*

Description

Adds a *rug* representation (1-d plot) of the data to the plot.

Usage

```
rug(x, ticksize=0.03, side=1, lwd=0.5, col,
    quiet = getOption("warn") < 0, ...)
```

Arguments

x	A numeric vector
ticksize	The length of the ticks making up the 'rug'. Positive lengths give inwards ticks.
side	On which side of the plot box the rug will be plotted. Normally 1 (bottom) or 3 (top).
lwd	The line width of the ticks.
col	The colour the ticks are plotted in, default is black.
quiet	logical indicating if there should be a warning about clipped values.
...	further arguments, passed to axis(...), such as line or pos for specifying the location of the rug.

Details

Because of the way rug is implemented, only values of x that fall within the plot region are included. There will be a warning if any finite values are omitted, but non-finite values are omitted silently.

Because of the way colours are done the axis itself is coloured the same as the ticks. You can always replot the box in black if you don't like this feature.

References

Chambers, J. M. and Hastie, T. J. (1992) *Statistical Models in S.* Wadsworth & Brooks/Cole.

See Also

jitter which you may want for ties in x.

Examples

```
data(faithful)
with(faithful, {
  plot(density(eruptions, bw=0.15))
  rug(eruptions)
  rug(jitter(eruptions, amount = .01), side = 3,
      col = "light blue")
})
```

| screen | Creating and Controlling Multiple Screens on a Single |
| Device | |

Description

`split.screen` defines a number of regions within the current device which can, to some extent, be treated as separate graphics devices. It is useful for generating multiple plots on a single device. Screens can themselves be split, allowing for quite complex arrangements of plots.

`screen` is used to select which screen to draw in.

`erase.screen` is used to clear a single screen, which it does by filling with the background colour.

`close.screen` removes the specified screen definition(s).

Usage

```
split.screen(figs, screen = , erase = TRUE)
screen(n = , new = TRUE)
erase.screen(n = )
close.screen(n, all.screens = FALSE)
```

Arguments

figs
: A two-element vector describing the number of rows and the number of columns in a screen matrix *or* a matrix with 4 columns. If a matrix, then each row describes a screen with values for the left, right, bottom, and top of the screen (in that order) in NDC units.

screen
: A number giving the screen to be split.

erase
: logical: should be selected screen be cleared?

n
: A number indicating which screen to prepare for drawing (`screen`), erase (`erase.screen`), or close (`close.screen`).

new
: A logical value indicating whether the screen should be erased as part of the preparation for drawing in the screen.

all.screens
: A logical value indicating whether all of the screens should be closed.

Details

The first call to `split.screen` places R into split-screen mode. The other split-screen functions only work within this mode. While in this mode, certain other commands should be avoided (see WARNINGS below). Split-screen mode is exited by the command `close.screen(all = TRUE)`

Value

`split.screen` returns a vector of screen numbers for the newly-created screens. With no arguments, `split.screen` returns a vector of valid screen numbers.

`screen` invisibly returns the number of the selected screen. With no arguments, `screen` returns the number of the current screen.

`close.screen` returns a vector of valid screen numbers.

`screen`, `erase.screen`, and `close.screen` all return `FALSE` if R is not in split-screen mode.

Warning

The recommended way to use these functions is to completely draw a plot and all additions (ie. points and lines) to the base plot, prior to selecting and plotting on another screen. The behavior associated with returning to a screen to add to an existing plot is unpredictable and may result in problems that are not readily visible.

These functions are totally incompatible with the other mechanisms for arranging plots on a device: `par(mfrow)`, `par(mfcol)`, and `layout()`.

The functions are also incompatible with some plotting functions, such as `coplot`, which make use of these other mechanisms.

The functions should not be used with multiple devices.

`erase.screen` will appear not to work if the background colour is transparent (as it is by default on most devices).

References

Chambers, J. M. and Hastie, T. J. (1992) *Statistical Models in S.* Wadsworth & Brooks/Cole.

See Also

`par`, `layout`, `Devices`, `dev.*`

Examples

```
if (interactive()) {
  par(bg = "white") # default is likely to be transparent
  split.screen(c(2,1)) # split display into two screens
  # now split the bottom half into 3
  split.screen(c(1,3), screen = 2)
  screen(1) # prepare screen 1 for output
  plot(10:1)
  screen(4) # prepare screen 4 for output
  plot(10:1)
  close.screen(all = TRUE) # exit split-screen mode

  split.screen(c(2,1))      # split display into two screens
  split.screen(c(1,2),2)    # split bottom half in two
  plot(1:10)                # screen 3 is active, draw plot
  erase.screen()            # forgot label, erase and redraw
  plot(1:10, ylab= "ylab 3")
  screen(1)                 # prepare screen 1 for output
  plot(1:10)
  screen(4)                 # prepare screen 4 for output
  plot(1:10, ylab="ylab 4")
  screen(1, FALSE) # return to screen 1, but do not clear
  # overlay second plot
  plot(10:1, axes=FALSE, lty=2, ylab="")
  axis(4)  # add tic marks to right-hand axis
  title("Plot 1")
  close.screen(all = TRUE) # exit split-screen mode
}
```

segments — Add Line Segments to a Plot

Description

Draw line segments between pairs of points.

Usage

```
segments(x0, y0, x1, y1,
  col = par("fg"), lty = par("lty"), lwd = par("lwd"), ...)
```

Arguments

x0,y0	coordinates of points from which to draw.
x1,y1	coordinates of points to which to draw.
col, lty, lwd	usual graphical parameters as in par.
...	further graphical parameters (from par).

Details

For each i, a line segment is drawn between the point (x0[i], y0[i]) and the point (x1[i],y1[i]).

The graphical parameters col and lty can be used to specify a color and line texture for the line segments (col may be a vector).

References

Becker, R. A., Chambers, J. M. and Wilks, A. R. (1988) *The New S Language*. Wadsworth & Brooks/Cole.

See Also

arrows, polygon for slightly easier and less flexible line drawing, and lines for the usual polygons.

Examples

```
x <- runif(12); y <- rnorm(12)
i <- order(x,y); x <- x[i]; y <- y[i]
plot(x,y, main="arrows(.) and segments(.)")
## draw arrows from point to point :
```

```
s <- seq(length(x)-1) # one shorter than data
arrows(x[s], y[s], x[s+1], y[s+1], col= 1:3)
s <- s[-length(s)]
segments(x[s], y[s], x[s+2], y[s+2], col= 'pink')
```

base — stars

stars	Star (Spider/Radar) Plots and Segment Diagrams

Description

Draw star plots or segment diagrams of a multivariate data set. With one single location, also draws "spider" (or "radar") plots.

Usage

```
stars(x, full = TRUE, scale = TRUE, radius = TRUE,
  labels = dimnames(x)[[1]], locations = NULL,
  nrow = NULL, ncol = NULL, len = 1, key.loc = NULL,
  key.labels = dimnames(x)[[2]], key.xpd = TRUE,
  xlim = NULL, ylim = NULL, flip.labels = NULL,
  draw.segments = FALSE, col.segments = 1:n.seg,
  col.stars = NA, axes = FALSE, frame.plot = axes,
  main = NULL, sub = NULL, xlab = "", ylab = "",
  cex = 0.8, lwd = 0.25, lty = par("lty"), xpd = FALSE,
  mar = pmin(par("mar"), 1.1+ c(2*axes+ (xlab != ""),
         2*axes+ (ylab != ""), 1,0)),
  add=FALSE, plot=TRUE, ...)
```

Arguments

x	matrix or data frame of data. One star or segment plot will be produced for each row of x. Missing values (NA) are allowed, but they are treated as if they were 0 (after scaling, if relevant).
full	logical flag: if TRUE, the segment plots will occupy a full circle. Otherwise, they occupy the (upper) semi-circle only.
scale	logical flag: if TRUE, the columns of the data matrix are scaled independently so that the maximum value in each column is 1 and the minimum is 0. If FALSE, the presumption is that the data have been scaled by some other algorithm to the range [0, 1].
radius	logical flag: in TRUE, the radii corresponding to each variable in the data will be drawn.
labels	vector of character strings for labeling the plots. Unlike the S function stars, no attempt is made to construct labels if labels = NULL.

locations	Either two column matrix with the x and y coordinates used to place each of the segment plots; or numeric of length 2 when all plots should be superimposed (for a "spider plot"). By default, locations = NULL, the segment plots will be placed in a rectangular grid.
nrow, ncol	integers giving the number of rows and columns to use when locations is NULL. By default, nrow == ncol, a square layout will be used.
len	scale factor for the length of radii or segments.
key.loc	vector with x and y coordinates of the unit key.
key.labels	vector of character strings for labeling the segments of the unit key. If omitted, the second component of dimnames(x) is used, if available.
key.xpd	clipping switch for the unit key (drawing and labeling), see par("xpd").
xlim	vector with the range of x coordinates to plot.
ylim	vector with the range of y coordinates to plot.
flip.labels	logical indicating if the label locations should flip up and down from diagram to diagram. Defaults to a somewhat smart heuristic.
draw.segments	logical. If TRUE draw a segment diagram.
col.segments	color vector (integer or character, see par), each specifying a color for one of the segments (variables). Ignored if draw.segments = FALSE.
col.stars	color vector (integer or character, see par), each specifying a color for one of the stars (cases). Ignored if draw.segments = TRUE.
axes	logical flag: if TRUE axes are added to the plot.
frame.plot	logical flag: if TRUE, the plot region is framed.
main	a main title for the plot.
sub	a sub title for the plot.
xlab	a label for the x axis.
ylab	a label for the y axis.
cex	character expansion factor for the labels.
lwd	line width used for drawing.
lty	line type used for drawing.

xpd	logical or NA indicating if clipping should be done, see par(xpd = .).
mar	argument to par(mar = *), typically choosing smaller margins than by default.
...	further arguments, passed to the first call of plot(), see plot.default and to box() if frame.plot is true.
add	logical, if TRUE *add* stars to current plot.
plot	logical, if FALSE, nothing is plotted.

Details

Missing values are treated as 0.

Each star plot or segment diagram represents one row of the input x. Variables (columns) start on the right and wind counterclockwise around the circle. The size of the (scaled) column is shown by the distance from the center to the point on the star or the radius of the segment representing the variable.

Only one page of output is produced.

Note

This code started life as spatial star plots by David A. Andrews.

Prior to 1.4.1, scaling only shifted the maximum to 1, although documented as here.

Author(s)

Thomas S. Dye

References

Becker, R. A., Chambers, J. M. and Wilks, A. R. (1988) *The New S Language.* Wadsworth & Brooks/Cole.

Examples

```
data(mtcars)
stars(mtcars[, 1:7], key.loc = c(14, 2),
  main = "Motor Trend Cars: stars(*, full=F)", full=FALSE)
stars(mtcars[, 1:7], key.loc = c(14, 1.5),
  main = "Motor Trend Cars: full stars()",
  flip.labels=FALSE)
```

```
## 'Spider' or 'Radar' plot:
stars(mtcars[, 1:7], locations = c(0,0), radius = FALSE,
      key.loc=c(0,0), main="Motor Trend Cars", lty = 2)

## Segment Diagrams:
palette(rainbow(12, s = 0.6, v = 0.75))
stars(mtcars[, 1:7], len = 0.8, key.loc = c(12, 1.5),
      main = "Motor Trend Cars", draw.segments = TRUE)
stars(mtcars[, 1:7], len = 0.6, key.loc = c(1.5, 0),
      main = "Motor Trend Cars", draw.segments = TRUE,
      frame.plot=TRUE, nrow = 4, cex = .7)

data(USJudgeRatings)
## scale linearly (not affinely) to [0, 1]
USJudge <- apply(USJudgeRatings, 2, function(x) x/max(x))
Jnam <- case.names(USJudgeRatings)
Snam <-
  abbreviate(substring(Jnam,1,regexpr("[,.]",Jnam) - 1), 7)
stars(USJudge, labels = Jnam, scale = FALSE,
  key.loc = c(13, 1.5), main = "Judge not ...", len = 0.8)
stars(USJudge, labels = Snam, scale = FALSE,
  key.loc = c(13, 1.5), radius = FALSE)

loc <- stars(USJudge, labels = NULL, scale = FALSE,
  radius = FALSE, frame.plot = TRUE,
  key.loc = c(13, 1.5), main = "Judge not ...", len = 1.2)
text(loc, Snam, col = "blue", cex = 0.8, xpd = TRUE)

## 'Segments':
stars(USJudge, draw.segments = TRUE, scale = FALSE,
  key.loc = c(13,1.5))

## 'Spider':
stars(USJudgeRatings, locations=c(0,0), scale=FALSE,
  radius = FALSE, col.stars=1:10, key.loc = c(0,0),
  main="US Judges rated")
## 'Radar-Segments'
stars(USJudgeRatings[1:10,], locations = 0:1, scale=FALSE,
  draw.segments = TRUE, col.segments=0, col.stars=1:10,
  key.loc= 0:1, main="US Judges 1-10 ")
palette("default")
stars(cbind(1:16,10*(16:1)),draw.segments=TRUE,
```

```
main = "A Joke -- do not use symbols on 2D data!")
```

stripchart   *1-D Scatter Plots*

Description

stripchart produces one dimensional scatter plots (or dot plots) of the given data. These plots are a good alternative to boxplots when sample sizes are small.

Usage

```
stripchart(x, method="overplot", jitter=0.1, offset=1/3,
   vertical=FALSE, group.names, add = FALSE, at = NULL,
   xlim=NULL, ylim=NULL, main="", ylab="", xlab="",
   log="", pch=0, col=par("fg"), cex=par("cex"))
```

Arguments

x	the data from which the plots are to be produced. The data can be specified as a single vector, or as list of vectors, each corresponding to a component plot. Alternatively a symbolic specification of the form x ~ g can be given, indicating the the observations in the vector x are to be grouped according to the levels of the factor g. NAs are allowed in the data.
method	the method to be used to separate coincident points. The default method "overplot" causes such points to be overplotted, but it is also possible to specify "jitter" to jitter the points, or "stack" have coincident points stacked. The last method only makes sense for very granular data.
jitter	when jittering is used, jitter gives the amount of jittering applied.
offset	when stacking is used, points are stacked this many line-heights (symbol widths) apart.
vertical	when vertical is TRUE the plots are drawn vertically rather than the default horizontal.
group.names	group labels which will be printed alongside (or underneath) each plot.
add	logical, if true *add* boxplot to current plot.

at numeric vector giving the locations where the boxplots should be drawn, particularly when add = TRUE; defaults to 1:n where n is the number of boxes.

xlim, ylim, main, ylab, xlab, log, pch, col, cex
Graphical parameters.

Details

Extensive examples of the use of this kind of plot can be found in Box, Hunter and Hunter or Seber and Wild.

Examples

```
x <- rnorm(50)
xr<- round(x, 1)
stripchart(x) ; m <- mean(par("usr")[1:2])
text(m, 1.04, "stripchart(x, \"overplot\")")
stripchart(xr, method = "stack", add = TRUE, at = 1.2)
text(m, 1.35, "stripchart(round(x,1), \"stack\")")
stripchart(xr, method = "jitter", add = TRUE, at = 0.7)
text(m, 0.85, "stripchart(round(x,1), \"jitter\")")

data(OrchardSprays)
with(OrchardSprays,
  stripchart(decrease ~ treatment,
    main = "stripchart(Orchardsprays)", ylab = "decrease",
    vertical = TRUE, log = "y"))

with(OrchardSprays,
  stripchart(decrease ~ treatment, at = c(1:8)^2,
    main = "stripchart(Orchardsprays)", ylab = "decrease",
    vertical = TRUE, log = "y"))
```

strwidth — *Plotting Dimensions of Character Strings and Math Expressions*

Description

These functions compute the width or height, respectively, of the given strings or mathematical expressions s[i] on the current plotting device in *user* coordinates, *inches* or as fraction of the figure width par("fin").

Usage

```
strwidth(s, units = "user", cex = NULL)
strheight(s, units = "user", cex = NULL)
```

Arguments

s
: character vector or expressions whose string widths in plotting units are to be determined. An attempt is made to coerce other vectors to character, and other language objects to expressions.

units
: character indicating in which units s is measured; should be one of "user", "inches", "figure"; partial matching is performed.

cex
: character expansion to which it applies. By default, the current par("cex") is used.

Value

Numeric vector with the same length as s, giving the width or height for each s[i]. NA strings are given width and height 0 (as they are not plotted).

See Also

text, nchar

Examples

```
str.ex <- c("W","w","I",".","WwI.")
op <- par(pty='s'); plot(1:100,1:100, type="n")
sw <- strwidth(str.ex); sw
```

```
# since the last string contains the others
all.equal(sum(sw[1:4]), sw[5])

# width in [mm]
sw.i <- strwidth(str.ex, "inches"); 25.4 * sw.i
unique(sw / sw.i)
# constant factor: 1 value
mean(sw.i / strwidth(str.ex, "fig")) / par('fin')[1]

## See how letters fall in classes -- depending on graphics
## device and font!
all.lett <- c(letters, LETTERS)
# 'big points'
shL <- strheight(all.lett, units = "inches") * 72
table(shL) # all have same heights ...
mean(shL)/par("cin")[2] # around 0.6

# 'big points'
(swL <- strwidth(all.lett, units="inches") * 72)
split(all.lett, factor(round(swL, 2)))

sumex <- expression(sum(x[i], i=1,n), e^{i * pi} == -1)
strwidth(sumex)
strheight(sumex)

par(op) # reset to previous setting
```

sunflowerplot   *Produce a Sunflower Scatter Plot*

Description

Multiple points are plotted as "sunflowers" with multiple leaves ("petals") such that overplotting is visualized instead of accidental and invisible.

Usage

```
sunflowerplot(x, y = NULL, number, log = "", digits = 6,
  xlab = NULL, ylab = NULL, xlim = NULL, ylim = NULL,
  add = FALSE, rotate = FALSE,
  pch = 16, cex = 0.8, cex.fact = 1.5,
  size = 1/8, seg.col = 2, seg.lwd = 1.5, ...)
```

Arguments

x	numeric vector of x-coordinates of length n, say, or another valid plotting structure, as for `plot.default`, see also `xy.coords`.
y	numeric vector of y-coordinates of length n.
number	integer vector of length n. `number[i]` = number of replicates for (`x[i]`,`y[i]`), may be 0. Default: compute the exact multiplicity of the points `x[]`,`y[]`.
log	character indicating log coordinate scale, see `plot.default`.
digits	when `number` is computed (i.e., not specified), x and y are rounded to `digits` significant digits before multiplicities are computed.
xlab,ylab	character label for x-, or y-axis, respectively.
xlim,ylim	numeric(2) limiting the extents of the x-, or y-axis.
add	logical; should the plot be added on a previous one? Default is `FALSE`.
rotate	logical; if `TRUE`, randomly rotate the sunflowers (preventing artefacts).
pch	plotting character to be used for points (`number[i]`==1) and center of sunflowers.

cex	numeric; character size expansion of center points (s. pch).
cex.fact	numeric *shrinking* factor to be used for the center points *when there are flower leaves*, i.e., cex / cex.fact is used for these.
size	of sunflower leaves in inches, 1[in] := 2.54[cm]. Default: 1/8; approximately 3.2mm.
seg.col	color to be used for the segments which make the sunflowers leaves, see par(col=); col = "gold" reminds of real sunflowers.
seg.lwd	numeric; the line width for the leaves' segments.
...	further arguments to plot [if add=FALSE].

Details

For number[i]==1, a (slightly enlarged) usual plotting symbol (pch) is drawn. For number[i] > 1, a small plotting symbol is drawn and number[i] equi-angular "rays" emanate from it.

If rotate=TRUE and number[i] >= 2, a random direction is chosen (instead of the y-axis) for the first ray. The goal is to jitter the orientations of the sunflowers in order to prevent artefactual visual impressions.

Value

A list with three components of same length,

x	x coordinates
y	y coordinates
number	number

Side Effects

A scatter plot is drawn with "sunflowers" as symbols.

Author(s)

Andreas Ruckstuhl, Werner Stahel, Martin Maechler, Tim Hesterberg, 1989–1993. Port to R by Martin Maechler.

References

Chambers, J. M., Cleveland, W. S., Kleiner, B. and Tukey, P. A. (1983) *Graphical Methods for Data Analysis.* Wadsworth.

Schilling, M. F. and Watkins, A. E. (1994) A suggestion for sunflower plots. *The American Statistician,* 48, 303–305.

See Also

density

Examples

```
data(iris)
## 'number' is computed automatically:
sunflowerplot(iris[, 3:4])
## Imitating Chambers et al., p.109, closely:
sunflowerplot(iris[, 3:4],cex=.2, cex.f=1, size=.035,
   seg.lwd=.8)

sunflowerplot(x=sort(2*round(rnorm(100))),
              y=round(rnorm(100),0),
              main = "Sunflower Plot of Rounded N(0,1)")

## A 'point process' {explicit 'number' argument}:
sunflowerplot(rnorm(100),rnorm(100),
              number=rpois(n=100,lambda=2),
              rotate=TRUE, main="Sunflower plot")
```

base — symbols

symbols *Draw symbols on a plot*

Description

This function draws symbols on a plot. One of six symbols; *circles, squares, rectangles, stars, thermometers,* and *boxplots,* can be plotted at a specified set of x and y coordinates. Specific aspects of the symbols, such as relative size, can be customized by additional parameters.

Usage

```
symbols(x, y = NULL, circles, squares, rectangles, stars,
    thermometers, boxplots, inches = TRUE, add = FALSE,
    fg = 1, bg = NA, xlab = NULL, ylab = NULL, main = NULL,
    xlim = NULL, ylim = NULL, ...)
```

Arguments

x, y
: the x and y co-ordinates for the symbols. They can be specified in any way which is accepted by xy.coords.

circles
: a vector giving the radii of the circles.

squares
: a vector giving the length of the sides of the squares.

rectangles
: a matrix with two columns. The first column gives widths and the second the heights of rectangle symbols.

stars
: a matrix with three or more columns giving the lengths of the rays from the center of the stars. NA values are replaced by zeroes.

thermometers
: a matrix with three or four columns. The first two columns give the width and height of the thermometer symbols. If there are three columns, the third is taken as a proportion. The thermometers are filled from their base to this proportion of their height. If there are four columns, the third and fourth columns are taken as proportions. The thermometers are filled between these two proportions of their heights.

boxplots
: a matrix with five columns. The first two columns give the width and height of the boxes, the next two columns give the lengths of the lower and upper whiskers and the fifth the proportion (with a warning

	if not in [0,1]) of the way up the box that the median line is drawn.
inches	If inches is FALSE, the units are taken to be those of the x axis. If inches is TRUE, the symbols are scaled so that the largest symbol is one inch in height. If a number is given the symbols are scaled to make the largest symbol this height in inches.
add	if add is TRUE, the symbols are added to an existing plot, otherwise a new plot is created.
fg	colors the symbols are to be drawn in (the default is the value of the col graphics parameter).
bg	if specified, the symbols are filled with this color. The default is to leave the symbols unfilled.
xlab	the x label of the plot if add is not true; this applies to the following arguments as well. Defaults to the deparsed expression used for x.
ylab	the y label of the plot.
main	a main title for the plot.
xlim	numeric of length 2 giving the x limits for the plot.
ylim	numeric of length 2 giving the y limits for the plot.
...	graphics parameters can also be passed to this function.

Details

Observations which have missing coordinates or missing size parameters are not plotted. The exception to this is *stars*. In that case, the length of any rays which are NA is reset to zero.

Circles of radius zero are plotted at radius one pixel (which is device-dependent).

References

Becker, R. A., Chambers, J. M. and Wilks, A. R. (1988) *The New S Language*. Wadsworth & Brooks/Cole.

W. S. Cleveland (1985) *The Elements of Graphing Data*. Monterey, California: Wadsworth.

See Also

stars for drawing *stars* with a bit more flexibility; sunflowerplot.

base — symbols

Examples

```
x <- 1:10
y <- sort(10*runif(10))
z <- runif(10)
z3 <- cbind(z, 2*runif(10), runif(10))
symbols(x, y, thermometers=cbind(.5, 1, z), inches=.5,
        fg = 1:10)
symbols(x, y, thermometers = z3, inches=FALSE)
text(x,y,
  apply(format(round(z3, dig=2)), 1, paste, collapse=","),
  adj = c(-.2,0), cex = .75, col = "purple", xpd=NA)

data(trees)
## Note that  example(trees) shows more sensible plots!
N <- nrow(trees)
attach(trees)
## Girth is diameter in inches
symbols(Height, Volume, circles=Girth/24, inches=FALSE,
        main="Trees' Girth") # xlab and ylab automatically
## Colors too:
palette(rainbow(N, end = 0.9))
symbols(Height, Volume, circles=Girth/16, inches=FALSE,
        bg = 1:N, fg="gray30",
        main = "symbols(*, circles=Girth/16, bg = 1:N)")
palette("default"); detach()
```

termplot   *Plot regression terms*

Description

Plots regression terms against their predictors, optionally with standard errors and partial residuals added.

Usage

```
termplot(model, data=NULL,
  envir=environment(formula(model)),
  partial.resid=FALSE, rug=FALSE,
  terms=NULL, se=FALSE, xlabs=NULL, ylabs=NULL, main=NULL,
  col.term = 2, lwd.term = 1.5,
  col.se = "orange", lty.se = 2, lwd.se = 1,
  col.res = "gray", cex = 1, pch = par("pch"),
  ask = interactive() && nb.fig < n.tms
                       && .Device != "postscript",
  use.factor.levels=TRUE, ...)
```

Arguments

model	fitted model object
data	data frame in which variables in model can be found
envir	environment in which variables in model can be found
partial.resid	logical; should partial residuals be plotted?
rug	add rugplots (jittered 1-d histograms) to the axes?
terms	which terms to plot (default NULL means all terms)
se	plot pointwise standard errors?
xlabs	vector of labels for the x axes
ylabs	vector of labels for the y axes
main	logical, or vector of main titles; if TRUE, the model's call is taken as main title, NULL or FALSE mean no titles.
col.term, lwd.term	color and line width for the "term curve", see lines.
col.se, lty.se, lwd.se	color, line type and line width for the "twice-standard-error curve" when se = TRUE.

col.res, cex, pch
: color, plotting character expansion and type for partial residuals, when `partial.resid = TRUE`, see points.

ask
: logical; if TRUE, the user is *ask*ed before each plot, see par(ask=.).

use.factor.levels
: Should x-axis ticks use factor levels or numbers for factor terms?

...
: other graphical parameters

Details

The model object must have a `predict` method that accepts `type=terms`, eg `glm` in the **base** package, `coxph` and `survreg` in the **survival** package.

For the `partial.resid=TRUE` option it must have a `residuals` method that accepts `type="partial"`, which `lm` and `glm` do.

The `data` argument should rarely be needed, but in some cases `termplot` may be unable to reconstruct the original data frame.

Nothing sensible happens for interaction terms.

See Also

For (generalized) linear models, `plot.lm` and `predict.glm`.

Examples

```
had.splines <- "package:splines" %in% search()
if(!had.splines) rs <- require(splines)
x <- 1:100
z <- factor(rep(LETTERS[1:4],25))
y <- rnorm(100,sin(x/10)+as.numeric(z))
model <- glm(y ~ ns(x,6) + z)

par(mfrow=c(2,2)) ## 2 x 2 plots for same model :
termplot(model,
  main = paste("termplot( ", deparse(model$call)," ...)"))
termplot(model, rug=TRUE)
termplot(model, partial=TRUE, rug= TRUE,
         main="termplot(..., partial = TRUE, rug = TRUE)")
termplot(model, partial=TRUE, se = TRUE, main = TRUE)
if(!had.splines && rs) detach("package:splines")
```

text *Add Text to a Plot*

Description

`text` draws the strings given in the vector `labels` at the coordinates given by `x` and `y`. `y` may be missing since `xy.coords(x,y)` is used for construction of the coordinates.

Usage

```
text(x, ...)

## Default S3 method:
text (x, y = NULL, labels = seq(along = x), adj = NULL,
      pos = NULL, offset = 0.5, vfont = NULL,
      cex = 1, col = NULL, font = NULL, xpd = NULL, ...)
```

Arguments

x, y
: numeric vectors of coordinates where the text `labels` should be written. If the length of `x` and `y` differs, the shorter one is recycled.

labels
: one or more character strings or expressions specifying the *text* to be written. An attempt is made to coerce other vectors to character, and other language objects to expressions.

adj
: one or two values in [0, 1] which specify the x (and optionally y) adjustment of the labels. On most devices values outside that interval will also work.

pos
: a position specifier for the text. If specified, this overrides any `adj` value given. Values of 1, 2, 3 and 4, respectively indicate positions below, to the left of, above and to the right of the specified coordinates.

offset
: when `pos` is specified, this value gives the offset of the label from the specified coordinate in fractions of a character width.

vfont
: if a character vector of length 2 is specified, then Hershey vector fonts are used. The first element of the vector selects a typeface and the second element selects a style.

base — text

`cex`	numeric character expansion factor; multiplied by `par("cex")` yields the final character size.
`col, font`	the color and font to be used; these default to the values of the global graphical parameters in `par()`.
`xpd`	(where) should clipping take place? Defaults to `par("xpd")`.
`...`	further graphical parameters (from `par`).

Details

`labels` must be of type `character` or `expression` (or be coercible to such a type). In the latter case, quite a bit of mathematical notation is available such as sub- and superscripts, greek letters, fractions, etc.

`adj` allows *adj*ustment of the text with respect to (x,y). Values of 0, 0.5, and 1 specify left/bottom, middle and right/top, respectively. The default is for centered text, i.e., `adj = c(0.5, 0.5)`. Accurate vertical centering needs character metric information on individual characters, which is only available on some devices.

The `pos` and `offset` arguments can be used in conjunction with values returned by `identify` to recreate an interactively labelled plot.

Text can be rotated by using graphical parameters `srt` (see `par`); this rotates about the centre set by `adj`.

Graphical parameters `col`, `cex` and `font` can be vectors and will then be applied cyclically to the `labels` (and extra values will be ignored).

Labels whose x, y, `labels`, `cex` or `col` value is `NA` are omitted from the plot.

References

Becker, R. A., Chambers, J. M. and Wilks, A. R. (1988) *The New S Language.* Wadsworth & Brooks/Cole.

See Also

`mtext`, `title`, `Hershey` for details on Hershey vector fonts, `plotmath` for details and more examples on mathematical annotation.

Examples

```
plot(-1:1,-1:1, type = "n", xlab = "Re", ylab = "Im")
K <- 16; text(exp(1i * 2 * pi * (1:K) / K), col = 2)
```

```
plot(1:10, 1:10,
     main = "text(...) examples\n~~~~~~~~~~~~~~~")
points(c(6,2), c(2,1), pch = 3, cex = 4, col = "red")
text(6, 2,
  "the text is CENTERED around (x,y) = (6,2) by default",
   cex = .8)
text(2, 1,
  "or Left/Bottom - JUSTIFIED at (2,1) by 'adj = c(0,0)'",
  adj = c(0,0))
text(4, 9,
  expression(hat(beta) == (X^t * X)^{-1} * X^t * y))
text(4, 8.4,
  "expression(hat(beta) == (X^t * X)^{-1} * X^t * y)",
  cex = .75)
text(4, 7,
  expression(bar(x) == sum(frac(x[i], n), i==1, n))
)
```

title *Plot Annotation*

Description

This function can be used to add labels to a plot. Its first four principal arguments can also be used as arguments in most high-level plotting functions. They must be of type `character` or `expression`. In the latter case, quite a bit of mathematical notation is available such as sub- and superscripts, greek letters, fractions, etc.

Usage

```
title(main = NULL, sub = NULL, xlab = NULL, ylab = NULL,
      line = NA, outer = FALSE, ...)
```

Arguments

main	The main title (on top) using font and size (character expansion) `par("font.main")` and color `par("col.main")`.
sub	Sub-title (at bottom) using font and size `par("font.sub")` and color `par("col.sub")`.
xlab	X axis label using font and character expansion `par("font.axis")` and color `par("col.axis")`.
ylab	Y axis label, same font attributes as `xlab`.
line	specifying a value for `line` overrides the default placement of labels, and places them this many lines from the plot.
outer	a logical value. If `TRUE`, the titles are placed in the outer margins of the plot.
...	further graphical parameters from `par`. Use e.g., `col.main` or `cex.sub` instead of just `col` or `cex`.

Details

The labels passed to title can be simple strings or expressions, or they can be a list containing the string to be plotted, and a selection of the optional modifying graphical parameters `cex=`, `col=`, `font=`.

References

Becker, R. A., Chambers, J. M. and Wilks, A. R. (1988) *The New S Language*. Wadsworth & Brooks/Cole.

See Also

mtext, text; plotmath for details on mathematical annotation.

Examples

```
data(cars)
plot(cars, main = "") # here, could use main directly
title(main = "Stopping Distance versus Speed")

plot(cars, main = "")
title(main = list("Stopping Distance versus Speed",
                cex=1.5, col="red", font=3))

## Specifying "..." :
plot(1, col.axis = "sky blue", col.lab = "thistle")
title("Main Title", sub = "sub title",
      cex.main = 2,   font.main= 4, col.main= "blue",
      cex.sub = 0.75, font.sub = 3, col.sub = "red")

x <- seq(-4, 4, len = 101)
y <- cbind(sin(x), cos(x))
matplot(x, y, type = "l", xaxt = "n",
  main = expression(paste(plain(sin) * phi, "  and  ",
                          plain(cos) * phi)),
  # only 1st is taken
  ylab = expression("sin" * phi, "cos" * phi),
  xlab = expression(paste("Phase Angle ", phi)),
  col.main = "blue")
axis(1, at = c(-pi, -pi/2, 0, pi/2, pi),
     lab = expression(-pi, -pi/2, 0, pi/2, pi))
abline(h = 0, v = pi/2 * c(-1,1), lty = 2, lwd = .1,
       col = "gray70")
```

units *Graphical Units*

Description

xinch and yinch convert the specified number of inches given as their arguments into the correct units for plotting with graphics functions. Usually, this only makes sense when normal coordinates are used, i.e., *no* log scale (see the log argument to par).

xyinch does the same for a pair of numbers xy, simultaneously.

cm translates inches in to cm (centimeters).

Usage

```
xinch(x = 1, warn.log = TRUE)
yinch(y = 1, warn.log = TRUE)
xyinch(xy = 1, warn.log = TRUE)
cm(x)
```

Arguments

x,y	numeric vector
xy	numeric of length 1 or 2.
warn.log	logical; if TRUE, a warning is printed in case of active log scale.

Examples

```
all(c(xinch(),yinch()) == xyinch()) # TRUE
xyinch()
xyinch # to see that is really delta{"usr"} / "pin"

cm(1) # = 2.54

## plot labels offset 0.12 inches to the right of plotted
## symbols in a plot
data(mtcars)
with(mtcars, {
    plot(mpg, disp, pch=19, main= "Motor Trend Cars")
    text(mpg + xinch(0.12), disp, row.names(mtcars),
         adj = 0, cex = .7, col = 'blue')
})
```

x11 X Window System Graphics

Description

X11 starts a graphics device driver for the X Window System (version 11). This can only be done on machines that run X. x11 is recognized as a synonym for X11.

Usage

```
X11(display = "", width = 7, height = 7, pointsize = 12,
    gamma = 1, colortype = getOption("X11colortype"),
    maxcubesize = 256, canvas = "white")
```

Arguments

display
: the display on which the graphics window will appear. The default is to use the value in the user's environment variable DISPLAY.

width
: the width of the plotting window in inches.

height
: the height of the plotting window in inches.

pointsize
: the default pointsize to be used.

gamma
: the gamma correction factor. This value is used to ensure that the colors displayed are linearly related to RGB values. A value of around 0.5 is appropriate for many PC displays. A value of 1.0 (no correction) is usually appropriate for high-end displays or Macintoshs.

colortype
: the kind of color model to be used. The possibilities are "mono", "gray", "pseudo", "pseudo.cube" and "true". Ignored if an X11 is already open.

maxcubesize
: can be used to limit the size of color cube allocated for pseudocolor devices.

canvas
: color. The color of the canvas, which is visible only when the background color is transparent.

Details

By default, an X11 device will use the best color rendering strategy that it can. The choice can be overriden with the `colortype` parameter. A value of "mono" results in black and white graphics, "gray" in grayscale and "true" in truecolor graphics (if this is possible). The values "pseudo" and "pseudo.cube" provide color strategies for pseudocolor displays. The first strategy provides on-demand color allocation which produces exact colors until the color resources of the display are exhausted. The second causes a standard color cube to be set up, and requested colors are approximated by the closest value in the cube. The default strategy for pseudocolor displays is "pseudo".

Note: All X11 devices share a `colortype` which is set by the first device to be opened. To change the `colortype` you need to close *all* open X11 devices then open one with the desired `colortype`.

With `colortype` equal to "pseudo.cube" or "gray" successively smaller palettes are tried until one is completely allocated. If allocation of the smallest attempt fails the device will revert to "mono".

See Also

`Devices`.

xfig   *XFig Graphics Device*

Description

xfig starts the graphics device driver for producing XFig (version 3.2) graphics.

The auxiliary function `ps.options` can be used to set and view (if called without arguments) default values for the arguments to `xfig` and `postscript`.

Usage

```
xfig(file = ifelse(onefile, "Rplots.fig", "Rplot%03d.fig"),
     onefile = FALSE, ...)
```

Arguments

file
: a character string giving the name of the file. If it is `""`, the output is piped to the command given by the argument command. For use with `onefile=FALSE` give a `printf` format such as `"Rplot%d.fig"` (the default in that case).

onefile
: logical: if true allow multiple figures in one file. If false, assume only one page per file and generate a file number containing the page number.

...
: further arguments to `ps.options` accepted by `xfig()`:

   paper the size of paper in the printer. The choices are `"A4"`, `"Letter"` and `"Legal"` (and these can be lowercase). A further choice is `"default"`, which is the default. If this is selected, the papersize is taken from the option `"papersize"` if that is set and to `"A4"` if it is unset or empty.

   horizontal the orientation of the printed image, a logical. Defaults to true, that is landscape orientation.

   width, height the width and height of the graphics region in inches. The default is to use the entire page less a 0.25 inch border.

> `family` the font family to be used. This must be one of `"AvantGarde"`, `"Bookman"`, `"Courier"`, `"Helvetica"`, `"Helvetica-Narrow"`, `"NewCenturySchoolbook"`, `"Palatino"` or `"Times"`.
>
> `pointsize` the default point size to be used.
>
> `bg` the default background color to be used.
>
> `fg` the default foreground color to be used.
>
> `pagecentre` logical: should the device region be centred on the page: defaults to `TRUE`.

Details

Although `xfig` can produce multiple plots in one file, the XFig format does not say how to separate or view them. So `onefile=FALSE` is the default.

Note

On some line textures (0 <= lty > 4) are used. Eventually this will be partially remedied, but the XFig file format does not allow as general line textures as the R model. Unimplemented line textures are displayed as *dash-double-dotted*.

There is a limit of 512 colours (plus white and black) per file.

See Also

`Devices`, `postscript`, `ps.options`.

xy.coords   *Extracting Plotting Structures*

Description

xy.coords is used by many functions to obtain x and y coordinates for plotting. The use of this common mechanism across all R functions produces a measure of consistency.

Usage

```
xy.coords(x, y, xlab = NULL, ylab = NULL, log = NULL,
         recycle = FALSE)
```

Arguments

x, y   the x and y coordinates of a set of points. Alternatively, a single argument x can be provided.

xlab, ylab   names for the x and y variables to be extracted.

log   character, "x", "y" or both, as for plot. Sets negative values to NA and gives a warning.

recycle   logical; if TRUE, recycle (rep) the shorter of x or y if their lengths differ.

Details

An attempt is made to interpret the arguments x and y in a way suitable for plotting.

If y is missing and x is a

formula: of the form yvar ~ xvar. xvar and yvar are used as x and y variables.

list: containing components x and y, these are used to define plotting coordinates.

time series: the x values are taken to be time(x) and the y values to be the time series.

matrix with two columns: the first is assumed to contain the x values and the second the y values.

In any other case, the x argument is coerced to a vector and returned as y component where the resulting x is just the index vector 1:n. In this case, the resulting xlab component is set to "Index".

If x (after transformation as above) inherits from class "POSIXt" it is coerced to class "POSIXct".

Value

A list with the components

x	numeric (i.e., "double") vector of abscissa values.
y	numeric vector of the same length as x.
xlab	character(1) or NULL, the 'label' of x.
ylab	character(1) or NULL, the 'label' of y.

See Also

plot.default, lines, points and lowess are examples of functions which use this mechanism.

Examples

```
xy.coords(fft(c(1:10)), NULL)
data(cars) ; attach(cars)
xy.coords(dist ~ speed, NULL)$xlab # = "speed"

str(xy.coords(1:3, 1:2, recycle=TRUE))
str(xy.coords(-2:10,NULL, log="y"))
## warning: 3 y values <=0 omitted ..
detach()
```

xyz.coords *Extracting Plotting Structures*

Description

Utility for obtaining consistent x, y and z coordinates and labels for three dimensional (3D) plots.

Usage

```
xyz.coords(x, y, z, xlab=NULL, ylab=NULL, zlab=NULL,
          log=NULL, recycle=FALSE)
```

Arguments

x, y, z   the x, y and z coordinates of a set of points. Alternatively, a single argument x can be provided. In this case, an attempt is made to interpret the argument in a way suitable for plotting.

If the argument is a formula zvar ~ xvar + yvar, xvar, yvar and zvar are used as x, y and z variables; if the argument is a list containing components x, y and z, these are assumed to define plotting coordinates; if the argument is a matrix with three columns, the first is assumed to contain the x values, etc.

Alternatively, two arguments x and y can be provided. One may be real, the other complex; in any other case, the arguments are coerced to vectors and the values plotted against their indices.

xlab, ylab, zlab
    names for the x, y and z variables to be extracted.

log   character, "x", "y", "z" or combinations. Sets negative values to NA and gives a warning.

recycle   logical; if TRUE, recycle (rep) the shorter ones of x, y or z if their lengths differ.

Value

A list with the components

x   numeric (i.e., double) vector of abscissa values.

y   numeric vector of the same length as x.

base — xyz.coords

z	numeric vector of the same length as x.
xlab	character(1) or NULL, the axis label of x.
ylab	character(1) or NULL, the axis label of y.
zlab	character(1) or NULL, the axis label of z.

Author(s)

Uwe Ligges and Martin Maechler

See Also

xy.coords for 2D.

Examples

```
str(xyz.coords(data.frame(10*1:9, -4),y=NULL,z=NULL))

str(xyz.coords(1:6, fft(1:6),z=NULL,xlab="X", ylab="Y"))

y <- 2 * (x2 <- 10 + (x1 <- 1:10))
str(xyz.coords(y ~ x1 + x2,y=NULL,z=NULL))

str(xyz.coords(data.frame(x=-1:9,y=2:12,z=3:13),
    y=NULL, z=NULL, log="xy"))
## Warning message: 2 x values <= 0 omitted ...
```

Chapter 2

Base package — math

abs	*Miscellaneous Mathematical Functions*

Description

These functions compute miscellaneous mathematical functions. The naming follows the standard for computer languages such as C or Fortran.

Usage

```
abs(x)
sqrt(x)
```

Arguments

x   a numeric vector

References

Becker, R. A., Chambers, J. M. and Wilks, A. R. (1988) *The New S Language.* Wadsworth & Brooks/Cole.

See Also

Arithmetic for simple, log for logarithmic, sin for trigonometric, and Special for special mathematical functions.

Examples

```
xx <- -9:9
plot(xx, sqrt(abs(xx)),  col = "red")
lines(spline(xx, sqrt(abs(xx)), n=101), col = "pink")
```

all.equal *Test if Two Objects are (Nearly) Equal*

Description

all.equal(x,y) is a utility to compare R objects x and y testing "near equality". If they are different, comparison is still made to some extent, and a report of the differences is returned. Don't use all.equal directly in if expressions—either use identical or combine the two, as shown in the documentation for identical.

Usage

```
all.equal(target, current, ...)

## S3 method for class 'numeric':
all.equal(target, current,
    tolerance= .Machine$double.eps ^ 0.5, scale=NULL, ...)
```

Arguments

target	R object.
current	other R object, to be compared with target.
...	Further arguments for different methods, notably the following two, for numerical comparison:
tolerance	numeric ≥ 0. Differences smaller than tolerance are not considered.
scale	numeric scalar > 0 (or NULL). See Details.

Details

There are several methods available, most of which are dispatched by the default method, see methods("all.equal"). all.equal.list and all.equal.language provide comparison of recursive objects.

Numerical comparisons for scale = NULL (the default) are done by first computing the mean absolute difference of the two numerical vectors. If this is smaller than tolerance or not finite, absolute differences are used, otherwise relative differences scaled by the mean absolute difference.

If scale is positive, absolute comparisons are after scaling (dividing) by scale.

For complex arguments, the modulus Mod of the difference is used.

attr.all.equal is used for comparing attributes, returning NULL or character.

Value

Either TRUE or a vector of mode "character" describing the differences between target and current.

Numerical differences are reported by relative error

References

Chambers, J. M. (1998) *Programming with Data. A Guide to the S Language.* Springer (for =).

See Also

==, and all for exact equality testing.

Examples

```
# not precise enough (default tol) > relative error
all.equal(pi, 355/113)

d45 <- pi*(1/4 + 1:10)
stopifnot(
all.equal(tan(d45), rep(1,10))) # TRUE, but
all       (tan(d45) == rep(1,10)) # FALSE, since not exactly
all.equal(tan(d45), rep(1,10), tol=0) # to see difference

all.equal(options(), .Options)
all.equal(options(), as.list(.Options)) # TRUE
.Options $ myopt <- TRUE
all.equal(options(), as.list(.Options))
rm(.Options)
```

base — approxfun 247

approxfun *Interpolation Functions*

Description

Return a list of points which linearly interpolate given data points, or a function performing the linear (or constant) interpolation.

Usage

```
approx   (x, y = NULL, xout, method="linear", n=50,
          yleft, yright, rule = 1, f=0, ties = mean)

approxfun(x, y = NULL,       method="linear",
          yleft, yright, rule = 1, f=0, ties = mean)
```

Arguments

x, y	vectors giving the coordinates of the points to be interpolated. Alternatively a single plotting structure can be specified: see xy.coords.
xout	an optional set of values specifying where interpolation is to take place.
method	specifies the interpolation method to be used. Choices are "linear" or "constant".
n	If xout is not specified, interpolation takes place at n equally spaced points spanning the interval [min(x), max(x)].
yleft	the value to be returned when input x values are less than min(x). The default is defined by the value of rule given below.
yright	the value to be returned when input x values are greater than max(x). The default is defined by the value of rule given below.
rule	an integer describing how interpolation is to take place outside the interval [min(x), max(x)]. If rule is 1 then NAs are returned for such points and if it is 2, the value at the closest data extreme is used.
f	For method="constant" a number between 0 and 1 inclusive, indicating a compromise between left- and

right-continuous step functions. If y0 and y1 are the values to the left and right of the point then the value is y0*(1-f)+y1*f so that f=0 is right-continuous and f=1 is left-continuous.

ties Handling of tied x values. Either a function with a single vector argument returning a single number result or the string "ordered".

Details

The inputs can contain missing values which are deleted, so at least two complete (x, y) pairs are required. If there are duplicated (tied) x values and ties is a function it is applied to the y values for each distinct x value. Useful functions in this context include mean, min, and max. If ties="ordered" the x values are assumed to be already ordered. The first y value will be used for interpolation to the left and the last one for interpolation to the right.

Value

approx returns a list with components x and y, containing n coordinates which interpolate the given data points according to the method (and rule) desired.

The function approxfun returns a function performing (linear or constant) interpolation of the given data points. For a given set of x values, this function will return the corresponding interpolated values. This is often more useful than approx.

References

Becker, R. A., Chambers, J. M. and Wilks, A. R. (1988) *The New S Language*. Wadsworth & Brooks/Cole.

See Also

spline and splinefun for spline interpolation.

Examples

```
x <- 1:10
y <- rnorm(10)
par(mfrow = c(2,1))
plot(x, y, main = "approx(.) and approxfun(.)")
points(approx(x, y), col = 2, pch = "*")
```

```
points(approx(x, y, method = "constant"), col=4, pch="*")

f <- approxfun(x, y)
curve(f(x), 0, 10, col = "green")
points(x, y)
is.function(fc <- approxfun(x, y, method = "const")) # TRUE
curve(fc(x), 0, 10, col = "darkblue", add = TRUE)

## Show treatment of 'ties' :
x <- c(2,2:4,4,4,5,5,7,7,7)
y <- c(1:6, 5:4, 3:1)
approx(x,y, xout=x)$y # warning
(ay <- approx(x,y, xout=x, ties = "ordered")$y)
stopifnot(ay == c(2,2,3,6,6,6,4,4,1,1,1))
approx(x,y, xout=x, ties = min)$y
approx(x,y, xout=x, ties = max)$y
```

Arithmetic	*Arithmetic Operators*

Description

These binary operators perform arithmetic on vector objects.

Usage

```
x + y
x - y
x * y
x / y
x ^ y
x %% y
x %/% y
```

Details

1 ^ y and y ^ 0 are 1, *always*. x ^ y should also give the proper "limit" result when either argument is infinite (i.e., ±Inf).

Objects such as arrays or time-series can be operated on this way provided they are conformable.

Value

They return numeric vectors containing the result of the element by element operations. The elements of shorter vectors are recycled as necessary (with a warning when they are recycled only *fractionally*). The operators are + for addition, - for subtraction, * for multiplication, / for division and ^ for exponentiation.

%% indicates x mod y and %/% indicates integer division. It is guaranteed that x == (x %% y) + y * (x %/% y) unless y == 0 where the result is NA or NaN (depending on the typeof of the arguments).

References

Becker, R. A., Chambers, J. M. and Wilks, A. R. (1988) *The New S Language.* Wadsworth & Brooks/Cole.

base — Arithmetic

See Also

sqrt for miscellaneous and **Special** for special mathematical functions.
Syntax for operator precedence.

Examples

```
x <- -1:12
x + 1
2 * x + 3
x %% 2 # is periodic
x %/% 5
```

backsolve   *Solve an Upper or Lower Triangular System*

Description

Solves a system of linear equations where the coefficient matrix is upper or lower triangular.

Usage

```
backsolve(r, x, k=ncol(r), upper.tri=TRUE, transpose=FALSE)
forwardsolve(l, x, k=ncol(l), upper.tri=FALSE,
   transpose=FALSE)
```

Arguments

r,l
: an upper (or lower) triangular matrix giving the coefficients for the system to be solved. Values below (above) the diagonal are ignored.

x
: a matrix whose columns give "right-hand sides" for the equations.

k
: The number of columns of r and rows of x to use.

upper.tri
: logical; if TRUE (default), the *upper tri*angular part of r is used. Otherwise, the lower one.

transpose
: logical; if TRUE, solve $r' * y = x$ for y, i.e., t(r) %*% y == x.

Value

The solution of the triangular system. The result will be a vector if x is a vector and a matrix if x is a matrix.

References

Becker, R. A., Chambers, J. M. and Wilks, A. R. (1988) *The New S Language*. Wadsworth & Brooks/Cole.

Dongarra, J. J., Bunch, J. R., Moler, C. B. and Stewart, G. W. (1978) *LINPACK Users Guide*. Philadelphia: SIAM Publications.

See Also

chol, qr, solve.

Examples

```
## upper triangular matrix 'r':
r <- rbind(c(1,2,3),
           c(0,1,1),
           c(0,0,2))
( y <- backsolve(r, x <- c(8,4,2)) ) # -1 3 1
r %*% y # == x = (8,4,2)
backsolve(r, x, transpose = TRUE) # 8 -12 -5
```

Bessel *Bessel Functions*

Description

Bessel Functions of integer and fractional order, of first and second kind, J_ν and Y_ν, and Modified Bessel functions (of first and third kind), I_ν and K_ν.

gammaCody is the (Γ) function as from the Specfun package and originally used in the Bessel code.

Usage

```
besselI(x, nu, expon.scaled = FALSE)
besselK(x, nu, expon.scaled = FALSE)
besselJ(x, nu)
besselY(x, nu)
gammaCody(x)
```

Arguments

x	numeric, ≥ 0.
nu	numeric; The *order* (maybe fractional!) of the corresponding Bessel function.
expon.scaled	logical; if TRUE, the results are exponentially scaled in order to avoid overflow (I_ν) or underflow (K_ν), respectively.

Details

The underlying C code stems from *Netlib* (http://www.netlib.org/specfun/r[ijky]besl).

If expon.scaled = TRUE, $e^{-x}I_\nu(x)$, or $e^x K_\nu(x)$ are returned.

gammaCody may be somewhat faster but less precise and/or robust than R's standard gamma. It is here for experimental purpose mainly, and *may be defunct very soon.*

For $\nu < 0$, formulae 9.1.2 and 9.6.2 from the reference below are applied (which is probably suboptimal), except for besselK which is symmetric in nu.

Value

Numeric vector of the same length of x with the (scaled, if expon.scale=TRUE) values of the corresponding Bessel function.

Author(s)

Original Fortran code: W. J. Cody, Argonne National Laboratory
Translation to C and adaption to R: Martin Maechler.

References

Abramowitz, M. and Stegun, I. A. (1972) *Handbook of Mathematical Functions*. Dover, New York; Chapter 9: Bessel Functions of Integer Order.

See Also

Other special mathematical functions, such as gamma, $\Gamma(x)$, and beta, $B(x)$.

Examples

```
nus <- c(0:5,10,20)

x <- seq(0,4, len= 501)
plot(x,x, ylim = c(0,6), ylab="",type='n',
     main = "Bessel Functions I_nu(x)")
for(nu in nus) lines(x,besselI(x,nu=nu), col = nu+2)
legend(0,6, leg=paste("nu=",nus), col = nus+2, lwd=1)

x <- seq(0,40,len=801); yl <- c(-.8,.8)
plot(x,x, ylim = yl, ylab="",type='n',
     main = "Bessel Functions J_nu(x)")
for(nu in nus) lines(x,besselJ(x,nu=nu), col = nu+2)
legend(32,-.18, leg=paste("nu=",nus), col = nus+2, lwd=1)

## Negative nu's :
xx <- 2:7
nu <- seq(-10,9, len = 2001)
op <- par(lab = c(16,5,7))
matplot(nu, t(outer(xx,nu, besselI)), type = 'l',
  ylim = c(-50,200),
  main = expression(paste("Bessel ",I[nu](x)," for fixed ",
```

```
                          x, ", as ",f(nu))),
    xlab = expression(nu))
abline(v=0, col = "light gray", lty = 3)
legend(5,200, leg = paste("x=",xx), col=seq(xx),
    lty=seq(xx))
par(op)


x0 <- 2^(-20:10)
plot(x0,x0^-8, log='xy', ylab="",type='n',
    main = "Bessel Functions J_nu(x) near 0\n log-log scale")
for(nu in sort(c(nus,nus+.5)))
    lines(x0,besselJ(x0,nu=nu), col = nu+2)
legend(3,1e50, leg=paste("nu=", paste(nus,nus+.5,sep=",")),
    col=nus+2, lwd=1)


plot(x0,x0^-8, log='xy', ylab="",type='n',
    main = "Bessel Functions K_nu(x) near 0\n log-log scale")
for(nu in sort(c(nus,nus+.5)))
    lines(x0,besselK(x0,nu=nu), col = nu+2)
legend(3,1e50,leg=paste("nu=", paste(nus,nus+.5, sep=",")),
    col=nus+2, lwd=1)


x <- x[x > 0]
plot(x,x, ylim=c(1e-18,1e11),log="y", ylab="",type='n',
      main = "Bessel Functions  K_nu(x)")
for(nu in nus) lines(x,besselK(x,nu=nu), col = nu+2)
legend(0,1e-5, leg=paste("nu=",nus), col = nus+2, lwd=1)


yl <- c(-1.6, .6)
plot(x,x, ylim = yl, ylab="",type='n',
      main = "Bessel Functions  Y_nu(x)")
for(nu in nus){
    xx <- x[x > .6*nu];
    lines(xx,besselY(xx,nu=nu), col = nu+2)
}
legend(25,-.5, leg=paste("nu=",nus), col = nus+2, lwd=1)
```

| chol | The Choleski Decomposition |

Description

Compute the Choleski factorization of a real symmetric positive-definite square matrix.

Usage

```
chol(x, pivot = FALSE,  LINPACK = pivot)
La.chol(x)
```

Arguments

x	a real symmetric, positive-definite matrix
pivot	Should pivoting be used?
LINPACK	logical. Should LINPACK be used (for compatibility with R $<$ 1.7.0)?

Details

chol(pivot = TRUE) provides an interface to the LINPACK routine DCHDC. La.chol provides an interface to the LAPACK routine DPOTRF.

Note that only the upper triangular part of x is used, so that $R'R = x$ when x is symmetric.

If pivot = FALSE and x is not non-negative definite an error occurs. If x is positive semi-definite (i.e., some zero eigenvalues) an error will also occur, as a numerical tolerance is used.

If pivot = TRUE, then the Choleski decomposition of a positive semi-definite x can be computed. The rank of x is returned as attr(Q, "rank"), subject to numerical errors. The pivot is returned as attr(Q, "pivot"). It is no longer the case that t(Q) %*% Q equals x. However, setting pivot <- attr(Q, "pivot") and oo <- order(pivot), it is true that t(Q[, oo]) %*% Q[, oo] equals x, or, alternatively, t(Q) %*% Q equals x[pivot, pivot]. See the examples.

Value

The upper triangular factor of the Choleski decomposition, i.e., the matrix R such that $R'R = x$ (see example).

If pivoting is used, then two additional attributes "pivot" and "rank" are also returned.

Warning

The code does not check for symmetry.

If pivot = TRUE and x is not non-negative definite then there will be no error message but a meaningless result will occur. So only use pivot = TRUE when x is non-negative definite by construction.

References

Becker, R. A., Chambers, J. M. and Wilks, A. R. (1988) *The New S Language.* Wadsworth & Brooks/Cole.

Dongarra, J. J., Bunch, J. R., Moler, C. B. and Stewart, G. W. (1978) *LINPACK Users Guide.* Philadelphia: SIAM Publications.

Anderson. E., et al. (1999) *LAPACK Users' Guide.* Third Edition. SIAM. ISBN 0-89871-447-8.

See Also

chol2inv for its *inverse* (without pivoting), backsolve for solving linear systems with upper triangular left sides.

qr, svd for related matrix factorizations.

Examples

```
( m <- matrix(c(5,1,1,3),2,2) )
( cm <- chol(m) )
t(cm) %*% cm   # = 'm'
crossprod(cm)  # = 'm'

# now for something positive semi-definite
x <- matrix(c(1:5, (1:5)^2), 5, 2)
x <- cbind(x, x[, 1] + 3*x[, 2])
m <- crossprod(x)
qr(m)$rank # is 2, as it should be

# chol() may fail, depending on numerical rounding:
```

```
# chol() unlike qr() does not use a tolerance.
try(chol(m))

# NB wrong rank here ... see Warning section.
(Q <- chol(m, pivot = TRUE))
## we can use this by
pivot <- attr(Q, "pivot")
oo <- order(pivot)
t(Q[, oo]) %*% Q[, oo] # recover m
```

chol2inv   *Inverse from Choleski Decomposition*

Description

Invert a symmetric, positive definite square matrix from its Choleski decomposition.

Usage

```
chol2inv(x, size = NCOL(x), LINPACK = FALSE)
La.chol2inv(x, size = ncol(x))
```

Arguments

x
: a matrix. The first nc columns of the upper triangle contain the Choleski decomposition of the matrix to be inverted.

size
: the number of columns of x containing the Choleski decomposition.

LINPACK
: logical. Should LINPACK be used (for compatibility with $R < 1.7.0$)?

Details

chol2inv(LINPACK=TRUE) provides an interface to the LINPACK routine DPODI. La.chol2inv provides an interface to the LAPACK routine DPOTRI.

Value

The inverse of the decomposed matrix.

References

Dongarra, J. J., Bunch, J. R., Moler, C. B. and Stewart, G. W. (1978) *LINPACK Users Guide.* Philadelphia: SIAM Publications.

Anderson. E., et al. (1999) *LAPACK Users' Guide.* Third Edition. SIAM. ISBN 0-89871-447-8.

See Also

chol, solve.

Examples

```
cma <- chol(ma  <- cbind(1, 1:3, c(1,3,7)))
ma %*% chol2inv(cma)
```

colSums   *Form Row and Column Sums and Means*

Description

Form row and column sums and means for numeric arrays.

Usage

```
colSums (x, na.rm = FALSE, dims = 1)
rowSums (x, na.rm = FALSE, dims = 1)
colMeans(x, na.rm = FALSE, dims = 1)
rowMeans(x, na.rm = FALSE, dims = 1)
```

Arguments

x
: an array of two or more dimensions, containing numeric, complex, integer or logical values, or a numeric data frame.

na.rm
: logical. Should missing values (including NaN) be omitted from the calculations?

dims
: Which dimensions are regarded as "rows" or "columns" to sum over. For row*, the sum or mean is over dimensions dims+1, ...; for col* it is over dimensions 1:dims.

Details

These functions are equivalent to the use of apply with FUN = mean or FUN = sum with appropriate margins, but are a lot faster. As they are written for speed, they blur over some of the subtleties of NaN and NA. If na.rm = FALSE and either NaN or NA appears in a sum, the result will be one of NaN or NA, but which might be platform-dependent.

Value

A numeric or complex array of suitable size, or a vector if the result is one-dimensional. The dimnames (or names for a vector result) are taken from the original array.

If there are no values in a range to be summed over (after removing missing values with na.rm = TRUE), that component of the output is set to 0 (*Sums) or NA (*Means), consistent with sum and mean.

See Also

apply, rowsum

Examples

```
## Compute row and column sums for a matrix:
x <- cbind(x1 = 3, x2 = c(4:1, 2:5))
rowSums(x); colSums(x)
dimnames(x)[[1]] <- letters[1:8]
rowSums(x); colSums(x); rowMeans(x); colMeans(x)
x[] <- as.integer(x)
rowSums(x); colSums(x)
x[] <- x < 3
rowSums(x); colSums(x)
x <- cbind(x1 = 3, x2 = c(4:1, 2:5))
x[3, ] <- NA; x[4, 2] <- NA
rowSums(x); colSums(x); rowMeans(x); colMeans(x)
rowSums(x, na.rm = TRUE); colSums(x, na.rm = TRUE)
rowMeans(x, na.rm = TRUE); colMeans(x, na.rm = TRUE)

## an array
data(UCBAdmissions)
dim(UCBAdmissions)
rowSums(UCBAdmissions); rowSums(UCBAdmissions, dims = 2)
colSums(UCBAdmissions); colSums(UCBAdmissions, dims = 2)

## complex case
x <- cbind(x1 = 3 + 2i, x2 = c(4:1, 2:5) - 5i)
x[3, ] <- NA; x[4, 2] <- NA
rowSums(x); colSums(x); rowMeans(x); colMeans(x)
rowSums(x, na.rm = TRUE); colSums(x, na.rm = TRUE)
rowMeans(x, na.rm = TRUE); colMeans(x, na.rm = TRUE)
```

convolve   *Fast Convolution*

Description

Use the Fast Fourier Transform to compute several kinds of convolutions of two sequences.

Usage

```
convolve(x, y, conj = TRUE, type = c("circular", "open",
        "filter"))
```

Arguments

x,y   numeric sequences *of the same length* to be convolved.

conj   logical; if TRUE, take the complex *conjugate* before back-transforming (default, and used for usual convolution).

type   character; one of "circular", "open", "filter" (beginning of word is ok). For circular, the two sequences are treated as *circular*, i.e., periodic.

For open and filter, the sequences are padded with 0s (from left and right) first; "filter" returns the middle sub-vector of "open", namely, the result of running a weighted mean of x with weights y.

Details

The Fast Fourier Transform, fft, is used for efficiency.

The input sequences x and y must have the same length if circular is true.

Note that the usual definition of convolution of two sequences x and y is given by convolve(x, rev(y), type = "o").

Value

If r <- convolve(x,y, type = "open") and n <- length(x), m <- length(y), then

$$r_k = \sum_i x_{k-m+i} y_i$$

where the sum is over all valid indices i, for $k = 1, \ldots, n + m - 1$

If type == "circular", $n = m$ is required, and the above is true for $i, k = 1, \ldots, n$ when $x_j := x_{n+j}$ for $j < 1$.

References

Brillinger, D. R. (1981) *Time Series: Data Analysis and Theory*, Second Edition. San Francisco: Holden-Day.

See Also

fft, nextn, and particularly filter (from the ts package) which may be more appropriate.

Examples

```
x <- c(0,0,0,100,0,0,0)
y <- c(0,0,1, 2 ,1,0,0)/4
zapsmall(convolve(x,y)) # NOT what you first thought.
zapsmall(convolve(x, y[3:5], type="f")) # rather
x <- rnorm(50)
y <- rnorm(50)
# Circular convolution has this symmetry:
all.equal(convolve(x,y, conj = FALSE),
          rev(convolve(rev(y),x)))

n <- length(x <- -20:24)
y <- (x-10)^2/1000 + rnorm(x)/8

Han <- function(y) # Hanning
       convolve(y, c(1,2,1)/4, type = "filter")

plot(x,y, main="Using  convolve(.) for Hanning filters")
lines(x[-c(1,n)],Han(y),col="red")
lines(x[-c(1:2,(n-1):n)],Han(Han(y)),lwd=2,col="dark blue")
```

crossprod   *Matrix Crossproduct*

Description

Given matrices x and y as arguments, crossprod returns their matrix cross-product. This is formally equivalent to, but faster than, the call t(x) %*% y.

Usage

crossprod(x, y = NULL)

Arguments

x, y   matrices: y = NULL is taken to be the same matrix as x.

References

Becker, R. A., Chambers, J. M. and Wilks, A. R. (1988) *The New S Language.* Wadsworth & Brooks/Cole.

See Also

%*% and outer product %o%.

Examples

```
(z <- crossprod(1:4))    # = sum(1 + 2^2 + 3^2 + 4^2)
drop(z)                  # scalar
```

| cumsum | *Cumulative Sums, Products, and Extremes* |

Description

Returns a vector whose elements are the cumulative sums, products, minima or maxima of the elements of the argument.

Usage

```
cumsum(x)
cumprod(x)
cummax(x)
cummin(x)
```

Arguments

x   a numeric object.

Details

An NA value in x causes the corresponding and following elements of the return value to be NA.

References

Becker, R. A., Chambers, J. M. and Wilks, A. R. (1988) *The New S Language.* Wadsworth & Brooks/Cole. (cumsum only.)

Examples

```
cumsum(1:10)
cumprod(1:10)
cummin(c(3:1, 2:0, 4:2))
cummax(c(3:1, 2:0, 4:2))
```

deriv	*Symbolic and Algorithmic Derivatives of Simple Expressions*

Description

Compute derivatives of simple expressions, symbolically.

Usage

```
    D (expr, name)
 deriv(expr, namevec, function.arg, tag = ".expr",
     hessian = FALSE)
 deriv3(expr, namevec, function.arg, tag = ".expr",
     hessian = TRUE)
```

Arguments

expr	`expression` or `call` to be differentiated.
name,namevec	character vector, giving the variable names (only one for `D()`) with respect to which derivatives will be computed.
function.arg	If specified, a character vector of arguments for a function return, or a function (with empty body) or `TRUE`, the latter indicating that a function with argument names `namevec` should be used.
tag	character; the prefix to be used for the locally created variables in result.
hessian	a logical value indicating whether the second derivatives should be calculated and incorporated in the return value.

Details

`D` is modelled after its S namesake for taking simple symbolic derivatives.

`deriv` is a *generic* function with a default and a `formula` method. It returns a `call` for computing the `expr` and its (partial) derivatives, simultaneously. It uses so-called "*algorithmic derivatives*". If `function.arg` is a function, its arguments can have default values, see the `fx` example below.

Currently, deriv.formula just calls deriv.default after extracting the expression to the right of ~.

deriv3 and its methods are equivalent to deriv and its methods except that hessian defaults to TRUE for deriv3.

Value

D returns a call and therefore can easily be iterated for higher derivatives.

deriv and deriv3 normally return an expression object whose evaluation returns the function values with a "gradient" attribute containing the gradient matrix. If hessian is TRUE the evaluation also returns a "hessian" attribute containing the Hessian array.

If function.arg is specified, deriv and deriv3 return a function with those arguments rather than an expression.

References

Griewank, A. and Corliss, G. F. (1991) *Automatic Differentiation of Algorithms: Theory, Implementation, and Application.* SIAM proceedings, Philadelphia.

Bates, D. M. and Chambers, J. M. (1992) *Nonlinear models.* Chapter 10 of *Statistical Models in S* eds J. M. Chambers and T. J. Hastie, Wadsworth & Brooks/Cole.

See Also

nlm and optim for numeric minimization which could make use of derivatives, nls in package **nls**.

Examples

```
## formula argument :
dx2x <- deriv(~ x^2, "x") ; dx2x
expression({
   .value <- x^2
   .grad <- array(0, c(length(.value),1), list(NULL,c("x")))
   .grad[, "x"] <- 2 * x
   attr(.value, "gradient") <- .grad
   .value
})
mode(dx2x)
x <- -1:2
eval(dx2x)
```

```
## Something 'tougher':
trig.exp <- expression(sin(cos(x + y^2)))
( D.sc <- D(trig.exp, "x") )
all.equal(D(trig.exp[[1]], "x"), D.sc)

( dxy <- deriv(trig.exp, c("x", "y")) )
y <- 1
eval(dxy)
eval(D.sc)

## function returned:
deriv((y ~ sin(cos(x) * y)), c("x","y"), func = TRUE)

## function with defaulted arguments:
(fx <- deriv(y ~ b0 + b1 * 2^(-x/th), c("b0", "b1", "th"),
             function(b0, b1, th, x = 1:7){} ) )
fx(2,3,4)

## Higher derivatives
deriv3(y ~ b0 + b1 * 2^(-x/th), c("b0", "b1", "th"),
    c("b0", "b1", "th", "x") )

## Higher derivatives:
DD <- function(expr,name, order = 1) {
   if(order < 1) stop("'order' must be >= 1")
   if(order == 1) D(expr,name)
   else DD(D(expr, name), name, order - 1)
}

DD(expression(sin(x^2)), "x", 3)
## showing the limits of the internal "simplify()" :
-sin(x^2) * (2 * x) * 2
+ ((cos(x^2) * (2 * x) * (2 * x) + sin(x^2) * 2) * (2 * x)
+ sin(x^2) * (2 * x) * 2)
```

eigen — *Spectral Decomposition of a Matrix*

Description

Computes eigenvalues and eigenvectors.

Usage

```
eigen(x, symmetric, only.values = FALSE, EISPACK = FALSE)
La.eigen(x, symmetric, only.values = FALSE,
         method = c("dsyevr", "dsyev"))
```

Arguments

x	a matrix whose spectral decomposition is to be computed.
symmetric	if TRUE, the matrix is assumed to be symmetric (or Hermitian if complex) and only its lower triangle is used. If symmetric is not specified, the matrix is inspected for symmetry.
only.values	if TRUE, only the eigenvalues are computed and returned, otherwise both eigenvalues and eigenvectors are returned.
EISPACK	logical. Should EISPACK be used (for compatibility with R < 1.7.0)?
method	The LAPACK routine to use in the real symmetric case.

Details

These functions use the LAPACK routines DSYEV/DSYEVR, DGEEV, ZHEEV and ZGEEV, and eigen(EISPACK=TRUE) provides an interface to the EISPACK routines RS, RG, CH and CG.

If symmetric is unspecified, the code attempts to determine if the matrix is symmetric up to plausible numerical inaccuracies. It is faster and surer to set the value yourself.

eigen is preferred to eigen(EISPACK=TRUE) for new projects, but its eigenvectors may differ in sign and (in the asymmetric case) in normalization. (They may also differ between methods and between platforms.)

The LAPACK routine DSYEVR is usually substantially faster than DSYEV. Most benefits are seen with an optimized BLAS system.

Using `method="dsyevr"` requires IEEE 754 arithmetic. Should this not be supported on your platform, `method="dsyev"` is used, with a warning.

Computing the eigenvectors is the slow part for large matrices.

Value

The spectral decomposition of x is returned as components of a list.

`values`   a vector containing the p eigenvalues of x, sorted in *decreasing* order, according to `Mod(values)` if they are complex.

`vectors`   a $p \times p$ matrix whose columns contain the eigenvectors of x, or NULL if `only.values` is TRUE.

For `eigen(..., symmetric=FALSE, EISPACK=TRUE)` the choice of length of the eigenvectors is not defined by EISPACK. In all other cases the vectors are normalized to unit length.

Recall that the eigenvectors are only defined up to a constant: even when the length is specified they are still only defined up to a scalar of modulus one (the sign for real matrices).

References

Becker, R. A., Chambers, J. M. and Wilks, A. R. (1988) *The New S Language.* Wadsworth & Brooks/Cole.

Smith, B. T, Boyle, J. M., Dongarra, J. J., Garbow, B. S., Ikebe,Y., Klema, V., and Moler, C. B. (1976). *Matrix Eigensystems Routines – EISPACK Guide.* Springer-Verlag Lecture Notes in Computer Science.

Anderson. E., et al. (1999) *LAPACK Users' Guide.* Third Edition. SIAM. ISBN 0-89871-447-8.

See Also

`svd`, a generalization of `eigen`; `qr`, and `chol` for related decompositions.

To compute the determinant of a matrix, the `qr` decomposition is much more efficient: `det`.

`capabilities` to test for IEEE 754 arithmetic.

Examples

```
eigen(cbind(c(1,-1),c(-1,1)))
# same (different algorithm)
eigen(cbind(c(1,-1),c(-1,1)), symmetric = FALSE)

eigen(cbind(1,c(1,-1)), only.values = TRUE)
eigen(cbind(-1,2:1)) # complex values
# Hermitian, real eigenvalues
eigen(print(cbind(c(0,1i), c(-1i,0))))
## 3 x 3:
eigen(cbind( 1,3:1,1:3))
eigen(cbind(-1,c(1:2,0),0:2)) # complex values
```

Extremes   *Maxima and Minima*

Description

Returns the (parallel) maxima and minima of the input values.

Usage

```
max(..., na.rm=FALSE)
min(..., na.rm=FALSE)

pmax(..., na.rm=FALSE)
pmin(..., na.rm=FALSE)
```

Arguments

...   numeric arguments.

na.rm   a logical indicating whether missing values should be removed.

Value

max and min return the maximum or minimum of all the values present in their arguments, as integer if all are integer, or as double otherwise.

The minimum and maximum of an empty set are +Inf and -Inf (in this order!) which ensures *transitivity*, e.g., min(x1, min(x2)) == min(x1,x2). In R versions before 1.5, min(integer(0)) == .Machine$integer.max, and analogously for max, preserving argument *type*, whereas from R version 1.5.0, max(x) == -Inf and min(x) == +Inf whenever length(x) == 0 (after removing missing values if requested).

If na.rm is FALSE an NA value in any of the arguments will cause a value of NA to be returned, otherwise NA values are ignored.

pmax and pmin take several vectors (or matrices) as arguments and return a single vector giving the parallel maxima (or minima) of the vectors. The first element of the result is the maximum (minimum) of the first elements of all the arguments, the second element of the result is the maximum (minimum) of the second elements of all the arguments and so on. Shorter vectors are recycled if necessary. If na.rm is FALSE, NA values in the input vectors will produce NA values in the output. If na.rm is TRUE, NA values are ignored. attributes (such as names or dim) are transferred from the first argument (if applicable).

References

Becker, R. A., Chambers, J. M. and Wilks, A. R. (1988) *The New S Language.* Wadsworth & Brooks/Cole.

See Also

range (*both* min and max) and which.min (which.max) for the *arg min*, i.e., the location where an extreme value occurs.

Examples

```
min(5:1,pi)
pmin(5:1, pi)
x <- sort(rnorm(100));   cH <- 1.35
pmin(cH, quantile(x)) # no names
pmin(quantile(x), cH) # has names
plot(x, pmin(cH, pmax(-cH, x)), type='b',
     main= "Huber's function")
```

| fft | *Fast Discrete Fourier Transform* |

Description

Performs the Fast Fourier Transform of an array.

Usage

```
fft(z, inverse = FALSE)
mvfft(z, inverse = FALSE)
```

Arguments

z
: a real or complex array containing the values to be transformed.

inverse
: if TRUE, the unnormalized inverse transform is computed (the inverse has a + in the exponent of *e*, but here, we do *not* divide by 1/length(x)).

Value

When z is a vector, the value computed and returned by fft is the unnormalized univariate Fourier transform of the sequence of values in z. When z contains an array, fft computes and returns the multivariate (spatial) transform. If inverse is TRUE, the (unnormalized) inverse Fourier transform is returned, i.e., if y <- fft(z), then z is fft(y, inverse = TRUE) / length(y).

By contrast, mvfft takes a real or complex matrix as argument, and returns a similar shaped matrix, but with each column replaced by its discrete Fourier transform. This is useful for analyzing vector-valued series.

The FFT is fastest when the length of the series being transformed is highly composite (i.e., has many factors). If this is not the case, the transform may take a long time to compute and will use a large amount of memory.

References

Becker, R. A., Chambers, J. M. and Wilks, A. R. (1988) *The New S Language*. Wadsworth & Brooks/Cole.

Singleton, R. C. (1979) Mixed Radix Fast Fourier Transforms, in *Programs for Digital Signal Processing*, IEEE Digital Signal Processing Committee eds. IEEE Press.

See Also

`convolve`, `nextn`.

Examples

```
x <- 1:4
fft(x)
fft(fft(x), inverse = TRUE)/length(x)
```

| findInterval | Find Interval Numbers or Indices |

Description

Find the indices of x in vec, where vec must be sorted (non-decreasingly); i.e., if i <- findInterval(x,v), we have $v_{i_j} \leq x_j < v_{i_j+1}$ where $v_0 := -\infty$, $v_{N+1} := +\infty$, and N <- length(vec). At the two boundaries, the returned index may differ by 1, depending on the optional arguments rightmost.closed and all.inside.

Usage

```
findInterval(x, vec, rightmost.closed = FALSE,
             all.inside = FALSE)
```

Arguments

x   numeric.

vec   numeric, sorted (weakly) increasingly, of length N, say.

rightmost.closed
            logical; if true, the rightmost interval, vec[N-1] .. vec[N] is treated as *closed*, see below.

all.inside logical; if true, the returned indices are coerced into $\{1, \ldots, N-1\}$, i.e., 0 is mapped to 1 and N to $N-1$.

Details

The function findInterval finds the index of one vector x in another, vec, where the latter must be non-decreasing. Where this is trivial, equivalent to apply(outer(x, vec, ">="), 1, sum), as a matter of fact, the internal algorithm uses interval search ensuring $O(n \log N)$ complexity where n <- length(x) (and N <- length(vec)). For (almost) sorted x, it will be even faster, basically $O(n)$.

This is the same computation as for the empirical distribution function, and indeed, findInterval(t, sort(X)) is *identical* to $nF_n(t; X_1, \ldots, X_n)$ where F_n is the empirical distribution function of $X_1, \ldots, X_n$.

When rightmost.closed = TRUE, the result for x[j] = vec[N] (= max(*vec*)), is N - 1 as for all other values in the last interval.

Value

vector of length length(x) with values in 0:N where N <- length(vec), or values coerced to 1:(N-1) if all.inside = TRUE (equivalently coercing all x values *inside* the intervals).

Author(s)

Martin Maechler

See Also

approx(*, method = "constant") which is a generalization of findInterval(), ecdf for computing the empirical distribution function which is (up to a factor of n) also basically the same as findInterval(.).

Examples

```
N <- 100
X <- sort(round(rt(N, df=2), 2))
tt <- c(-100, seq(-2,2, len=201), +100)
it <- findInterval(tt, X)
# only first and last are outside range(X)
tt[it < 1 | it >= N]
```

gl *Generate Factor Levels*

Description

Generate factors by specifying the pattern of their levels.

Usage

```
gl(n, k, length = n*k, labels = 1:n, ordered = FALSE)
```

Arguments

n	an integer giving the number of levels.
k	an integer giving the number of replications.
length	an integer giving the length of the result.
labels	an optional vector of labels for the resulting factor levels.
ordered	a logical indicating whether the result should be ordered or not.

Value

The result has levels from 1 to n with each value replicated in groups of length k out to a total length of length.

gl is modelled on the *GLIM* function of the same name.

See Also

The underlying factor().

Examples

```
## First control, then treatment:
gl(2, 8, label = c("Control", "Treat"))
## 20 alternating 1s and 2s
gl(2, 1, 20)
## alternating pairs of 1s and 2s
gl(2, 2, 20)
```

Hyperbolic   *Hyperbolic Functions*

Description

These functions give the obvious hyperbolic functions. They respectively compute the hyperbolic cosine, sine, tangent, arc-cosine, arc-sine, arc-tangent.

Usage

```
cosh(x)
sinh(x)
tanh(x)
acosh(x)
asinh(x)
atanh(x)
```

Arguments

x   a numeric vector

See Also

cos, sin, tan, acos, asin, atan.

`integrate` *Integration of One-Dimensional Functions*

Description

Adaptive quadrature of functions of one variable over a finite or infinite interval.

Usage

```
integrate(f, lower, upper, subdivisions=100,
   rel.tol = .Machine$double.eps^0.25, abs.tol = rel.tol,
   stop.on.error = TRUE, keep.xy = FALSE, aux = NULL, ...)
```

Arguments

`f` an R function taking a numeric first argument and returning a numeric vector of the same length. Returning a non-finite element will generate an error.

`lower, upper` the limits of integration. Can be infinite.

`subdivisions` the maximum number of subintervals.

`rel.tol` relative accuracy requested.

`abs.tol` absolute accuracy requested.

`stop.on.error` logical. If true (the default) an error stops the function. If false some errors will give a result with a warning in the `message` component.

`keep.xy` unused. For compatibility with S.

`aux` unused. For compatibility with S.

`...` additional arguments to be passed to `f`. Remember to use argument names *not* matching those of `integrate(.)`!

Details

If one or both limits are infinite, the infinite range is mapped onto a finite interval.

For a finite interval, globally adaptive interval subdivision is used in connection with extrapolation by the Epsilon algorithm.

`rel.tol` cannot be less than `max(50*.Machine$double.eps, 0.5e-28)` if `abs.tol <= 0`.

Value

A list of class "integrate" with components

value
: the final estimate of the integral.

abs.error
: estimate of the modulus of the absolute error.

subdivisions
: the number of subintervals produced in the subdivision process.

message
: "OK" or a character string giving the error message.

call
: the matched call.

Note

Like all numerical integration routines, these evaluate the function on a finite set of points. If the function is approximately constant (in particular, zero) over nearly all its range it is possible that the result and error estimate may be seriously wrong.

When integrating over infinite intervals do so explicitly, rather than just using a large number as the endpoint. This increases the chance of a correct answer – any function whose integral over an infinite interval is finite must be near zero for most of that interval.

References

Based on QUADPACK routines dqags and dqagi by R. Piessens and E. deDoncker-Kapenga, available from Netlib.
See
R. Piessens, E. deDoncker-Kapenga, C. Uberhuber, D. Kahaner (1983) *Quadpack: a Subroutine Package for Automatic Integration*; Springer Verlag.

See Also

The function adapt in the **adapt** package on CRAN, for multivariate integration.

Examples

```
integrate(dnorm, -1.96, 1.96)
integrate(dnorm, -Inf, Inf)

## a slowly-convergent integral
integrand <- function(x) {1/((x+1)*sqrt(x))}
```

```r
integrate(integrand, lower = 0, upper = Inf)

## don't do this if you really want the integral from 0 to
## Inf
integrate(integrand, lower = 0, upper = 10)
integrate(integrand, lower = 0, upper = 100000)
integrate(integrand, lower = 0, upper = 1000000,
          stop.on.error = FALSE)

## no vectorizable function
try(integrate(function(x) 2, 0, 1))
integrate(function(x) rep(2, length(x)), 0, 1)   ## correct

## integrate can fail if misused
integrate(dnorm,0,2)
integrate(dnorm,0,20)
integrate(dnorm,0,200)
integrate(dnorm,0,2000)
integrate(dnorm,0,20000) ## fails on many systems
integrate(dnorm,0,Inf)   ## works
```

kappa — *Estimate the Condition Number*

Description

An estimate of the condition number of a matrix or of the R matrix of a QR decomposition, perhaps of a linear fit. The condition number is defined as the ratio of the largest to the smallest *non-zero* singular value of the matrix.

Usage

```
kappa(z, ...)
## S3 method for class 'lm':
kappa(z, ...)
## Default S3 method:
kappa(z, exact = FALSE, ...)
## S3 method for class 'qr':
kappa(z, ...)

kappa.tri(z, exact = FALSE, ...)
```

Arguments

z   A matrix or a the result of qr or a fit from a class inheriting from "lm".

exact   logical. Should the result be exact?

...   further arguments passed to or from other methods.

Details

If exact = FALSE (the default) the condition number is estimated by a cheap approximation. Following S, this uses the LINPACK routine 'dtrco.f'. However, in R (or S) the exact calculation is also likely to be quick enough.

kappa.tri is an internal function called by kappa.qr.

Value

The condition number, *kappa*, or an approximation if exact = FALSE.

Author(s)

The design was inspired by (but differs considerably from) the S function of the same name described in Chambers (1992).

References

Chambers, J. M. (1992) *Linear models.* Chapter 4 of *Statistical Models in S* eds J. M. Chambers and T. J. Hastie, Wadsworth & Brooks/Cole.

See Also

svd for the singular value decomposition and qr for the QR one.

Examples

```
kappa(x1 <- cbind(1,1:10)) # 15.71
kappa(x1, exact = TRUE)         # 13.68
kappa(x2 <- cbind(x1,2:11)) # high! [x2 is singular!]

hilbert <- function(n) {
   i <- 1:n; 1 / outer(i - 1, i, "+")
}
sv9 <- svd(h9 <- hilbert(9))$ d
kappa(h9) # pretty high!
kappa(h9, exact = TRUE) == max(sv9) / min(sv9)
# .677 (i.e., rel.error = 32%)
kappa(h9, exact = TRUE) / kappa(h9)
```

log *Logarithms and Exponentials*

Description

log computes natural logarithms, log10 computes common (i.e., base 10) logarithms, and log2 computes binary (i.e., base 2) logarithms. The general form logb(x, base) computes logarithms with base base (log10 and log2 are only special cases).

log1p(x) computes $\log(1+x)$ accurately also for $|x| \ll 1$ (and less accurately when $x \approx -1$).

exp computes the exponential function.

expm1(x) computes $\exp(x) - 1$ accurately also for $|x| \ll 1$.

Usage

```
log(x, base = exp(1))
logb(x, base = exp(1))
log10(x)
log2(x)
exp(x)
expm1(x)
log1p(x)
```

Arguments

x	a numeric or complex vector.
base	positive number. The base with respect to which logarithms are computed. Defaults to e=exp(1).

Value

A vector of the same length as x containing the transformed values. log(0) gives -Inf (when available).

Note

log and logb are the same thing in R, but logb is preferred if base is specified, for S-PLUS compatibility.

References

Becker, R. A., Chambers, J. M. and Wilks, A. R. (1988) *The New S Language*. Wadsworth & Brooks/Cole. (for log, log10 and exp.)

Chambers, J. M. (1998) *Programming with Data. A Guide to the S Language*. Springer. (for logb.)

See Also

Trig, sqrt, Arithmetic.

Examples

```
log(exp(3))
log10(1e7) # = 7

x <- 10^-(1+2*1:9)
cbind(x, log(1+x), log1p(x), exp(x)-1, expm1(x))
```

matmult *Matrix Multiplication*

Description

Multiplies two matrices, if they are conformable. If one argument is a vector, it will be coerced to either a row or column matrix to make the two arguments conformable. If both are vectors it will return the inner product.

Usage

```
a %*% b
```

Arguments

a, b   numeric or complex matrices or vectors.

Value

The matrix product. Use `drop` to get rid of dimensions which have only one level.

References

Becker, R. A., Chambers, J. M. and Wilks, A. R. (1988) *The New S Language.* Wadsworth & Brooks/Cole.

See Also

`matrix`, `Arithmetic`, `diag`.

Examples

```
x <- 1:4
(z <- x %*% x)     # scalar ("inner") product (1 x 1 matrix)
drop(z)            # as scalar

y <- diag(x)
z <- matrix(1:12, ncol = 3, nrow = 4)
y %*% z
y %*% x
x %*% z
```

`matrix`   *Matrices*

Description

`matrix` creates a matrix from the given set of values.

`as.matrix` attempts to turn its argument into a matrix.

`is.matrix` tests if its argument is a (strict) matrix. It is generic: you can write methods to handle specific classes of objects, see InternalMethods.

Usage

```
matrix(data = NA, nrow = 1, ncol = 1, byrow = FALSE,
       dimnames = NULL)
as.matrix(x)
is.matrix(x)
```

Arguments

`data`	an optional data vector.
`nrow`	the desired number of rows
`ncol`	the desired number of columns
`byrow`	logical. If `FALSE` (the default) the matrix is filled by columns, otherwise the matrix is filled by rows.
`dimnames`	A dimnames attribute for the matrix: a `list` of length 2.
`x`	an R object.

Details

If either of `nrow` or `ncol` is not given, an attempt is made to infer it from the length of `data` and the other parameter.

`is.matrix` returns TRUE if x is a matrix (i.e., it is *not* a `data.frame` and has a `dim` attribute of length 2) and `FALSE` otherwise.

`as.matrix` is a generic function. The method for data frames will convert any non-numeric column into a character vector using `format` and so return a character matrix.

References

Becker, R. A., Chambers, J. M. and Wilks, A. R. (1988) *The New S Language*. Wadsworth & Brooks/Cole.

See Also

data.matrix, which attempts to convert to a numeric matrix.

Examples

```
is.matrix(as.matrix(1:10))
data(warpbreaks)
!is.matrix(warpbreaks) # data.frame, NOT matrix!
str(warpbreaks)
# using as.matrix.data.frame(.) method
str(as.matrix(warpbreaks))
```

| nextn | *Highly Composite Numbers* |

Description

nextn returns the smallest integer, greater than or equal to n, which can be obtained as a product of powers of the values contained in factors. nextn is intended to be used to find a suitable length to zero-pad the argument of fft to so that the transform is computed quickly. The default value for factors ensures this.

Usage

```
nextn(n, factors=c(2,3,5))
```

Arguments

n an integer.

factors a vector of positive integer factors.

See Also

convolve, fft.

Examples

```
nextn(1001) # 1024
table(sapply(599:630, nextn))
```

poly   *Compute Orthogonal Polynomials*

Description

Returns or evaluates orthogonal polynomials of degree 1 to `degree` over the specified set of points x. These are all orthogonal to the constant polynomial of degree 0.

Usage

```
poly(x, ..., degree = 1, coefs = NULL)
polym(..., degree = 1)

## S3 method for class 'poly':
predict(object, newdata, ...)
```

Arguments

x, newdata	a numeric vector at which to evaluate the polynomial. x can also be a matrix.
degree	the degree of the polynomial
coefs	for prediction, coefficients from a previous fit.
object	an object inheriting from class "poly", normally the result of a call to poly with a single vector argument.
...	poly, polym: further vectors. predict.poly: arguments to be passed to or from other methods.

Details

Although formally `degree` should be named (as it follows ...), an unnamed second argument of length 1 will be interpreted as the degree.

The orthogonal polynomial is summarized by the coefficients, which can be used to evaluate it via the three-term recursion given in Kennedy & Gentle (1980, pp. 343-4), and used in the "predict" part of the code.

Value

For `poly` with a single vector argument:
A matrix with rows corresponding to points in x and columns corresponding to the degree, with attributes `"degree"` specifying the degrees

of the columns and "coefs" which contains the centering and normalization constants used in constructing the orthogonal polynomials. The matrix is given class c("poly", "matrix") as from R 1.5.0.

Other cases of poly and polym, and predict.poly: a matrix.

Note

This routine is intended for statistical purposes such as contr.poly: it does not attempt to orthogonalize to machine accuracy.

References

Chambers, J. M. and Hastie, T. J. (1992) *Statistical Models in S*. Wadsworth & Brooks/Cole.

Kennedy, W. J. Jr and Gentle, J. E. (1980) *Statistical Computing* Marcel Dekker.

See Also

contr.poly

Examples

```
(z <- poly(1:10, 3))
predict(z, seq(2, 4, 0.5))
poly(seq(4, 6, 0.5), 3, coefs = attr(z, "coefs"))

polym(1:4, c(1, 4:6), degree=3) # or just poly()
poly(cbind(1:4, c(1, 4:6)), degree=3)
```

polyroot — Find Zeros of a Real or Complex Polynomial

Description

Find zeros of a real or complex polynomial.

Usage

```
polyroot(z)
```

Arguments

z   the vector of polynomial coefficients in increasing order.

Details

A polynomial of degree $n - 1$,

$$p(x) = z_1 + z_2 x + \cdots + z_n x^{n-1}$$

is given by its coefficient vector `z[1:n]`. polyroot returns the $n - 1$ complex zeros of $p(x)$ using the Jenkins-Traub algorithm.

If the coefficient vector z has zeroes for the highest powers, these are discarded.

Value

A complex vector of length $n - 1$, where n is the position of the largest non-zero element of z.

References

Jenkins and Traub (1972) TOMS Algorithm 419. *Comm. ACM*, **15**, 97–99.

See Also

uniroot for numerical root finding of arbitrary functions; complex and the zero example in the demos directory.

Examples

```
polyroot(c(1, 2, 1))
round(polyroot(choose(8, 0:8)), 11) # guess what!
for (n1 in 1:4) print(polyroot(1:n1), digits = 4)
polyroot(c(1, 2, 1, 0, 0)) # same as the first
```

prod *Product of Vector Elements*

Description

`prod` returns the product of all the values present in its arguments.

Usage

`prod(..., na.rm = FALSE)`

Arguments

...	numeric vectors.
na.rm	logical. Should missing values be removed?

Details

If `na.rm` is `FALSE` an `NA` value in any of the arguments will cause a value of `NA` to be returned, otherwise `NA` values are ignored.

References

Becker, R. A., Chambers, J. M. and Wilks, A. R. (1988) *The New S Language.* Wadsworth & Brooks/Cole.

See Also

`sum`, `cumprod`, `cumsum`.

Examples

`print(prod(1:7)) == print(gamma(8))`

qr The QR Decomposition of a Matrix

Description

qr computes the QR decomposition of a matrix. It provides an interface to the techniques used in the LINPACK routine DQRDC or the LAPACK routines DGEQP3 and (for complex matrices) ZGEQP3.

Usage

```
qr(x, tol = 1e-07 , LAPACK = FALSE)
qr.coef(qr, y)
qr.qy(qr, y)
qr.qty(qr, y)
qr.resid(qr, y)
qr.fitted(qr, y, k = qr$rank)
qr.solve(a, b, tol = 1e-7)
## S3 method for class 'qr':
solve(a, b, ...)

is.qr(x)
as.qr(x)
```

Arguments

x	a matrix whose QR decomposition is to be computed.
tol	the tolerance for detecting linear dependencies in the columns of x. Only used by LINPACK.
qr	a QR decomposition of the type computed by qr.
y, b	a vector or matrix of right-hand sides of equations.
a	A QR decomposition or (qr.solve only) a rectangular matrix.
k	effective rank.
LAPACK	logical. For real x, if true use LAPACK otherwise use LINPACK.
...	further arguments passed to or from other methods

Details

The QR decomposition plays an important role in many statistical techniques. In particular it can be used to solve the equation $Ax = b$ for given matrix A, and vector b. It is useful for computing regression coefficients and in applying the Newton-Raphson algorithm.

The functions qr.coef, qr.resid, and qr.fitted return the coefficients, residuals and fitted values obtained when fitting y to the matrix with QR decomposition qr. qr.qy and qr.qty return Q %*% y and t(Q) %*% y, where Q is the Q matrix.

All the above functions keep dimnames (and names) of x and y if there are.

solve.qr is the method for solve for qr objects. qr.solve solves systems of equations via the QR decomposition: if a is a QR decomposition it is the same as solve.qr, but if a is a rectangular matrix the QR decomposition is computed first. Either will handle over- and under-determined systems, providing a minimal-length solution or a least-squares fit if appropriate.

is.qr returns TRUE if x is a list with components named qr, rank and qraux and FALSE otherwise.

It is not possible to coerce objects to mode "qr". Objects either are QR decompositions or they are not.

Value

The QR decomposition of the matrix as computed by LINPACK or LAPACK. The components in the returned value correspond directly to the values returned by DQRDC/DGEQP3/ZGEQP3.

qr	a matrix with the same dimensions as x. The upper triangle contains the R of the decomposition and the lower triangle contains information on the Q of the decomposition (stored in compact form). Note that the storage used by DQRDC and DGEQP3 differs.
qraux	a vector of length ncol(x) which contains additional information on Q.
rank	the rank of x as computed by the decomposition: always full rank in the LAPACK case.
pivot	information on the pivoting strategy used during the decomposition.

Non-complex QR objects computed by LAPACK have the attribute "useLAPACK" with value TRUE.

Note

To compute the determinant of a matrix (do you *really* need it?), the QR decomposition is much more efficient than using Eigen values (eigen). See det.

Using LAPACK (including in the complex case) uses column pivoting and does not attempt to detect rank-deficient matrices.

References

Becker, R. A., Chambers, J. M. and Wilks, A. R. (1988) *The New S Language*. Wadsworth & Brooks/Cole.

Dongarra, J. J., Bunch, J. R., Moler, C. B. and Stewart, G. W. (1978) *LINPACK Users Guide*. Philadelphia: SIAM Publications.

Anderson. E., et al. (1999) *LAPACK Users' Guide*. Third Edition. SIAM. ISBN 0-89871-447-8.

See Also

qr.Q, qr.R, qr.X for reconstruction of the matrices. solve.qr, lsfit, eigen, svd.

det (using qr) to compute the determinant of a matrix.

Examples

```
hilbert <- function(n) {
   i <- 1:n; 1 / outer(i - 1, i, "+")
}
h9 <- hilbert(9); h9
qr(h9)$rank              # only 7
qrh9 <- qr(h9, tol = 1e-10)
qrh9$rank                # 9
## Solve linear equation system H %*% x = y :
y <- 1:9/10
x <- qr.solve(h9, y, tol = 1e-10) # or equivalently :
x <- qr.coef(qrh9, y) # is == but much better than
                         # solve(h9) %*% y
h9 %*% x                 # = y
```

QR.Auxiliaries — Reconstruct the Q, R, or X Matrices from a QR Object

Description

Returns the original matrix from which the object was constructed or the components of the decomposition.

Usage

```
qr.X(qr, complete = FALSE, ncol =)
qr.Q(qr, complete = FALSE, Dvec =)
qr.R(qr, complete = FALSE)
```

Arguments

qr
: object representing a QR decomposition. This will typically have come from a previous call to qr or lsfit.

complete
: logical expression of length 1. Indicates whether an arbitrary orthogonal completion of the Q or X matrices is to be made, or whether the R matrix is to be completed by binding zero-value rows beneath the square upper triangle.

ncol
: integer in the range 1:nrow(qr$qr). The number of columns to be in the reconstructed X. The default when complete is FALSE is the first min(ncol(X), nrow(X)) columns of the original X from which the qr object was constructed. The default when complete is TRUE is a square matrix with the original X in the first ncol(X) columns and an arbitrary orthogonal completion (unitary completion in the complex case) in the remaining columns.

Dvec
: vector (not matrix) of diagonal values. Each column of the returned Q will be multiplied by the corresponding diagonal value. Defaults to all 1s.

Value

qr.X returns X, the original matrix from which the qr object was constructed, provided ncol(X) <= nrow(X). If complete is TRUE or the argument ncol is greater than ncol(X), additional columns from an arbitrary orthogonal (unitary) completion of X are returned.

qr.Q returns **Q**, the order-nrow(X) orthogonal (unitary) transformation represented by qr. If complete is TRUE, **Q** has nrow(X) columns. If complete is FALSE, **Q** has ncol(X) columns. When Dvec is specified, each column of **Q** is multiplied by the corresponding value in Dvec.

qr.R returns **R**, the upper triangular matrix such that X == Q %*% R. The number of rows of **R** is nrow(X) or ncol(X), depending on whether complete is TRUE or FALSE.

See Also

qr, qr.qy.

Examples

```
data(LifeCycleSavings)
p <- ncol(x <- LifeCycleSavings[,-1]) # not the 'sr'
qrstr <- qr(x)    # dim(x) == c(n,p)
qrstr $ rank # = 4 = p
Q <- qr.Q(qrstr) # dim(Q) == dim(x)
R <- qr.R(qrstr) # dim(R) == ncol(x)
X <- qr.X(qrstr) # X == x
range(X - as.matrix(x)) # ~ < 6e-12
## X == Q %*% R :
Q %*% R
```

range   *Range of Values*

Description

range returns a vector containing the minimum and maximum of all the given arguments.

Usage

```
range(..., na.rm = FALSE)

## Default S3 method:
range(..., na.rm = FALSE, finite = FALSE)
```

Arguments

...	any numeric objects.
na.rm	logical, indicating if NA's should be omitted.
finite	logical, indicating if all non-finite elements should be omitted.

Details

This is a generic function; currently, it has only a default method (range.default).

It is also a member of the Summary group of functions, see Methods.

If na.rm is FALSE, NA and NaN values in any of the arguments will cause NA values to be returned, otherwise NA values are ignored.

If finite is TRUE, the minimum and maximum of all finite values is computed, i.e., finite=TRUE *includes* na.rm=TRUE.

A special situation occurs when there is no (after omission of NAs) nonempty argument left, see min.

References

Becker, R. A., Chambers, J. M. and Wilks, A. R. (1988) *The New S Language.* Wadsworth & Brooks/Cole.

See Also

min, max, Methods.

Examples

```
print(r.x <- range(rnorm(100)))
diff(r.x) # the SAMPLE range

x <- c(NA, 1:3, -1:1/0); x
range(x)
range(x, na.rm = TRUE)
range(x, finite = TRUE)
```

Round *Rounding of Numbers*

Description

ceiling takes a single numeric argument x and returns a numeric vector containing the smallest integers not less than the corresponding elements of x.

floor takes a single numeric argument x and returns a numeric vector containing the largest integers not greater than the corresponding elements of x.

round rounds the values in its first argument to the specified number of decimal places (default 0). Note that for rounding off a 5, the IEEE standard is used, *"go to the even digit"*. Therefore round(0.5) is 0 and round(-1.5) is -2.

signif rounds the values in its first argument to the specified number of significant digits.

trunc takes a single numeric argument x and returns a numeric vector containing the integers by truncating the values in x toward 0.

zapsmall determines a digits argument dr for calling round(x, digits = dr) such that values "close to zero" (compared with the maximal absolute one) are "zapped", i.e., treated as 0.

Usage

```
ceiling(x)
floor(x)
round(x, digits = 0)
signif(x, digits = 6)
trunc(x)
zapsmall(x, digits= getOption("digits"))
```

Arguments

x	a numeric vector.
digits	integer indicating the precision to be used.

References

Becker, R. A., Chambers, J. M. and Wilks, A. R. (1988) *The New S Language.* Wadsworth & Brooks/Cole. (except zapsmall.)

Chambers, J. M. (1998) *Programming with Data. A Guide to the S Language.* Springer. (zapsmall.)

See Also

as.integer.

Examples

```
round(.5 + -2:4) # IEEE rounding: -2  0  0  2  2  4  4
( x1 <- seq(-2, 4, by = .5) )
round(x1) # IEEE rounding !
x1[trunc(x1) != floor(x1)]
x1[round(x1) != floor(x1 + .5)]
(non.int <- ceiling(x1) != floor(x1))

x2 <- pi * 100^(-1:3)
round(x2, 3)
signif(x2, 3)

print   (x2 / 1000, digits=4)
zapsmall(x2 / 1000, digits=4)
zapsmall(exp(1i*0:4*pi/2))
```

| sign | *Sign Function* |

Description

sign returns a vector with the signs of the corresponding elements of x (the sign of a real number is 1, 0, or −1 if the number is positive, zero, or negative, respectively).

Note that sign does not operate on complex vectors.

Usage

```
sign(x)
```

Arguments

x   a numeric vector

See Also

abs

Examples

```
sign(pi)   # == 1
sign(-2:3) # -1 -1 0 1 1 1
```

solve   Solve a System of Equations

Description

This generic function solves the equation a %*% x = b for x, where b can be either a vector or a matrix.

Usage

```
solve(a, b, ...)

## Default S3 method:
solve(a, b, tol, LINPACK = FALSE, ...)
```

Arguments

a	a square numeric or complex matrix containing the coefficients of the linear system.
b	a numeric or complex vector or matrix giving the right-hand side(s) of the linear system. If missing, b is taken to be an identity matrix and solve will return the inverse of a.
tol	the tolerance for detecting linear dependencies in the columns of a. If LINPACK is TRUE the default is 1e-7, otherwise it is .Machine$double.eps. Future versions of R may use a tighter tolerance. Not presently used with complex matrices a.
LINPACK	logical. Should LINPACK be used (for compatibility with R < 1.7.0)?
...	further arguments passed to or from other methods

Details

As from R 1.3.0, a or b can be complex, in which case LAPACK routine ZESV is used. This uses double complex arithmetic which might not be available on all platforms.

The row and column names of the result are taken from the column names of a and of b respectively. As from R 1.7.0 if b is missing the column names of the result are the row names of a. No check is made that the column names of a and the row names of b are equal.

For back-compatibility a can be a (real) QR decomposition, although qr.solve should be called in that case. qr.solve can handle non-square systems.

References

Becker, R. A., Chambers, J. M. and Wilks, A. R. (1988) *The New S Language.* Wadsworth & Brooks/Cole.

See Also

solve.qr for the qr method, backsolve, qr.solve.

Examples

```
hilbert <- function(n) {
   i <- 1:n; 1 / outer(i - 1, i, "+")
}
h8 <- hilbert(8); h8
sh8 <- solve(h8)
round(sh8 %*% h8, 3)

A <- hilbert(4)
A[] <- as.complex(A)
## might not be supported on all platforms
try(solve(A))
```

sort   *Sorting or Ordering Vectors*

Description

Sort (or *order*) a numeric or complex vector (partially) into ascending (or descending) order.

Usage

```
sort(x, partial = NULL, na.last = NA, decreasing = FALSE,
     method = c("shell", "quick"), index.return = FALSE)
is.unsorted(x, na.rm = FALSE)
```

Arguments

x	a numeric or complex vector.
partial	a vector of indices for partial sorting.
na.last	for controlling the treatment of NAs. If TRUE, missing values in the data are put last; if FALSE, they are put first; if NA, they are removed.
decreasing	logical. Should the sort be increasing or decreasing?
method	character specifying the algorithm used.
index.return	logical indicating if the ordering index vector should be returned as well; this is only available for the default na.last = NA.
na.rm	logical. Should missing values be removed?

Details

If `partial` is not NULL, it is taken to contain indices of elements of x which are to be placed in their correct positions by partial sorting. After the sort, the values specified in `partial` are in their correct position in the sorted array. Any values smaller than these values are guaranteed to have a smaller index in the sorted array and any values which are greater are guaranteed to have a bigger index in the sorted array.

The sort order for character vectors will depend on the collating sequence of the locale in use: see Comparison.

`is.unsorted` returns a logical indicating if x is sorted increasingly, i.e., `is.unsorted(x)` is true if `any(x != sort(x))` (and there are no NAs).

method = "shell" uses Shellsort (an $O(n^{4/3})$ variant from Sedgewick (1996)). If x has names a stable sort is used, so ties are not reordered. (This only matters if names are present.)

Method "quick" uses Singleton's Quicksort implementation and is only available when x is numeric (double or integer) and partial is NULL. It is normally somewhat faster than Shellsort (perhaps twice as fast on vectors of length a million) but has poor performance in the rare worst case. (Peto's modification using a pseudo-random midpoint is used to make the worst case rarer.) This is not a stable sort, and ties may be reordered.

Value

For sort the sorted vector unless index.return is true, when the result is a list with components named x and ix containing the sorted numbers and the ordering index vector. In the latter case, if method == "quick" ties may be reversed in the ordering, unlike sort.list, as quicksort is not stable.

References

Becker, R. A., Chambers, J. M. and Wilks, A. R. (1988) *The New S Language.* Wadsworth & Brooks/Cole.

Sedgewick, R. (1986) A new upper bound for Shell sort. *J. Algorithms* **7**, 159–173.

Singleton, R. C. (1969) An efficient algorithm for sorting with minimal storage: Algorithm 347. *Communications of the ACM* **12**, 185–187.

See Also

order, rank.

Examples

```
data(swiss)
x <- swiss$Education[1:25]
x; sort(x); sort(x, partial = c(10, 15))
median # shows you another example for 'partial'

## illustrate 'stable' sorting (of ties):
sort(c(10:3,2:12), method = "sh", index=TRUE) # is stable
## $x : 2  3  3  4  4  5  5  6  6  7  7  8  8  9  9 10 10
## 11 12
```

```
## $ix: 9  8 10  7 11  6 12  5 13  4 14  3 15  2 16
## 1 17 18 19
sort(c(10:3,2:12), method = "qu", index=TRUE) # is not
## $x : 2  3  3  4  4  5  5  6  6  7  7  8  8  9  9 10 10
## 11 12
## $ix: 9 10  8  7 11  6 12  5 13  4 14  3 15 16  2
## 17  1 18 19

## Small speed comparison simulation:
N <- 2000
Sim <- 20
rep <- 50 # << adjust to your CPU
c1 <- c2 <- numeric(Sim)
for(is in 1:Sim) {
  x <- rnorm(N)
  gc()   ## sort should not have to pay for gc
  c1[is] <- system.time(for(i in 1:rep)
                          sort(x, method = "shell"))[1]
  c2[is] <- system.time(for(i in 1:rep)
                          sort(x, method = "quick"))[1]
  stopifnot(sort(x, meth = "s") == sort(x, meth = "q"))
}
100 * rbind(ShellSort = c1, QuickSort = c2)
cat("Speedup factor of quick sort():\n")
summary({qq <- c1 / c2; qq[is.finite(qq)]})

## A larger test
x <- rnorm(1e6)
gc()
system.time(x1 <- sort(x, method = "shell"))
gc()
system.time(x2 <- sort(x, method = "quick"))
stopifnot(identical(x1, x2))
```

Special — Special Functions of Mathematics

Description

Special mathematical functions related to the beta and gamma functions.

Usage

```
beta(a, b)
lbeta(a, b)
gamma(x)
lgamma(x)
digamma(x)
trigamma(x)
tetragamma(x)
pentagamma(x)
choose(n, k)
lchoose(n, k)
```

Arguments

a, b, x   numeric vectors.

n, k   integer vectors.

Details

The functions beta and lbeta return the beta function and the natural logarithm of the beta function,

$$B(a,b) = \frac{\Gamma(a)\Gamma(b)}{\Gamma(a+b)}.$$

The functions gamma and lgamma return the gamma function $\Gamma(x)$ and the natural logarithm of the absolute value of the gamma function.

The functions digamma, trigamma, tetragamma and pentagamma return the first, second, third and fourth derivatives of the logarithm of the gamma function.

$$\texttt{digamma(x)} = \psi(x) = \frac{d}{dx}\ln\Gamma(x) = \frac{\Gamma'(x)}{\Gamma(x)}$$

The functions choose and lchoose return binomial coefficients and their logarithms.

References

Becker, R. A., Chambers, J. M. and Wilks, A. R. (1988) *The New S Language.* Wadsworth & Brooks/Cole. (for gamma and lgamma.)

Abramowitz, M. and Stegun, I. A. (1972) *Handbook of Mathematical Functions.* New York: Dover. Chapter 6: Gamma and Related Functions.

See Also

Arithmetic for simple, sqrt for miscellaneous mathematical functions and Bessel for the real Bessel functions.

Examples

```
choose(5, 2)
for (n in 0:10) print(choose(n, k = 0:n))

## gamma has discontinuities are 0, -1, -2, ... use plots
## of points to show this.
curve(gamma(x),-3,4, n=1001, ylim=c(-10,100),
      col="red", lwd=2, main="gamma(x)")
abline(h=0,v=0, lty=3, col="midnightblue")

x <- seq(.1, 4, length = 201); dx <- diff(x)[1]
par(mfrow = c(2, 3))
for (ch in c("", "l","di","tri","tetra","penta")) {
  is.deriv <- nchar(ch) >= 2
  if (is.deriv) dy <- diff(y) / dx
  nm <- paste(ch, "gamma", sep = "")
  y <- get(nm)(x)
  plot(x, y, type = "l", main = nm, col = "red")
  abline(h = 0, col = "lightgray")
  if (is.deriv) lines(x[-1], dy, col = "blue", lty = 2)
}
```

splinefun   *Interpolating Splines*

Description

Perform cubic spline interpolation of given data points, returning either a list of points obtained by the interpolation or a function performing the interpolation.

Usage

```
splinefun(x, y = NULL, method = "fmm")

spline(x, y = NULL, n = 3*length(x), method = "fmm",
       xmin = min(x), xmax = max(x))
```

Arguments

x,y	vectors giving the coordinates of the points to be interpolated. Alternatively a single plotting structure can be specified: see xy.coords.
method	specifies the type of spline to be used. Possible values are "fmm", "natural" and "periodic".
n	interpolation takes place at n equally spaced points spanning the interval [xmin, xmax].
xmin	left-hand endpoint of the interpolation interval.
xmax	right-hand endpoint of the interpolation interval.

Details

If method = "fmm", the spline used is that of Forsythe, Malcolm and Moler (an exact cubic is fitted through the four points at each end of the data, and this is used to determine the end conditions). Natural splines are used when method = "natural", and periodic splines when method = "periodic".

These interpolation splines can also be used for extrapolation, that is prediction at points outside the range of x. Extrapolation makes little sense for method = "fmm"; for natural splines it is linear using the slope of the interpolating curve at the nearest data point.

Value

`spline` returns a list containing components `x` and `y` which give the ordinates where interpolation took place and the interpolated values.

`splinefun` returns a function which will perform cubic spline interpolation of the given data points. This is often more useful than `spline`.

References

Becker, R. A., Chambers, J. M. and Wilks, A. R. (1988) *The New S Language.* Wadsworth & Brooks/Cole.

Forsythe, G. E., Malcolm, M. A. and Moler, C. B. (1977) *Computer Methods for Mathematical Computations.*

See Also

`approx` and `approxfun` for constant and linear interpolation.

Package **splines**, especially `interpSpline` and `periodicSpline` for interpolation splines. That package also generates spline bases that can be used for regression splines.

`smooth.spline` in package **modreg** for smoothing splines.

Examples

```
op <- par(mfrow = c(2,1), mgp = c(2,.8,0),
          mar = .1+c(3,3,3,1))
n <- 9
x <- 1:n
y <- rnorm(n)
plot(x, y,
     main = paste("spline[fun](.) through", n, "points"))
lines(spline(x, y))
lines(spline(x, y, n = 201), col = 2)

y <- (x-6)^2
plot(x, y, main = "spline(.) -- 3 methods")
lines(spline(x, y, n = 201), col = 2)
lines(spline(x, y, n = 201, method = "natural"), col = 3)
lines(spline(x, y, n = 201, method = "periodic"), col = 4)
legend(6,25, c("fmm","natural","periodic"), col=2:4, lty=1)

f <- splinefun(x, y)
ls(envir = environment(f))
```

```
splinecoef <- eval(expression(z), envir = environment(f))
curve(f(x), 1, 10, col = "green", lwd = 1.5)
points(splinecoef, col = "purple", cex = 2)
par(op)
```

| sum | *Sum of Vector Elements* |

Description

sum returns the sum of all the values present in its arguments. If na.rm is FALSE an NA value in any of the arguments will cause a value of NA to be returned, otherwise NA values are ignored.

Usage

sum(..., na.rm=FALSE)

Arguments

...   numeric or complex vectors.
na.rm   logical. Should missing values be removed?

Value

The sum. If all of ... are of type integer, then so is the sum, and in that case the result will be NA (with a warning) if integer overflow occurs.

NB: the sum of an empty set is zero, by definition.

References

Becker, R. A., Chambers, J. M. and Wilks, A. R. (1988) *The New S Language.* Wadsworth & Brooks/Cole.

svd　　　*Singular Value Decomposition of a Matrix*

Description

Compute the singular-value decomposition of a rectangular matrix.

Usage

```
svd(x, nu = min(n, p), nv = min(n, p), LINPACK = FALSE)
La.svd(x, nu = min(n, p), nv = min(n, p),
       method = c("dgesdd", "dgesvd"))
```

Arguments

x　　　　　　a matrix whose SVD decomposition is to be computed.

nu　　　　　 the number of left singular vectors to be computed. This must be one of 0, nrow(x) and ncol(x), except for method = "dgesdd".

nv　　　　　 the number of right singular vectors to be computed. This must be one of 0 and ncol(x).

LINPACK　　 logical. Should LINPACK be used (for compatibility with R < 1.7.0)?

method　　　The LAPACK routine to use in the real case.

Details

The singular value decomposition plays an important role in many statistical techniques. svd and La.svd provide two slightly different interfaces. The main functions used are the LAPACK routines DGESDD and ZGESVD; svd(LINPACK=TRUE) provides an interface to the LINPACK routine DSVDC, purely for backwards compatibility.

La.svd provides an interface to both the LAPACK routines DGESVD and DGESDD. The latter is usually substantially faster if singular vectors are required. Most benefit is seen with an optimized BLAS system. Using method="dgesdd" requires IEEE 754 arithmetic. Should this not be supported on your platform, method="dgesvd" is used, with a warning.

Computing the singular vectors is the slow part for large matrices.

Value

The SVD decomposition of the matrix as computed by LINPACK,

$$X = UDV',$$

where U and V are orthogonal, V' means V *transposed*, and D is a diagonal matrix with the singular values D_{ii}. Equivalently, $D = U'XV$, which is verified in the examples, below.

The returned value is a list with components

d   a vector containing the singular values of x.

u   a matrix whose columns contain the left singular vectors of x, present if nu > 0

v   a matrix whose columns contain the right singular vectors of x, present if nv > 0.

For La.svd the return value replaces v by vt, the (conjugated if complex) transpose of v.

References

Becker, R. A., Chambers, J. M. and Wilks, A. R. (1988) *The New S Language*. Wadsworth & Brooks/Cole.

Dongarra, J. J., Bunch, J. R., Moler, C. B. and Stewart, G. W. (1978) *LINPACK Users Guide*. Philadelphia: SIAM Publications.

Anderson. E., et al. (1999) *LAPACK Users' Guide*. Third Edition. SIAM. ISBN 0-89871-447-8.

See Also

eigen, qr.

capabilities to test for IEEE 754 arithmetic.

Examples

```
hilbert <- function(n) {
   i <- 1:n; 1 / outer(i - 1, i, "+")
}
str(X <- hilbert(9)[,1:6])
str(s <- svd(X))
D <- diag(s$d)
s$u %*% D %*% t(s$v) #   X = U D V'
t(s$u) %*% X %*% s$v #   D = U' X V
```

tabulate   *Tabulation for Vectors*

Description

tabulate takes the integer valued vector bin and counts the number of times each integer occurs in it. tabulate is used as the basis of the table function.

Usage

```
tabulate(bin, nbins = max(1, bin))
```

Arguments

bin   a vector of integers, or a factor.
nbins   the number of bins to be used.

Details

If bin is a factor, its internal integer representation is tabulated. If the elements of bin are not integers, they are rounded to the nearest integer. Elements outside the range 1,..., nbin are (silently) ignored in the tabulation.

See Also

factor, table.

Examples

```
tabulate(c(2,3,5))
tabulate(c(2,3,3,5), nb = 10)
tabulate(c(-2,0,2,3,3,5), nb = 3)
tabulate(factor(letters[1:10]))
```

| Trig | *Trigonometric Functions* |

Description

These functions give the obvious trigonometric functions. They respectively compute the cosine, sine, tangent, arc-cosine, arc-sine, arc-tangent, and the two-argument arc-tangent.

Usage

```
cos(x)
sin(x)
tan(x)
acos(x)
asin(x)
atan(x)
atan2(y, x)
```

Arguments

x, y   numeric vector

Details

The arc-tangent of two arguments `atan2(y,x)` returns the angle between the x-axis and the vector from the origin to (x, y), i.e., for positive arguments `atan2(y,x) == atan(y/x)`.

Angles are in radians, not degrees (i.e., a right angle is $\pi/2$).

References

Becker, R. A., Chambers, J. M. and Wilks, A. R. (1988) *The New S Language.* Wadsworth & Brooks/Cole.

Chapter 3

Base package — distributions and random numbers

bandwidth — Bandwidth Selectors for Kernel Density Estimation

Description

Bandwidth selectors for gaussian windows in `density`.

Usage

```
bw.nrd0(x)
bw.nrd(x)
bw.ucv(x, nb = 1000, lower, upper)
bw.bcv(x, nb = 1000, lower, upper)
bw.SJ(x, nb = 1000, lower, upper, method = c("ste", "dpi"))
```

Arguments

x	A data vector.
nb	number of bins to use.
lower, upper	Range over which to minimize. The default is almost always satisfactory.
method	Either "ste" ("solve-the-equation") or "dpi" ("direct plug-in").

Details

`bw.nrd0` implements a rule-of-thumb for choosing the bandwidth of a Gaussian kernel density estimator. It defaults to 0.9 times the minimum of the standard deviation and the interquartile range divided by 1.34 times the sample size to the negative one-fifth power (= Silverman's "rule of thumb", Silverman (1986, page 48, eqn (3.31)) *unless* the quartiles coincide when a positive result will be guaranteed.

`bw.nrd` is the more common variation given by Scott (1992), using factor 1.06.

`bw.ucv` and `bw.bcv` implement unbiased and biased cross-validation respectively.

`bw.SJ` implements the methods of Sheather & Jones (1991) to select the bandwidth using pilot estimation of derivatives.

Value

A bandwidth on a scale suitable for the `bw` argument of `density`.

References

Scott, D. W. (1992) *Multivariate Density Estimation: Theory, Practice, and Visualization.* Wiley.

Sheather, S. J. and Jones, M. C. (1991) A reliable data-based bandwidth selection method for kernel density estimation. *Journal of the Royal Statistical Society series B*, **53**, 683–690.

Silverman, B. W. (1986) *Density Estimation.* London: Chapman and Hall.

Venables, W. N. and Ripley, B. D. (2002) *Modern Applied Statistics with S.* Springer.

See Also

density.

bandwidth.nrd, ucv, bcv and width.SJ in package **MASS**, which are all scaled to the width argument of density and so give answers four times as large.

Examples

```
data(precip)
plot(density(precip, n = 1000))
rug(precip)
lines(density(precip, bw="nrd"), col = 2)
lines(density(precip, bw="ucv"), col = 3)
lines(density(precip, bw="bcv"), col = 4)
lines(density(precip, bw="SJ-ste"), col = 5)
lines(density(precip, bw="SJ-dpi"), col = 6)
legend(55, 0.035,
   legend = c("nrd0","nrd","ucv","bcv","SJ-ste","SJ-dpi"),
   col = 1:6, lty = 1)
```

Beta   *The Beta Distribution*

Description

Density, distribution function, quantile function and random generation for the Beta distribution with parameters shape1 and shape2 (and optional non-centrality parameter ncp).

Usage

```
dbeta(x,shape1,shape2,ncp=0,log=FALSE)
pbeta(q,shape1,shape2,ncp=0,lower.tail=TRUE,log.p=FALSE)
qbeta(p,shape1,shape2,lower.tail=TRUE,log.p=FALSE)
rbeta(n,shape1,shape2)
```

Arguments

x, q	vector of quantiles.
p	vector of probabilities.
n	number of observations. If length(n) > 1, the length is taken to be the number required.
shape1, shape2	positive parameters of the Beta distribution.
ncp	non-centrality parameter.
log, log.p	logical; if TRUE, probabilities p are given as log(p).
lower.tail	logical; if TRUE (default), probabilities are $P[X \leq x]$, otherwise, $P[X > x]$.

Details

The Beta distribution with parameters shape1 = a and shape2 = b has density

$$f(x) = \frac{\Gamma(a+b)}{\Gamma(a)\Gamma(b)} x^a (1-x)^b$$

for $a > 0$, $b > 0$ and $0 \leq x \leq 1$ where the boundary values at $x = 0$ or $x = 1$ are defined as by continuity (as limits).

Value

dbeta gives the density, pbeta the distribution function, qbeta the quantile function, and rbeta generates random deviates.

References

Becker, R. A., Chambers, J. M. and Wilks, A. R. (1988) *The New S Language.* Wadsworth & Brooks/Cole.

See Also

beta for the Beta function, and dgamma for the Gamma distribution.

Examples

```
x <- seq(0, 1, length=21)
dbeta(x, 1, 1)
pbeta(x, 1, 1)
```

Binomial — The Binomial Distribution

Description

Density, distribution function, quantile function and random generation for the binomial distribution with parameters `size` and `prob`.

Usage

```
dbinom(x, size, prob, log = FALSE)
pbinom(q, size, prob, lower.tail = TRUE, log.p = FALSE)
qbinom(p, size, prob, lower.tail = TRUE, log.p = FALSE)
rbinom(n, size, prob)
```

Arguments

x, q	vector of quantiles.
p	vector of probabilities.
n	number of observations. If `length(n) > 1`, the length is taken to be the number required.
size	number of trials.
prob	probability of success on each trial.
log, log.p	logical; if TRUE, probabilities p are given as log(p).
lower.tail	logical; if TRUE (default), probabilities are $P[X \leq x]$, otherwise, $P[X > x]$.

Details

The binomial distribution with `size` $= n$ and `prob` $= p$ has density

$$p(x) = \binom{n}{x} p^x (1-p)^{n-x}$$

for $x = 0, \ldots, n$.

If an element of x is not integer, the result of `dbinom` is zero, with a warning. $p(x)$ is computed using Loader's algorithm, see the reference below.

The quantile is defined as the smallest value x such that $F(x) \geq p$, where F is the distribution function.

Value

dbinom gives the density, pbinom gives the distribution function, qbinom gives the quantile function and rbinom generates random deviates.

If size is not an integer, NaN is returned.

References

Catherine Loader (2000). *Fast and Accurate Computation of Binomial Probabilities*

See Also

dnbinom for the negative binomial, and dpois for the Poisson distribution.

Examples

```
# Compute P(45 < X < 55) for X Binomial(100,0.5)
sum(dbinom(46:54, 100, 0.5))

## Using "log = TRUE" for an extended range :
n <- 2000
k <- seq(0, n, by = 20)
plot (k, dbinom(k, n, pi/10, log=TRUE), type='l',
 ylab="log density",
 main="dbinom(*, log=TRUE) is better than log(dbinom(*))")
lines(k, log(dbinom(k, n, pi/10)), col='red', lwd=2)
## extreme points are omitted since dbinom gives 0.
mtext("dbinom(k, log=TRUE)", adj=0)
mtext("extended range", adj=0, line = -1, font=4)
mtext("log(dbinom(k))", col="red", adj=1)
```

birthday *Probability of coincidences*

Description

Computes approximate answers to a generalised "birthday paradox" problem. pbirthday computes the probability of a coincidence and qbirthday computes the number of observations needed to have a specified probability of coincidence.

Usage

```
qbirthday(prob = 0.5, classes = 365, coincident = 2)
pbirthday(n, classes = 365, coincident = 2)
```

Arguments

classes   How many distinct categories the people could fall into
prob   The desired probability of coincidence
n   The number of people
coincident The number of people to fall in the same category

Details

The birthday paradox is that a very small number of people, 23, suffices to have a 50-50 chance that two of them have the same birthday. This function generalises the calculation to probabilities other than 0.5, numbers of coincident events other than 2, and numbers of classes other than 365.

This formula is approximate, as the example below shows. For coincident=2 the exact computation is straightforward and may be preferable.

Value

qbirthday   Number of people needed for a probability prob that k of them have the same one out of classes equiprobable labels.

pbirthday   Probability of the specified coincidence

References

Diaconis P, Mosteller F., "Methods for studying coincidences". JASA 84:853-861

Examples

```
## the standard version
qbirthday()
## same 4-digit PIN number
qbirthday(classes=10^4)
## 0.9 probability of three coincident birthdays
qbirthday(coincident=3,prob=0.9)
## Chance of 4 coincident birthdays in 150 people
pbirthday(150,coincident=4)
## Accuracy compared to exact calculation
x1<- sapply(10:100, pbirthday)
x2<-1-sapply(10:100, function(n)
                    prod((365:(365-n+1))/rep(365,n)))
par(mfrow=c(2,2))
plot(x1,x2,xlab="approximate",ylab="exact")
abline(0,1)
plot(x1,x1-x2,xlab="approximate",ylab="error")
abline(h=0)
plot(x1,x2,log="xy",xlab="approximate",ylab="exact")
abline(0,1)
plot(1-x1,1-x2,log="xy",xlab="approximate",ylab="exact")
abline(0,1)
```

Cauchy   *The Cauchy Distribution*

Description

Density, distribution function, quantile function and random generation for the Cauchy distribution with location parameter `location` and scale parameter `scale`.

Usage

```
dcauchy(x,location=0,scale=1,log=FALSE)
pcauchy(q,location=0,scale=1,lower.tail=TRUE,log.p=FALSE)
qcauchy(p,location=0,scale=1,lower.tail=TRUE,log.p=FALSE)
rcauchy(n,location=0,scale=1)
```

Arguments

`x, q`	vector of quantiles.
`p`	vector of probabilities.
`n`	number of observations. If `length(n) > 1`, the length is taken to be the number required.
`location, scale`	location and scale parameters.
`log, log.p`	logical; if TRUE, probabilities p are given as log(p).
`lower.tail`	logical; if TRUE (default), probabilities are $P[X \leq x]$, otherwise, $P[X > x]$.

Details

If `location` or `scale` are not specified, they assume the default values of 0 and 1 respectively.

The Cauchy distribution with location l and scale s has density

$$f(x) = \frac{1}{\pi s} \left(1 + \left(\frac{x-l}{s} \right)^2 \right)^{-1}$$

for all x.

Value

dcauchy, pcauchy, and qcauchy are respectively the density, distribution function and quantile function of the Cauchy distribution. rcauchy generates random deviates from the Cauchy.

References

Becker, R. A., Chambers, J. M. and Wilks, A. R. (1988) *The New S Language.* Wadsworth & Brooks/Cole.

See Also

dt for the t distribution which generalizes dcauchy(*, l = 0, s = 1).

Examples

dcauchy(-1:4)

Chisquare   *The (non-central) Chi-Squared Distribution*

Description

Density, distribution function, quantile function and random generation for the chi-squared (χ^2) distribution with df degrees of freedom and optional non-centrality parameter ncp.

Usage

```
dchisq(x, df, ncp=0, log = FALSE)
pchisq(q, df, ncp=0, lower.tail = TRUE, log.p = FALSE)
qchisq(p, df, ncp=0, lower.tail = TRUE, log.p = FALSE)
rchisq(n, df, ncp=0)
```

Arguments

x, q	vector of quantiles.
p	vector of probabilities.
n	number of observations. If length(n) > 1, the length is taken to be the number required.
df	degrees of freedom (non-negative, but can be non-integer).
ncp	non-centrality parameter (non-negative). Note that ncp values larger than about 1417 are not allowed currently for pchisq and qchisq.
log, log.p	logical; if TRUE, probabilities p are given as log(p).
lower.tail	logical; if TRUE (default), probabilities are $P[X \leq x]$, otherwise, $P[X > x]$.

Details

The chi-squared distribution with df$= n$ degrees of freedom has density

$$f_n(x) = \frac{1}{2^{n/2}\Gamma(n/2)} x^{n/2-1} e^{-x/2}$$

for $x > 0$. The mean and variance are n and $2n$.

The non-central chi-squared distribution with df= n degrees of freedom and non-centrality parameter ncp = λ has density

$$f(x) = e^{-\lambda/2} \sum_{r=0}^{\infty} \frac{(\lambda/2)^r}{r!} f_{n+2r}(x)$$

for $x \geq 0$. For integer n, this is the distribution of the sum of squares of n normals each with variance one, λ being the sum of squares of the normal means. Note that the degrees of freedom df= n, can be non-integer, and for non-centrality $\lambda > 0$, even $n = 0$; see the reference, chapter 29.

Value

dchisq gives the density, pchisq gives the distribution function, qchisq gives the quantile function, and rchisq generates random deviates.

References

Becker, R. A., Chambers, J. M. and Wilks, A. R. (1988) *The New S Language*. Wadsworth & Brooks/Cole.

Johnson, Kotz and Balakrishnan (1995). *Continuous Univariate Distributions*, Vol 2; Wiley NY;

See Also

dgamma for the Gamma distribution which generalizes the chi-squared one.

Examples

```
dchisq(1, df=1:3)
pchisq(1, df= 3)
pchisq(1, df= 3, ncp = 0:4) # includes the above

x <- 1:10
## Chi-squared(df = 2) is a special exponential
## distribution
all.equal(dchisq(x, df=2), dexp(x, 1/2))
all.equal(pchisq(x, df=2), pexp(x, 1/2))

## non-central RNG -- df=0 is ok for ncp > 0: Z0 has point
## mass at 0!
Z0 <- rchisq(100, df = 0, ncp = 2.)
```

```
stem(Z0)

## visual testing do P-P plots for 1000 points at various
## degrees of freedom
L <- 1.2; n <- 1000; pp <- ppoints(n)
op <- par(mfrow = c(3,3), mar = c(3,3,1,1)+.1,
          mgp = c(1.5,.6,0), oma = c(0,0,3,0))
for(df in 2^(4*rnorm(9))) {
  plot(pp, sort(pchisq(rr <- rchisq(n,df=df, ncp=L),
       df=df, ncp=L)), ylab="pchisq(rchisq(.),.)", pch=".")
  mtext(paste("df = ",formatC(df, digits = 4)),
        line= -2, adj=0.05)
  abline(0,1,col=2)
}
mtext(expression("P-P plots : Noncentral "*
                 chi^2 *"(n=1000, df=X, ncp= 1.2)"),
      cex = 1.5, font = 2, outer=TRUE)
par(op)
```

density *Kernel Density Estimation*

Description

The function `density` computes kernel density estimates with the given kernel and bandwidth.

Usage

```
density(x, bw = "nrd0", adjust = 1,
        kernel = c("gaussian", "epanechnikov",
                   "rectangular", "triangular",
                   "biweight", "cosine", "optcosine"),
        window = kernel, width,
        give.Rkern = FALSE,
        n = 512, from, to, cut = 3, na.rm = FALSE)
```

Arguments

x
: the data from which the estimate is to be computed.

bw
: the smoothing bandwidth to be used. The kernels are scaled such that this is the standard deviation of the smoothing kernel. (Note this differs from the reference books cited below, and from S-PLUS.)

    bw can also be a character string giving a rule to choose the bandwidth. See `bw.nrd`.

    The specified (or computed) value of bw is multiplied by adjust.

adjust
: the bandwidth used is actually adjust*bw. This makes it easy to specify values like "half the default" bandwidth.

kernel, window
: a character string giving the smoothing kernel to be used. This must be one of "gaussian", "rectangular", "triangular", "epanechnikov", "biweight", "cosine" or "optcosine", with default "gaussian", and may be abbreviated to a unique prefix (single letter).

    "cosine" is smoother than "optcosine", which is the usual "cosine" kernel in the literature and almost

	MSE-efficient. However, `"cosine"` is the version used by S.
`width`	this exists for compatibility with S; if given, and `bw` is not, will set `bw` to `width` if this is a character string, or to a kernel-dependent multiple of `width` if this is numeric.
`give.Rkern`	logical; if true, *no* density is estimated, and the "canonical bandwidth" of the chosen `kernel` is returned instead.
`n`	the number of equally spaced points at which the density is to be estimated. When n > 512, it is rounded up to the next power of 2 for efficiency reasons (`fft`).
`from,to`	the left and right-most points of the grid at which the density is to be estimated.
`cut`	by default, the values of `left` and `right` are cut bandwidths beyond the extremes of the data. This allows the estimated density to drop to approximately zero at the extremes.
`na.rm`	logical; if `TRUE`, missing values are removed from x. If `FALSE` any missing values cause an error.

Details

The algorithm used in `density` disperses the mass of the empirical distribution function over a regular grid of at least 512 points and then uses the fast Fourier transform to convolve this approximation with a discretized version of the kernel and then uses linear approximation to evaluate the density at the specified points.

The statistical properties of a kernel are determined by $\sigma_K^2 = \int t^2 K(t) dt$ which is always $= 1$ for our kernels (and hence the bandwidth `bw` is the standard deviation of the kernel) and $R(K) = \int K^2(t) dt$.

MSE-equivalent bandwidths (for different kernels) are proportional to $\sigma_K R(K)$ which is scale invariant and for our kernels equal to $R(K)$. This value is returned when `give.Rkern = TRUE`. See the examples for using exact equivalent bandwidths.

Infinite values in x are assumed to correspond to a point mass at +/-`Inf` and the density estimate is of the sub-density on (-`Inf`, +`Inf`).

Value

If `give.Rkern` is true, the number $R(K)$, otherwise an object with class `"density"` whose underlying structure is a list containing the following components.

x   the n coordinates of the points where the density is estimated.
y   the estimated density values.
bw   the bandwidth used.
N   the sample size after elimination of missing values.
call the call which produced the result.
data.name the deparsed name of the x argument.
has.na   logical, for compatibility (always FALSE).

References

Becker, R. A., Chambers, J. M. and Wilks, A. R. (1988) *The New S Language.* Wadsworth & Brooks/Cole (for S version).

Scott, D. W. (1992) *Multivariate Density Estimation. Theory, Practice and Visualization.* New York: Wiley.

Sheather, S. J. and Jones M. C. (1991) A reliable data-based bandwidth selection method for kernel density estimation. *J. Roy. Statist. Soc.* **B**, 683–690.

Silverman, B. W. (1986) *Density Estimation.* London: Chapman and Hall.

Venables, W. N. and Ripley, B. D. (1999) *Modern Applied Statistics with S-PLUS.* New York: Springer.

See Also

bw.nrd, plot.density, hist.

Examples

```
plot(density(c(-20,rep(0,98),20)), xlim = c(-4,4)) # IQR = 0

# The Old Faithful geyser data
data(faithful)
d <- density(faithful$eruptions, bw = "sj")
d
plot(d)

plot(d, type = "n")
polygon(d, col = "wheat")
```

```
## Missing values:
x <- xx <- faithful$eruptions
x[i.out <- sample(length(x), 10)] <- NA
doR <- density(x, bw = 0.15, na.rm = TRUE)
lines(doR, col = "blue")
points(xx[i.out], rep(0.01, 10))

(kernels <- eval(formals(density)$kernel))

## show the kernels in the R parametrization
plot (density(0, bw = 1), xlab = "",
      main="R's density() kernels with bw = 1")
for(i in 2:length(kernels))
   lines(density(0, bw = 1, kern = kernels[i]), col = i)
legend(1.5,.4, legend = kernels, col = seq(kernels),
       lty = 1, cex = .8, y.int = 1)

## show the kernels in the S parametrization
plot(density(0, from=-1.2, to=1.2, width=2,
     kern="gaussian"), type="l", ylim = c(0, 1), xlab="",
     main="R's density() kernels with width = 1")
for(i in 2:length(kernels))
   lines(density(0, width=2, kern = kernels[i]), col = i)
legend(0.6, 1.0, legend = kernels, col = seq(kernels),
       lty = 1)

(RKs <- cbind(sapply(kernels, function(k)
                  density(kern = k, give.Rkern = TRUE))))
100*round(RKs["epanechnikov",]/RKs, 4) ## Efficiencies

if(interactive()) {
data(precip)
bw <- bw.SJ(precip) ## sensible automatic choice
plot(density(precip, bw = bw, n = 2^13),
     main = "same sd bandwidths, 7 different kernels")
for(i in 2:length(kernels))
   lines(density(precip, bw = bw, kern = kernels[i],
                 n = 2^13), col = i)

## Bandwidth Adjustment for "Exactly Equivalent Kernels"
h.f <- sapply(kernels, function(k)
                  density(kern = k, give.Rkern = TRUE))
```

```
(h.f <- (h.f["gaussian"] / h.f)^ .2)
## -> 1, 1.01, .995, 1.007,... close to 1 => adjustment
## barely visible..

plot(density(precip, bw = bw, n = 2^13),
     main = "equivalent bandwidths, 7 different kernels")
for(i in 2:length(kernels))
    lines(density(precip, bw = bw, adjust = h.f[i],
                  kern = kernels[i], n = 2^13), col = i)
legend(55, 0.035, legend = kernels, col = seq(kernels),
       lty = 1)
}
```

Exponential — *The Exponential Distribution*

Description

Density, distribution function, quantile function and random generation for the exponential distribution with rate `rate` (i.e., mean 1/rate).

Usage

```
dexp(x, rate = 1, log = FALSE)
pexp(q, rate = 1, lower.tail = TRUE, log.p = FALSE)
qexp(p, rate = 1, lower.tail = TRUE, log.p = FALSE)
rexp(n, rate = 1)
```

Arguments

x, q	vector of quantiles.
p	vector of probabilities.
n	number of observations. If length(n) > 1, the length is taken to be the number required.
rate	vector of rates.
log, log.p	logical; if TRUE, probabilities p are given as log(p).
lower.tail	logical; if TRUE (default), probabilities are $P[X \leq x]$, otherwise, $P[X > x]$.

Details

If `rate` is not specified, it assumes the default value of 1.

The exponential distribution with rate λ has density

$$f(x) = \lambda e^{-\lambda x}$$

for $x \geq 0$.

Value

`dexp` gives the density, `pexp` gives the distribution function, `qexp` gives the quantile function, and `rexp` generates random deviates.

Note

The cumulative hazard $H(t) = -\log(1 - F(t))$ is -pexp(t, r, lower = FALSE, log = TRUE).

References

Becker, R. A., Chambers, J. M. and Wilks, A. R. (1988) *The New S Language*. Wadsworth & Brooks/Cole.

See Also

exp for the exponential function, dgamma for the gamma distribution and dweibull for the Weibull distribution, both of which generalize the exponential.

Examples

dexp(1) - exp(-1) # 0

FDist   *The F Distribution*

Description

Density, distribution function, quantile function and random generation for the F distribution with `df1` and `df2` degrees of freedom (and optional non-centrality parameter `ncp`).

Usage

```
df(x, df1, df2, log = FALSE)
pf(q, df1, df2, ncp=0, lower.tail = TRUE, log.p = FALSE)
qf(p, df1, df2,        lower.tail = TRUE, log.p = FALSE)
rf(n, df1, df2)
```

Arguments

`x, q`	vector of quantiles.
`p`	vector of probabilities.
`n`	number of observations. If `length(n) > 1`, the length is taken to be the number required.
`df1, df2`	degrees of freedom.
`ncp`	non-centrality parameter.
`log, log.p`	logical; if TRUE, probabilities p are given as log(p).
`lower.tail`	logical; if TRUE (default), probabilities are $P[X \leq x]$, otherwise, $P[X > x]$.

Details

The F distribution with `df1` = n_1 and `df2` = n_2 degrees of freedom has density

$$f(x) = \frac{\Gamma(n_1/2 + n_2/2)}{\Gamma(n_1/2)\Gamma(n_2/2)} \left(\frac{n_1}{n_2}\right)^{n_1/2} x^{n_1/2 - 1} \left(1 + \frac{n_1 x}{n_2}\right)^{-(n_1+n_2)/2}$$

for $x > 0$.

It is the distribution of the ratio of the mean squares of n_1 and n_2 independent standard normals, and hence of the ratio of two independent chi-squared variates each divided by its degrees of freedom. Since the ratio of a normal and the root mean-square of m independent normals has a Student's t_m distribution, the square of a t_m variate has a F distribution on 1 and m degrees of freedom.

The non-central F distribution is again the ratio of mean squares of independent normals of unit variance, but those in the numerator are allowed to have non-zero means and ncp is the sum of squares of the means. See Chisquare for further details on non-central distributions.

Value

df gives the density, pf gives the distribution function qf gives the quantile function, and rf generates random deviates.

References

Becker, R. A., Chambers, J. M. and Wilks, A. R. (1988) *The New S Language.* Wadsworth & Brooks/Cole.

See Also

dchisq for chi-squared and dt for Student's t distributions.

Examples

```
## the density of the square of a t_m is 2*dt(x, m)/(2*x)
# check this is the same as the density of F_{1,m}
x <- seq(0.001, 5, len=100)
all.equal(df(x^2, 1, 5), dt(x, 5)/x)

## Identity:   qf(2*p - 1, 1, df)) == qt(p, df)^2) for p
## >= 1/2
p <- seq(1/2, .99, length=50); df <- 10
rel.err <- function(x,y)
   ifelse(x==y, 0, abs(x-y)/mean(abs(c(x,y))))
quantile(rel.err(qf(2*p - 1, df1=1, df2=df), qt(p, df)^2),
  .90) # ~= 7e-9
```

GammaDist   *The Gamma Distribution*

Description

Density, distribution function, quantile function and random generation for the Gamma distribution with parameters shape and scale.

Usage

```
dgamma(x,shape,rate=1,scale=1/rate,log=FALSE)
pgamma(q,shape,rate=1,scale=1/rate,lower.tail=TRUE,
      log.p=FALSE)
qgamma(p,shape,rate=1,scale=1/rate,lower.tail=TRUE,log.
      p=FALSE)
rgamma(n,shape,rate=1,scale=1/rate)
```

Arguments

x, q	vector of quantiles.
p	vector of probabilities.
n	number of observations. If length(n) > 1, the length is taken to be the number required.
rate	an alternative way to specify the scale.
shape, scale	shape and scale parameters.
log, log.p	logical; if TRUE, probabilities p are given as log(p).
lower.tail	logical; if TRUE (default), probabilities are $P[X \leq x]$, otherwise, $P[X > x]$.

Details

If scale is omitted, it assumes the default value of 1.

The Gamma distribution with parameters shape $= \alpha$ and scale $= \sigma$ has density

$$f(x) = \frac{1}{\sigma^\alpha \Gamma(\alpha)} x^{\alpha-1} e^{-x/\sigma}$$

for $x > 0$, $\alpha > 0$ and $\sigma > 0$. The mean and variance are $E(X) = \alpha\sigma$ and $Var(X) = \alpha\sigma^2$.

Value

dgamma gives the density, **pgamma** gives the distribution function **qgamma** gives the quantile function, and **rgamma** generates random deviates.

Note

The S parametrization is via **shape** and **rate**: S has no **scale** parameter. Prior to 1.4.0 R only had **scale**.

The cumulative hazard $H(t) = -\log(1 - F(t))$ is -pgamma(t, ..., lower = FALSE, log = TRUE).

References

Becker, R. A., Chambers, J. M. and Wilks, A. R. (1988) *The New S Language.* Wadsworth & Brooks/Cole.

See Also

gamma for the Gamma function, **dbeta** for the Beta distribution and **dchisq** for the chi-squared distribution which is a special case of the Gamma distribution.

Examples

```
-log(dgamma(1:4, shape=1))
p <- (1:9)/10
pgamma(qgamma(p,shape=2), shape=2)
1 - 1/exp(qgamma(p, shape=1))
```

Geometric — The Geometric Distribution

Description

Density, distribution function, quantile function and random generation for the geometric distribution with parameter `prob`.

Usage

```
dgeom(x, prob, log = FALSE)
pgeom(q, prob, lower.tail = TRUE, log.p = FALSE)
qgeom(p, prob, lower.tail = TRUE, log.p = FALSE)
rgeom(n, prob)
```

Arguments

`x, q`	vector of quantiles representing the number of failures in a sequence of Bernoulli trials before success occurs.
`p`	vector of probabilities.
`n`	number of observations. If `length(n) > 1`, the length is taken to be the number required.
`prob`	probability of success in each trial.
`log, log.p`	logical; if TRUE, probabilities p are given as log(p).
`lower.tail`	logical; if TRUE (default), probabilities are $P[X \leq x]$, otherwise, $P[X > x]$.

Details

The geometric distribution with `prob` $= p$ has density

$$p(x) = p(1-p)^x$$

for $x = 0, 1, 2, \ldots$

If an element of x is not integer, the result of `pgeom` is zero, with a warning.

The quantile is defined as the smallest value x such that $F(x) \geq p$, where F is the distribution function.

Value

`dgeom` gives the density, `pgeom` gives the distribution function, `qgeom` gives the quantile function, and `rgeom` generates random deviates.

See Also

dnbinom for the negative binomial which generalizes the geometric distribution.

Examples

```
qgeom((1:9)/10, prob = .2)
Ni <- rgeom(20, prob = 1/4); table(factor(Ni, 0:max(Ni)))
```

Hypergeometric — The Hypergeometric Distribution

Description

Density, distribution function, quantile function and random generation for the hypergeometric distribution.

Usage

```
dhyper(x, m, n, k, log = FALSE)
phyper(q, m, n, k, lower.tail = TRUE, log.p = FALSE)
qhyper(p, m, n, k, lower.tail = TRUE, log.p = FALSE)
rhyper(nn, m, n, k)
```

Arguments

x, q	vector of quantiles representing the number of white balls drawn without replacement from an urn which contains both black and white balls.
m	the number of white balls in the urn.
n	the number of black balls in the urn.
k	the number of balls drawn from the urn.
p	probability, it must be between 0 and 1.
nn	number of observations. If length(nn) > 1, the length is taken to be the number required.
log, log.p	logical; if TRUE, probabilities p are given as log(p).
lower.tail	logical; if TRUE (default), probabilities are $P[X \leq x]$, otherwise, $P[X > x]$.

Details

The hypergeometric distribution is used for sampling *without* replacement. The density of this distribution with parameters m, n and k (named Np, $N - Np$, and n, respectively in the reference below) is given by

$$p(x) = \binom{m}{x}\binom{n}{k-x} \bigg/ \binom{m+n}{k}$$

for $x = 0, \ldots, k$.

Value

dhyper gives the density, **phyper** gives the distribution function, **qhyper** gives the quantile function, and **rhyper** generates random deviates.

References

Johnson, N. L., Kotz, S., and Kemp, A. W. (1992) *Univariate Discrete Distributions*, Second Edition. New York: Wiley.

Examples

```
m <- 10; n <- 7; k <- 8
x <- 0:(k+1)
rbind(phyper(x, m, n, k), dhyper(x, m, n, k))
# FALSE
all(phyper(x, m, n, k) == cumsum(dhyper(x, m, n, k)))
## but error is very small:
signif(phyper(x, m, n, k) - cumsum(dhyper(x, m, n, k)),
       dig=3)
```

Logistic   *The Logistic Distribution*

Description

Density, distribution function, quantile function and random generation for the logistic distribution with parameters `location` and `scale`.

Usage

```
dlogis(x,location=0,scale=1,log=FALSE)
plogis(q,location=0,scale=1,lower.tail=TRUE,log.p=FALSE)
qlogis(p,location=0,scale=1,lower.tail=TRUE,log.p=FALSE)
rlogis(n,location=0,scale=1)
```

Arguments

x, q	vector of quantiles.
p	vector of probabilities.
n	number of observations. If length(n) > 1, the length is taken to be the number required.
location, scale	
	location and scale parameters.
log, log.p	logical; if TRUE, probabilities p are given as log(p).
lower.tail	logical; if TRUE (default), probabilities are $P[X \leq x]$, otherwise, $P[X > x]$.

Details

If `location` or `scale` are omitted, they assume the default values of 0 and 1 respectively.

The Logistic distribution with `location` $= \mu$ and `scale` $= \sigma$ has distribution function

$$F(x) = \frac{1}{1 + e^{-(x-\mu)/\sigma}}$$

and density

$$f(x) = \frac{1}{\sigma} \frac{e^{(x-\mu)/\sigma}}{(1 + e^{(x-\mu)/\sigma})^2}$$

It is a long-tailed distribution with mean μ and variance $\pi^2/3\sigma^2$.

Value

`dlogis` gives the density, `plogis` gives the distribution function, `qlogis` gives the quantile function, and `rlogis` generates random deviates.

References

Becker, R. A., Chambers, J. M. and Wilks, A. R. (1988) *The New S Language.* Wadsworth & Brooks/Cole.

Examples

```
var(rlogis(4000, 0, s = 5))  # approximately (+/- 3)
pi^2/3 * 5^2
```

Lognormal — *The Log Normal Distribution*

Description

Density, distribution function, quantile function and random generation for the log normal distribution whose logarithm has mean equal to meanlog and standard deviation equal to sdlog.

Usage

```
dlnorm(x, meanlog=0, sdlog=1, log=false)
plnorm(q, meanlog=0, sdlog=1, lower.tail=true, log.p=false)
qlnorm(p, meanlog=0, sdlog=1, lower.tail=true, log.p=false)
rlnorm(n, meanlog=0, sdlog=1)
```

Arguments

x, q	vector of quantiles.
p	vector of probabilities.
n	number of observations. If length(n) > 1, the length is taken to be the number required.
meanlog, sdlog	mean and standard deviation of the distribution on the log scale with default values of 0 and 1 respectively.
log, log.p	logical; if TRUE, probabilities p are given as log(p).
lower.tail	logical; if TRUE (default), probabilities are $P[X \leq x]$, otherwise, $P[X > x]$.

Details

The log normal distribution has density

$$f(x) = \frac{1}{\sqrt{2\pi}\sigma x} e^{-(\log(x)-\mu)^2/2\sigma^2}$$

where μ and σ are the mean and standard deviation of the logarithm. The mean is $E(X) = exp(\mu + 1/2\sigma^2)$, and the variance $Var(X) = exp(2\mu + \sigma^2)(exp(\sigma^2) - 1)$ and hence the coefficient of variation is $\sqrt{exp(\sigma^2) - 1}$ which is approximately σ when that is small (e.g., $\sigma < 1/2$).

Value

dlnorm gives the density, plnorm gives the distribution function, qlnorm gives the quantile function, and rlnorm generates random deviates.

Note

The cumulative hazard $H(t) = -\log(1-F(t))$ is -plnorm(t, r, lower = FALSE, log = TRUE).

References

Becker, R. A., Chambers, J. M. and Wilks, A. R. (1988) *The New S Language*. Wadsworth & Brooks/Cole.

See Also

dnorm for the normal distribution.

Examples

dlnorm(1) == dnorm(0)

Multinomial   *The Multinomial Distribution*

Description

Generate multinomially distributed random number vectors and compute multinomial "density" probabilities.

Usage

```
rmultinom(n, size, prob)
dmultinom(x, size = NULL, prob, log = FALSE)
```

Arguments

x	vector of length K of integers in 0:size.
n	number of random vectors to draw.
size	integer, say N, specifying the total number of objects that are put into K boxes in the typical multinomial experiment. For dmultinom, it defaults to sum(x).
prob	numeric non-negative vector of length K, specifying the probability for the K classes; is internally normalized to sum 1.
log	logical; if TRUE, log probabilities are computed.

Details

If x is a K-component vector, dmultinom(x, prob) is the probability

$$P(X_1 = x_1, \ldots, X_K = x_k) = C \times \prod_{j=1}^{K} \pi_j^{x_j}$$

where C is the "multinomial coefficient" $C = N!/(x_1! \cdots x_K!)$ and $N = \sum_{j=1}^{K} x_j$.
By definition, each component X_j is binomially distributed as Bin(size, prob[j]) for $j = 1, \ldots, K$.

The rmultinom() algorithm draws binomials from $Bin(n_j, P_j)$ sequentially, where $n_1 = N$ (N := size), $P_1 = \pi_1$ (π is prob scaled to sum 1), and for $j \geq 2$, recursively $n_j = N - \sum_{k=1}^{j-1} n_k$ and $P_j = \pi_j/(1 - \sum_{k=1}^{j-1} \pi_k)$.

Value

For rmultinom(), an integer K x n matrix where each column is a random vector generated according to the desired multinomial law, and hence summing to size. Whereas the *transposed* result would seem more natural at first, the returned matrix is more efficient because of columnwise storage.

Note

dmultinom is currently *not vectorized* at all and has no C interface (API); this may be amended in the future.

See Also

rbinom which is a special case conceptually.

Examples

```
rmultinom(10, size = 12, prob=c(0.1,0.2,0.8))

pr <-c(1,3,6,10) # normalization unnecessary for generation
rmultinom(10, 20, prob = pr)

## all possible outcomes of Multinom(N = 3, K = 3)
x <- t(as.matrix(expand.grid(0:3, 0:3)));
x <- x[, colsums(x) <= 3]
x <- rbind(x, 3:3 - colsums(x));
dimnames(x) <- list(letters[1:3], null)
X
round(apply(X, 2, function(x)
                     dmultinom(x, prob = c(1,2,5))), 3)
```

NegBinomial *The Negative Binomial Distribution*

Description

Density, distribution function, quantile function and random generation for the negative binomial distribution with parameters size and prob.

Usage

```
dnbinom(x, size, prob, mu, log=FALSE)
pnbinom(q, size, prob, mu, lower.tail=TRUE, log.p=FALSE)
qnbinom(p, size, prob, mu, lower.tail=TRUE, log.p=FALSE)
rnbinom(n, size, prob, mu)
```

Arguments

x	vector of (non-negative integer) quantiles.
q	vector of quantiles.
p	vector of probabilities.
n	number of observations. If length(n) > 1, the length is taken to be the number required.
size	target for number of successful trials, or dispersion parameter (the shape parameter of the gamma mixing distribution).
prob	probability of success in each trial.
mu	alternative parametrization via mean: see Details
log, log.p	logical; if TRUE, probabilities p are given as log(p).
lower.tail	logical; if TRUE (default), probabilities are $P[X \leq x]$, otherwise, $P[X > x]$.

Details

The negative binomial distribution with size $= n$ and prob $= p$ has density

$$p(x) = \frac{\Gamma(x+n)}{\Gamma(n)x!} p^n (1-p)^x$$

for $x = 0, 1, 2, \ldots$

This represents the number of failures which occur in a sequence of Bernoulli trials before a target number of successes is reached.

A negative binomial distribution can arise as a mixture of Poisson distributions with mean distributed as a gamma (pgamma) distribution with scale parameter (1 - prob)/prob and shape parameter size. (This definition allows non-integer values of size.) In this model prob = scale/(1+scale), and the mean is size * (1 - prob)/prob

The alternative parametrization (often used in ecology) is by the *mean* mu, and size, the *dispersion parameter*, where prob = size/(size+mu). In this parametrization the variance is mu + mu^2/size.

If an element of x is not integer, the result of dnbinom is zero, with a warning.

The quantile is defined as the smallest value x such that $F(x) \geq p$, where F is the distribution function.

Value

dnbinom gives the density, pnbinom gives the distribution function, qnbinom gives the quantile function, and rnbinom generates random deviates.

See Also

dbinom for the binomial, dpois for the Poisson and dgeom for the geometric distribution, which is a special case of the negative binomial.

Examples

```
x <- 0:11
dnbinom(x, size = 1, prob = 1/2) * 2^(1 + x) # == 1
# theoretically integer
126 /  dnbinom(0:8, size  = 2, prob  = 1/2)

## Cumulative ('p') = Sum of discrete prob.s ('d');
## Relative error :
summary(1 - cumsum(dnbinom(x, size = 2, prob = 1/2)) /
             pnbinom(x, size  = 2, prob = 1/2))

x <- 0:15
size <- (1:20)/4
persp(x,size,
   dnb <- outer(x, size, function(x,s) dnbinom(x,s,pr=0.4)),
   xlab = "x", ylab = "s", zlab="density", theta = 150)
title(
   tit <- "negative binomial density(x,s,pr=0.4) vs. x,s"
```

```
)

image  (x,size, log10(dnb), main= paste("log [",tit,"]"))
contour(x,size, log10(dnb),add=TRUE)

## Alternative parametrization
x1 <- rnbinom(500, mu = 4, size = 1)
x2 <- rnbinom(500, mu = 4, size = 10)
x3 <- rnbinom(500, mu = 4, size = 100)
h1 <- hist(x1, breaks = 20, plot = FALSE)
h2 <- hist(x2, breaks = h1$breaks, plot = FALSE)
h3 <- hist(x3, breaks = h1$breaks, plot = FALSE)
barplot(rbind(h1$counts, h2$counts, h3$counts),
        beside = TRUE, col = c("red","blue","cyan"),
        names.arg = round(h1$breaks[-length(h1$breaks)]))
```

Normal *The Normal Distribution*

Description

Density, distribution function, quantile function and random generation for the normal distribution with mean equal to mean and standard deviation equal to sd.

Usage

```
dnorm(x, mean=0, sd=1, log = FALSE)
pnorm(q, mean=0, sd=1, lower.tail = TRUE, log.p = FALSE)
qnorm(p, mean=0, sd=1, lower.tail = TRUE, log.p = FALSE)
rnorm(n, mean=0, sd=1)
```

Arguments

x,q	vector of quantiles.
p	vector of probabilities.
n	number of observations. If length(n) > 1, the length is taken to be the number required.
mean	vector of means.
sd	vector of standard deviations.
log, log.p	logical; if TRUE, probabilities p are given as log(p).
lower.tail	logical; if TRUE (default), probabilities are $P[X \leq x]$, otherwise, $P[X > x]$.

Details

If mean or sd are not specified they assume the default values of 0 and 1, respectively.

The normal distribution has density

$$f(x) = \frac{1}{\sqrt{2\pi}\sigma} e^{-(x-\mu)^2/2\sigma^2}$$

where μ is the mean of the distribution and σ the standard deviation.

qnorm is based on Wichura's algorithm AS 241 which provides precise results up to about 16 digits.

Value

dnorm gives the density, pnorm gives the distribution function, qnorm gives the quantile function, and rnorm generates random deviates.

References

Becker, R. A., Chambers, J. M. and Wilks, A. R. (1988) *The New S Language*. Wadsworth & Brooks/Cole.

Wichura, M. J. (1988) Algorithm AS 241: The Percentage Points of the Normal Distribution. *Applied Statistics*, **37**, 477–484.

See Also

runif and .Random.seed about random number generation, and dlnorm for the *Log*normal distribution.

Examples

```
dnorm(0) == 1/ sqrt(2*pi)
dnorm(1) == exp(-1/2)/ sqrt(2*pi)
dnorm(1) == 1/ sqrt(2*pi*exp(1))

## Using "log = TRUE" for an extended range :
par(mfrow=c(2,1))
plot(function(x)dnorm(x, log=TRUE), -60, 50,
     main = "log { Normal density }")
curve(log(dnorm(x)), add=TRUE, col="red",lwd=2)
mtext("dnorm(x, log=TRUE)", adj=0);
mtext("log(dnorm(x))", col="red", adj=1)

plot(function(x)pnorm(x, log=TRUE), -50, 10,
     main = "log { Normal Cumulative }")
curve(log(pnorm(x)), add=TRUE, col="red",lwd=2)
mtext("pnorm(x, log=TRUE)", adj=0);
mtext("log(pnorm(x))", col="red", adj=1)

## if you want the so-called 'error function'
erf <- function(x) 2 * pnorm(x * sqrt(2)) - 1
## and the so-called 'complementary error function'
erfc <- function(x) 2 * pnorm(x * sqrt(2), lower=FALSE)
```

Poisson — The Poisson Distribution

Description

Density, distribution function, quantile function and random generation for the Poisson distribution with parameter `lambda`.

Usage

```
dpois(x, lambda, log = FALSE)
ppois(q, lambda, lower.tail = TRUE, log.p = FALSE)
qpois(p, lambda, lower.tail = TRUE, log.p = FALSE)
rpois(n, lambda)
```

Arguments

x	vector of (non-negative integer) quantiles.
q	vector of quantiles.
p	vector of probabilities.
n	number of random values to return.
lambda	vector of positive means.
log, log.p	logical; if TRUE, probabilities p are given as log(p).
lower.tail	logical; if TRUE (default), probabilities are $P[X \leq x]$, otherwise, $P[X > x]$.

Details

The Poisson distribution has density

$$p(x) = \frac{\lambda^x e^{-\lambda}}{x!}$$

for $x = 0, 1, 2, \ldots$. The mean and variance are $E(X) = Var(X) = \lambda$.

If an element of x is not integer, the result of dpois is zero, with a warning. $p(x)$ is computed using Loader's algorithm, see the reference in dbinom.

The quantile is left continuous: `qgeom(q, prob)` is the largest integer x such that $P(X \leq x) < q$.

Setting `lower.tail = FALSE` allows to get much more precise results when the default, `lower.tail = TRUE` would return 1, see the example below.

Value

dpois gives the (log) density, ppois gives the (log) distribution function, qpois gives the quantile function, and rpois generates random deviates.

See Also

dbinom for the binomial and dnbinom for the negative binomial distribution.

Examples

```
-log(dpois(0:7, lambda=1) * gamma(1+ 0:7)) # == 1
Ni <- rpois(50, lam= 4); table(factor(Ni, 0:max(Ni)))

# becomes 0 (cancellation)
1 - ppois(10*(15:25), lambda=100)
# no cancellation
ppois(10*(15:25), lambda=100, lower=FALSE)

par(mfrow = c(2, 1))
x <- seq(-0.01, 5, 0.01)
plot(x, ppois(x, 1), type="s", ylab="F(x)",
     main="Poisson(1) CDF")
plot(x, pbinom(x, 100, 0.01),type="s", ylab="F(x)",
     main="Binomial(100, 0.01) CDF")
```

r2dtable   *Random 2-way Tables with Given Marginals*

Description

Generate random 2-way tables with given marginals using Patefield's algorithm.

Usage

```
r2dtable(n, r, c)
```

Arguments

n
: a non-negative numeric giving the number of tables to be drawn.

r
: a non-negative vector of length at least 2 giving the row totals, to be coerced to `integer`. Must sum to the same as c.

c
: a non-negative vector of length at least 2 giving the column totals, to be coerced to `integer`.

Value

A list of length n containing the generated tables as its components.

References

Patefield, W. M. (1981) Algorithm AS159. An efficient method of generating r x c tables with given row and column totals. *Applied Statistics* **30**, 91–97.

Examples

```
## Fisher's Tea Drinker data.
TeaTasting <-
matrix(c(3, 1, 1, 3),
       nr = 2,
       dimnames = list(Guess = c("Milk", "Tea"),
                       Truth = c("Milk", "Tea")))
## Simulate permutation test for independence based on the
## maximum Pearson residuals (rather than their sum).
rowTotals <- rowSums(TeaTasting)
```

```
colTotals <- colSums(TeaTasting)
nOfCases <- sum(rowTotals)
expected <- outer(rowTotals, colTotals, "*") / nOfCases
maxSqResid <-
  function(x) max((x - expected) ^ 2 / expected)
simMaxSqResid <-
  sapply(r2dtable(1000, rowTotals, colTotals), maxSqResid)
sum(simMaxSqResid >= maxSqResid(TeaTasting)) / 1000
## Fisher's exact test gives p = 0.4857 ...
```

Random *Random Number Generation*

Description

`.Random.seed` is an integer vector, containing the random number generator (RNG) state for random number generation in R. It can be saved and restored, but should not be altered by the user.

`RNGkind` is a more friendly interface to query or set the kind of RNG in use.

`RNGversion` can be used to set the random generators as they were in an earlier R version (for reproducibility).

`set.seed` is the recommended way to specify seeds.

Usage

```
.Random.seed <- c(rng.kind, n1, n2, ...)
save.seed <- .Random.seed

RNGkind(kind = NULL, normal.kind = NULL)
RNGversion(vstr)
set.seed(seed, kind = NULL)
```

Arguments

`kind`	character or `NULL`. If `kind` is a character string, set R's RNG to the kind desired. If it is `NULL`, return the currently used RNG. Use `"default"` to return to the R default.
`normal.kind`	character string or `NULL`. If it is a character string, set the method of Normal generation. Use `"default"` to return to the R default.
`seed`	a single value, interpreted as an integer.
`vstr`	a character string containing a version number, e.g., `"1.6.2"`
`rng.kind`	integer code in `0:k` for the above `kind`.
`n1, n2, ...`	integers. See the details for how many are required (which depends on `rng.kind`).

Details

The currently available RNG kinds are given below. kind is partially matched to this list. The default is "Mersenne-Twister".

"Wichmann-Hill" The seed, .Random.seed[-1] == r[1:3] is an integer vector of length 3, where each r[i] is in 1:(p[i] - 1), where p is the length 3 vector of primes, p = (30269, 30307, 30323). The Wichmann–Hill generator has a cycle length of 6.9536×10^{12} (= prod(p-1)/4, see *Applied Statistics* (1984) **33**, 123 which corrects the original article).

"Marsaglia-Multicarry": A *multiply-with-carry* RNG is used, as recommended by George Marsaglia in his post to the mailing list 'sci.stat.math'. It has a period of more than 2^{60} and has passed all tests (according to Marsaglia). The seed is two integers (all values allowed).

"Super-Duper": Marsaglia's famous Super-Duper from the 70's. This is the original version which does *not* pass the MTUPLE test of the Diehard battery. It has a period of $\approx 4.6 \times 10^{18}$ for most initial seeds. The seed is two integers (all values allowed for the first seed: the second must be odd).

We use the implementation by Reeds et al. (1982–84).

The two seeds are the Tausworthe and congruence long integers, respectively. A one-to-one mapping to S's .Random.seed[1:12] is possible but we will not publish one, not least as this generator is not exactly the same as that in recent versions of S-PLUS.

"Mersenne-Twister": From Matsumoto and Nishimura (1998). A twisted GFSR with period $2^{19937} - 1$ and equidistribution in 623 consecutive dimensions (over the whole period). The "seed" is a 624-dimensional set of 32-bit integers plus a current position in that set.

"Knuth-TAOCP": From Knuth (1997). A GFSR using lagged Fibonacci sequences with subtraction. That is, the recurrence used is

$$X_j = (X_{j-100} - X_{j-37}) \bmod 2^{30}$$

and the "seed" is the set of the 100 last numbers (actually recorded as 101 numbers, the last being a cyclic shift of the buffer). The period is around 2^{129}.

"Knuth-TAOCP-2002": The 2002 version which not backwards compatible with the earlier version: the initialization of the GFSR from the seed was altered. R did not allow you to choose consecutive seeds, the reported 'weakness', and already scrambled the seeds.

"user-supplied": Use a user-supplied generator. See Random.user for details.

normal.kind can be "Kinderman-Ramage", "Buggy Kinderman-Ramage", "Ahrens-Dieter", "Box-Muller", "Inversion" (the default), or "user-supplied". (For inversion, see the reference in qnorm.) The Kinderman-Ramage generator used in versions prior to 1.7.1 had several approximation errors and should only be used for reproduction of older results.

set.seed uses its single integer argument to set as many seeds as are required. It is intended as a simple way to get quite different seeds by specifying small integer arguments, and also as a way to get valid seed sets for the more complicated methods (especially "Mersenne-Twister" and "Knuth-TAOCP").

Value

.Random.seed is an integer vector whose first element *codes* the kind of RNG and normal generator. The lowest two decimal digits are in 0:(k-1) where k is the number of available RNGs. The hundreds represent the type of normal generator (starting at 0).

In the underlying C, .Random.seed[-1] is unsigned; therefore in R .Random.seed[-1] can be negative.

RNGkind returns a two-element character vector of the RNG and normal kinds in use *before* the call, invisibly if either argument is not NULL. RNGversion returns the same information.

set.seed returns NULL, invisibly.

Note

Initially, there is no seed; a new one is created from the current time when one is required. Hence, different sessions will give different simulation results, by default.

.Random.seed saves the seed set for the uniform random-number generator, at least for the system generators. It does not necessarily save the state of other generators, and in particular does not save the state of the Box–Muller normal generator. If you want to reproduce work later, call set.seed rather than set .Random.seed.

As from R 1.8.0, .Random.seed is only looked for in the user's workspace.

Author(s)

of RNGkind: Martin Maechler. Current implementation, B. D. Ripley

References

Becker, R. A., Chambers, J. M. and Wilks, A. R. (1988) *The New S Language.* Wadsworth & Brooks/Cole. (set.seed, storing in .Random.seed.)

Wichmann, B. A. and Hill, I. D. (1982) *Algorithm AS 183: An Efficient and Portable Pseudo-random Number Generator*, Applied Statistics, **31**, 188–190; Remarks: **34**, 198 and **35**, 89.

De Matteis, A. and Pagnutti, S. (1993) *Long-range Correlation Analysis of the Wichmann-Hill Random Number Generator*, Statist. Comput., **3**, 67–70.

Marsaglia, G. (1997) *A random number generator for C.* Discussion paper, posting on Usenet newsgroup sci.stat.math on September 29, 1997.

Reeds, J., Hubert, S. and Abrahams, M. (1982-4) C implementation of SuperDuper, University of California at Berkeley. (Personal communication from Jim Reeds to Ross Ihaka.)

Marsaglia, G. and Zaman, A. (1994) Some portable very-long-period random number generators. *Computers in Physics*, **8**, 117–121.

Matsumoto, M. and Nishimura, T. (1998) Mersenne Twister: A 623-dimensionally equidistributed uniform pseudo-random number generator, *ACM Transactions on Modeling and Computer Simulation*, **8**, 3–30. Source code at http://www.math.keio.ac.jp/~matumoto/emt.html.

Knuth, D. E. (1997) *The Art of Computer Programming.* Volume 2, third edition.
Source code at http://www-cs-faculty.stanford.edu/~knuth/taocp.html.

Knuth, D. E. (2002) *The Art of Computer Programming.* Volume 2, third edition, ninth printing.

Kinderman, A. J. and Ramage, J. G. (1976) Computer generation of normal random variables. *Journal of the American Statistical Association* **71**, 893-896.

Ahrens, J.H. and Dieter, U. (1973) Extensions of Forsythe's method for random sampling from the normal distribution. *Mathematics of Computation* **27**, 927-937.

Box, G.E.P. and Muller, M.E. (1958) A note on the generation of normal random deviates. *Annals of Mathmatical Statistics* **29**, 610–611.

See Also

runif, rnorm, ...

Examples

```
runif(1); .Random.seed; runif(1); .Random.seed
## If there is no seed, a "random" new one is created:
rm(.Random.seed); runif(1); .Random.seed

RNGkind("Wich") # (partial string matching on 'kind')

## This shows how 'runif(.)' works for Wichmann-Hill, using
## only R functions:
p.WH <- c(30269, 30307, 30323)
a.WH <- c(  171,   172,   170)
next.WHseed <- function(i.seed = .Random.seed[-1])
  { (a.WH * i.seed) %% p.WH }
my.runif1 <- function(i.seed = .Random.seed)
  { ns <- next.WHseed(i.seed[-1]); sum(ns / p.WH) %% 1 }
rs <- .Random.seed
(WHs <- next.WHseed(rs[-1]))
u <- runif(1)
stopifnot(
 next.WHseed(rs[-1]) == .Random.seed[-1],
 all.equal(u, my.runif1(rs))
)

## ----
.Random.seed
ok <- RNGkind()
RNGkind("Super") # matches "Super-Duper"
RNGkind()
.Random.seed # new, corresponding to Super-Duper

## Reset:
RNGkind(ok[1])
```

Random.user   *User-supplied Random Number Generation*

Description

Function RNGkind allows user-coded uniform and normal random number generators to be supplied. The details are given here.

Details

A user-specified uniform RNG is called from entry points in dynamically-loaded compiled code. The user must supply the entry point user_unif_rand, which takes no arguments and returns a *pointer to* a double. The example below will show the general pattern.

Optionally, the user can supply the entry point user_unif_init, which is called with an unsigned int argument when RNGkind (or set.seed) is called, and is intended to be used to initialize the user's RNG code. The argument is intended to be used to set the "seeds"; it is the seed argument to set.seed or an essentially random seed if RNGkind is called.

If only these functions are supplied, no information about the generator's state is recorded in .Random.seed. Optionally, functions user_unif_nseed and user_unif_seedloc can be supplied which are called with no arguments and should return pointers to the number of "seeds" and to an integer array of "seeds". Calls to GetRNGstate and PutRNGstate will then copy this array to and from .Random.seed.

A user-specified normal RNG is specified by a single entry point user_norm_rand, which takes no arguments and returns a *pointer to* a double.

Warning

As with all compiled code, mis-specifying these functions can crash R. Do include the 'R_ext/Random.h' header file for type checking.

Examples

```
## Marsaglia's congruential PRNG
#include <R_ext/Random.h>

static Int32 seed;
static double res;
static int nseed = 1;
```

base — Random.user

```c
double * user_unif_rand()
{
    seed = 69069 * seed + 1;
    res = seed * 2.32830643653869e-10;
    return &res;
}

void  user_unif_init(Int32 seed_in) { seed = seed_in; }
int * user_unif_nseed() { return &nseed; }
int * user_unif_seedloc() { return (int *) &seed; }

/*  ratio-of-uniforms for normal  */
#include <math.h>
static double x;

double * user_norm_rand()
{
    double u, v, z;
    do {
        u = unif_rand();
        v = 0.857764 * (2. * unif_rand() - 1);
        x = v/u; z = 0.25 * x * x;
        if (z < 1. - u) break;
        if (z > 0.259/u + 0.35) continue;
    } while (z > -log(u));
    return &x;
}

## Use under Unix:
R SHLIB urand.c
R
> dyn.load("urand.so")
> RNGkind("user")
> runif(10)
> .Random.seed
> RNGkind(, "user")
> rnorm(10)
> RNGkind()
[1] "user-supplied" "user-supplied"
```

sample   *Random Samples and Permutations*

Description

sample takes a sample of the specified size from the elements of x using either with or without replacement.

Usage

```
sample(x, size, replace = FALSE, prob = NULL)
```

Arguments

x	Either a (numeric, complex, character or logical) vector of more than one element from which to choose, or a positive integer.
size	non-negative integer giving the number of items to choose.
replace	Should sampling be with replacement?
prob	A vector of probability weights for obtaining the elements of the vector being sampled.

Details

If x has length 1, sampling takes place from 1:x. *Note* that this convenience feature may lead to undesired behaviour when x is of varying length sample(x). See the resample() example below.

By default size is equal to length(x) so that sample(x) generates a random permutation of the elements of x (or 1:x).

The optional prob argument can be used to give a vector of weights for obtaining the elements of the vector being sampled. They need not sum to one, but they should be nonnegative and not all zero. If replace is false, these probabilities are applied sequentially, that is the probability of choosing the next item is proportional to the probabilities amongst the remaining items. The number of nonzero weights must be at least size in this case.

References

Becker, R. A., Chambers, J. M. and Wilks, A. R. (1988) *The New S Language.* Wadsworth & Brooks/Cole.

Examples

```
x <- 1:12
# a random permutation
sample(x)
# bootstrap sampling -- only if length(x) > 1 !
sample(x,replace=TRUE)

# 100 Bernoulli trials
sample(c(0,1), 100, replace = TRUE)

## More careful bootstrapping -- Consider this when using
## sample() programmatically (i.e., in your function or
## simulation)!

# sample()'s surprise -- example
x <- 1:10
    sample(x[x >  8]) # length 2
    sample(x[x >  9]) # oops -- length 10!
try(sample(x[x > 10])) # error!

## This is safer:
resample <- function(x, size, ...)
  if(length(x) <= 1) {
    if(!missing(size) && size == 0) x[FALSE] else x
  } else sample(x, size, ...)

resample(x[x >  8]) # length 2
resample(x[x >  9]) # length 1
resample(x[x > 10]) # length 0
```

SignRank — *Distribution of the Wilcoxon Signed Rank Statistic*

Description

Density, distribution function, quantile function and random generation for the distribution of the Wilcoxon Signed Rank statistic obtained from a sample with size n.

Usage

```
dsignrank(x, n, log = FALSE)
psignrank(q, n, lower.tail = TRUE, log.p = FALSE)
qsignrank(p, n, lower.tail = TRUE, log.p = FALSE)
rsignrank(nn, n)
```

Arguments

x,q	vector of quantiles.
p	vector of probabilities.
nn	number of observations. If length(nn) > 1, the length is taken to be the number required.
n	numbers of observations in the sample. Must be positive integers less than 50.
log, log.p	logical; if TRUE, probabilities p are given as log(p).
lower.tail	logical; if TRUE (default), probabilities are $P[X \leq x]$, otherwise, $P[X > x]$.

Details

This distribution is obtained as follows. Let x be a sample of size n from a continuous distribution symmetric about the origin. Then the Wilcoxon signed rank statistic is the sum of the ranks of the absolute values x[i] for which x[i] is positive. This statistic takes values between 0 and $n(n+1)/2$, and its mean and variance are $n(n+1)/4$ and $n(n+1)(2n+1)/24$, respectively.

Value

dsignrank gives the density, psignrank gives the distribution function, qsignrank gives the quantile function, and rsignrank generates random deviates.

Author(s)

Kurt Hornik

See Also

dwilcox etc, for the *two-sample* Wilcoxon rank sum statistic.

Examples

```
par(mfrow=c(2,2))
for(n in c(4:5,10,40)) {
  x <- seq(0, n*(n+1)/2, length=501)
  plot(x, dsignrank(x,n=n), type='l',
       main=paste("dsignrank(x,n=",n,")"))
}
```

TDist	The Student t Distribution

Description

Density, distribution function, quantile function and random generation for the t distribution with df degrees of freedom (and optional noncentrality parameter ncp).

Usage

```
dt(x, df, ncp=0, log = FALSE)
pt(q, df, ncp=0, lower.tail = TRUE, log.p = FALSE)
qt(p, df,        lower.tail = TRUE, log.p = FALSE)
rt(n, df)
```

Arguments

x, q	vector of quantiles.
p	vector of probabilities.
n	number of observations. If length(n) > 1, the length is taken to be the number required.
df	degrees of freedom (> 0, maybe non-integer).
ncp	non-centrality parameter δ; currently for pt() and dt(), only for ncp <= 37.62.
log, log.p	logical; if TRUE, probabilities p are given as log(p).
lower.tail	logical; if TRUE (default), probabilities are $P[X \leq x]$, otherwise, $P[X > x]$.

Details

The t distribution with df $= \nu$ degrees of freedom has density

$$f(x) = \frac{\Gamma((\nu+1)/2)}{\sqrt{\pi\nu}\Gamma(\nu/2)}(1 + x^2/\nu)^{-(\nu+1)/2}$$

for all real x. It has mean 0 (for $\nu > 1$) and variance $\frac{\nu}{\nu-2}$ (for $\nu > 2$).

The general *non-central t* with parameters (ν, δ) = (df, ncp) is defined as a the distribution of $T_\nu(\delta) := \frac{U+\delta}{\chi_\nu/\sqrt{\nu}}$ where U and χ_ν are independent random variables, $U \sim \mathcal{N}(0,1)$, and χ_ν^2 is chi-squared, see pchisq.

The most used applications are power calculations for t-tests:
Let $T = \frac{\bar{X}-\mu_0}{S/\sqrt{n}}$ where $\bar{X}$ is the mean and S the sample standard deviation (sd) of $X_1, X_2, \ldots, X_n$ which are i.i.d. $N(\mu, \sigma^2)$. Then T is distributed as non-centrally t with df$= n-1$ degrees of freedom and non-centrality parameter ncp$= (\mu - \mu_0)\sqrt{n}/\sigma$.

Value

dt gives the density, pt gives the distribution function, qt gives the quantile function, and rt generates random deviates.

References

Becker, R. A., Chambers, J. M. and Wilks, A. R. (1988) *The New S Language.* Wadsworth & Brooks/Cole. (except non-central versions.)

Lenth, R. V. (1989). *Algorithm AS 243* — Cumulative distribution function of the non-central t distribution, *Appl. Statist.* **38**, 185–189.

See Also

df for the F distribution.

Examples

```
1 - pt(1:5, df = 1)
qt(.975, df = c(1:10,20,50,100,1000))

tt <- seq(0,10, len=21)
ncp <- seq(0,6, len=31)
ptn <- outer(tt,ncp, function(t,d) pt(t, df = 3, ncp=d))
image(tt,ncp,ptn, zlim=c(0,1),
      main=t.tit <- "Non-central t - Probabilities")
persp(tt,ncp,ptn, zlim=0:1, r=2, phi=20, theta=200,
      main=t.tit, xlab = "t",
      ylab = "noncentrality parameter",
      zlab = "Pr(T <= t)")

op <- par(yaxs="i")
plot(function(x) dt(x, df = 3, ncp = 2), -3, 11,
     ylim = c(0, 0.32), main="Non-central t - Density")
par(op)
```

| Tukey | *The Studentized Range Distribution* |

Description

Functions on the distribution of the studentized range, R/s, where R is the range of a standard normal sample of size n and s^2 is independently distributed as chi-squared with *df* degrees of freedom, see pchisq.

Usage

```
ptukey(q,nmeans,df,nranges=1,lower.tail=TRUE,log.p=FALSE)
qtukey(p,nmeans,df,nranges=1,lower.tail=TRUE,log.p=FALSE)
```

Arguments

q	vector of quantiles.
p	vector of probabilities.
nmeans	sample size for range (same for each group).
df	degrees of freedom for s (see below).
nranges	number of *groups* whose **maximum** range is considered.
log.p	logical; if TRUE, probabilities p are given as log(p).
lower.tail	logical; if TRUE (default), probabilities are $P[X \leq x]$, otherwise, $P[X > x]$.

Details

If n_g =nranges is greater than one, R is the *maximum* of n_g groups of nmeans observations each.

Value

ptukey gives the distribution function and qtukey its inverse, the quantile function.

Note

A Legendre 16-point formula is used for the integral of ptukey. The computations are relatively expensive, especially for qtukey which uses a simple secant method for finding the inverse of ptukey. qtukey will be accurate to the 4th decimal place.

References

Copenhaver, Margaret Diponzio and Holland, Burt S. (1988) Multiple comparisons of simple effects in the two-way analysis of variance with fixed effects. *Journal of Statistical Computation and Simulation,* **30**, 1–15.

See Also

pnorm and qnorm for the corresponding functions for the normal distribution.

Examples

```
if(interactive())
   curve(ptukey(x, nm=6, df=5), from=-1, to=8, n=101)
(ptt <- ptukey(0:10, 2, df= 5))
(qtt <- qtukey(.95, 2, df= 2:11))
## The precision may be not much more than about 8 digits:
summary(abs(.95 - ptukey(qtt,2, df = 2:11)))
```

Uniform — The Uniform Distribution

Description

These functions provide information about the uniform distribution on the interval from min to max. dunif gives the density, punif gives the distribution function qunif gives the quantile function and runif generates random deviates.

Usage

```
dunif(x, min=0, max=1, log = FALSE)
punif(q, min=0, max=1, lower.tail = TRUE, log.p = FALSE)
qunif(p, min=0, max=1, lower.tail = TRUE, log.p = FALSE)
runif(n, min=0, max=1)
```

Arguments

x,q	vector of quantiles.
p	vector of probabilities.
n	number of observations. If length(n) > 1, the length is taken to be the number required.
min,max	lower and upper limits of the distribution.
log, log.p	logical; if TRUE, probabilities p are given as log(p).
lower.tail	logical; if TRUE (default), probabilities are $P[X \leq x]$, otherwise, $P[X > x]$.

Details

If min or max are not specified they assume the default values of 0 and 1 respectively.

The uniform distribution has density

$$f(x) = \frac{1}{max - min}$$

for $min \leq x \leq max$.

For the case of $u := min == max$, the limit case of $X \equiv u$ is assumed.

References

Becker, R. A., Chambers, J. M. and Wilks, A. R. (1988) *The New S Language*. Wadsworth & Brooks/Cole.

See Also

.Random.seed about random number generation, rnorm, etc for other distributions.

Examples

```
u <- runif(20)

## The following relations always hold :
punif(u) == u
dunif(u) == 1

var(runif(10000)) # ~ = 1/12 = .08333
```

Weibull — *The Weibull Distribution*

Description

Density, distribution function, quantile function and random generation for the Weibull distribution with parameters shape and scale.

Usage

```
dweibull(x, shape, scale=1, log=FALSE)
pweibull(q, shape, scale=1, lower.tail=TRUE, log.p=FALSE)
qweibull(p, shape, scale=1, lower.tail=TRUE, log.p=FALSE)
rweibull(n, shape, scale=1)
```

Arguments

x, q	vector of quantiles.
p	vector of probabilities.
n	number of observations. If length(n) > 1, the length is taken to be the number required.
shape, scale	shape and scale parameters, the latter defaulting to 1.
log, log.p	logical; if TRUE, probabilities p are given as log(p).
lower.tail	logical; if TRUE (default), probabilities are $P[X \leq x]$, otherwise, $P[X > x]$.

Details

The Weibull distribution with shape parameter a and scale parameter σ has density given by

$$f(x) = (a/\sigma)(x/\sigma)^{a-1} \exp(-(x/\sigma)^a)$$

for $x > 0$. The cumulative is $F(x) = 1 - \exp(-(x/\sigma)^a)$, the mean is $E(X) = \sigma\Gamma(1+1/a)$, and the $Var(X) = \sigma^2(\Gamma(1+2/a) - (\Gamma(1+1/a))^2)$.

Value

dweibull gives the density, pweibull gives the distribution function, qweibull gives the quantile function, and rweibull generates random deviates.

Note

The cumulative hazard $H(t) = -\log(1 - F(t))$ is -pweibull(t, a, b, lower = FALSE, log = TRUE) which is just $H(t) = (t/b)^a$.

See Also

dexp for the Exponential which is a special case of a Weibull distribution.

Examples

```
x <- c(0,rlnorm(50))
all.equal(dweibull(x, shape = 1), dexp(x))
all.equal(pweibull(x, shape = 1, scale = pi),
          pexp(x, rate = 1/pi))
## Cumulative hazard H():
all.equal(pweibull(x, 2.5, pi, lower=FALSE, log=TRUE),
          -(x/pi)^2.5, tol=1e-15)
all.equal(qweibull(x/11, shape = 1, scale = pi),
          qexp(x/11, rate = 1/pi))
```

| Wilcoxon | *Distribution of the Wilcoxon Rank Sum Statistic* |

Description

Density, distribution function, quantile function and random generation for the distribution of the Wilcoxon rank sum statistic obtained from samples with size m and n, respectively.

Usage

```
dwilcox(x, m, n, log = FALSE)
pwilcox(q, m, n, lower.tail = TRUE, log.p = FALSE)
qwilcox(p, m, n, lower.tail = TRUE, log.p = FALSE)
rwilcox(nn, m, n)
```

Arguments

x, q	vector of quantiles.
p	vector of probabilities.
nn	number of observations. If length(nn) > 1, the length is taken to be the number required.
m, n	numbers of observations in the first and second sample, respectively.
log, log.p	logical; if TRUE, probabilities p are given as log(p).
lower.tail	logical; if TRUE (default), probabilities are $P[X \leq x]$, otherwise, $P[X > x]$.

Details

This distribution is obtained as follows. Let x and y be two random, independent samples of size m and n. Then the Wilcoxon rank sum statistic is the number of all pairs (x[i], y[j]) for which y[j] is not greater than x[i]. This statistic takes values between 0 and m * n, and its mean and variance are m * n / 2 and m * n * (m + n + 1) / 12, respectively.

Value

dwilcox gives the density, pwilcox gives the distribution function, qwilcox gives the quantile function, and rwilcox generates random deviates.

Note

S-PLUS uses a different (but equivalent) definition of the Wilcoxon statistic.

Author(s)

Kurt Hornik

See Also

dsignrank etc, for the *one-sample* Wilcoxon rank statistic.

Examples

```
x <- -1:(4*6 + 1)
fx <- dwilcox(x, 4, 6)
Fx <- pwilcox(x, 4, 6)

layout(rbind(1,2),width=1,heights=c(3,2))
plot(x, fx,type='h', col="violet",
     main= "Prob (density) of Wilcoxon-Statist.(n=6,m=4)")
plot(x, Fx,type="s", col="blue",
     main= "Distribution of Wilcoxon-Statist.(n=6,m=4)")
abline(h=0:1, col="gray20",lty=2)
layout(1) # set back

N <- 200
hist(U <- rwilcox(N, m=4,n=6), breaks=0:25 - 1/2,
     border="red", col="pink", sub = paste("N =",N))
mtext("N * f(x),   f() = true \"density\"", side=3,
      col="blue")
lines(x, N*fx, type='h', col='blue', lwd=2)
points(x, N*fx, cex=2)

## Better is a Quantile-Quantile Plot
qqplot(U, qw <- qwilcox((1:N - 1/2)/N, m=4,n=6),
   main = paste("QQ-Plot of empirical and theor. quantiles",
     "Wilcoxon Statistic,  (m=4, n=6)",sep="\n"))
n <- as.numeric(names(print(tU <- table(U))))
text(n+.2, n+.5, labels=tU, col="red")
```

Chapter 4

Base package — models

add1 — *Add or Drop All Possible Single Terms to a Model*

Description

Compute all the single terms in the scope argument that can be added to or dropped from the model, fit those models and compute a table of the changes in fit.

Usage

```
add1(object, scope, ...)

## Default S3 method:
add1(object, scope, scale = 0, test = c("none", "Chisq"),
     k = 2, trace = FALSE, ...)

## S3 method for class 'lm':
add1(object, scope, scale = 0,
     test = c("none", "Chisq", "F"),
     x = NULL, k = 2, ...)

## S3 method for class 'glm':
add1(object, scope, scale = 0,
     test = c("none", "Chisq", "F"),
     x = NULL, k = 2, ...)

drop1(object, scope, ...)

## Default S3 method:
drop1(object, scope, scale = 0, test = c("none", "Chisq"),
      k = 2, trace = FALSE, ...)

## S3 method for class 'lm':
drop1(object, scope, scale = 0, all.cols = TRUE,
      test=c("none", "Chisq", "F"),k = 2, ...)

## S3 method for class 'glm':
drop1(object, scope, scale = 0,
      test = c("none", "Chisq", "F"),
      k = 2, ...)
```

Arguments

object a fitted model object.

scope a formula giving the terms to be considered for adding or dropping.

scale an estimate of the residual mean square to be used in computing C_p. Ignored if 0 or NULL.

test should the results include a test statistic relative to the original model? The F test is only appropriate for lm and aov models or perhaps for glm fits with estimated dispersion. The χ^2 test can be an exact test (lm models with known scale) or a likelihood-ratio test or a test of the reduction in scaled deviance depending on the method.

k the penalty constant in AIC / C_p.

trace if TRUE, print out progress reports.

x a model matrix containing columns for the fitted model and all terms in the upper scope. Useful if add1 is to be called repeatedly.

all.cols (Provided for compatibility with S.) Logical to specify whether all columns of the design matrix should be used. If FALSE then non-estimable columns are dropped, but the result is not usually statistically meaningful.

... further arguments passed to or from other methods.

Details

For drop1 methods, a missing scope is taken to be all terms in the model. The hierarchy is respected when considering terms to be added or dropped: all main effects contained in a second-order interaction must remain, and so on.

The methods for lm and glm are more efficient in that they do not recompute the model matrix and call the fit methods directly.

The default output table gives AIC, defined as minus twice log likelihood plus $2p$ where p is the rank of the model (the number of effective parameters). This is only defined up to an additive constant (like log-likelihoods). For linear Gaussian models with fixed scale, the constant is chosen to give Mallows' C_p, $RSS/scale + 2p - n$. Where C_p is used, the column is labelled as Cp rather than AIC.

Value

An object of class "anova" summarizing the differences in fit between the models.

Warning

The model fitting must apply the models to the same dataset. Most methods will attempt to use a subset of the data with no missing values for any of the variables if na.action=na.omit, but this may give biased results. Only use these functions with data containing missing values with great care.

Note

These are not fully equivalent to the functions in S. There is no keep argument, and the methods used are not quite so computationally efficient.

Their authors' definitions of Mallows' C_p and Akaike's AIC are used, not those of the authors of the models chapter of S.

Author(s)

The design was inspired by the S functions of the same names described in Chambers (1992).

References

Chambers, J. M. (1992) *Linear models.* Chapter 4 of *Statistical Models in S* eds J. M. Chambers and T. J. Hastie, Wadsworth & Brooks/Cole.

See Also

step, aov, lm, extractAIC, anova

Examples

```
example(step) # swiss
add1(lm1, ~ I(Education^2) + .^2)
drop1(lm1, test="F")   # So called 'type II' anova

example(glm)
drop1(glm.D93, test="Chisq")
drop1(glm.D93, test="F")
```

AIC   *Akaike's An Information Criterion*

Description

Generic function calculating the Akaike information criterion for one or several fitted model objects for which a log-likelihood value can be obtained, according to the formula $-2\text{log-likelihood} + kn_{par}$, where n_{par} represents the number of parameters in the fitted model, and $k = 2$ for the usual AIC, or $k = \log(n)$ (n the number of observations) for the so-called BIC or SBC (Schwarz's Bayesian criterion).

Usage

```
AIC(object, ..., k = 2)
```

Arguments

object   a fitted model object, for which there exists a logLik method to extract the corresponding log-likelihood, or an object inheriting from class logLik.

...   optionally more fitted model objects.

k   numeric, the "penalty" per parameter to be used; the default k = 2 is the classical AIC.

Details

The default method for AIC, AIC.default() entirely relies on the existence of a logLik method computing the log-likelihood for the given class.

When comparing fitted objects, the smaller the AIC, the better the fit.

Value

If just one object is provided, returns a numeric value with the corresponding AIC (or BIC, or ..., depending on k); if more than one object is provided, returns a data.frame with rows corresponding to the objects and columns representing the number of parameters in the model (df) and the AIC.

Author(s)

Jose Pinheiro and Douglas Bates

References

Sakamoto, Y., Ishiguro, M., and Kitagawa G. (1986). *Akaike Information Criterion Statistics*. D. Reidel Publishing Company.

See Also

extractAIC, logLik.

Examples

```
data(swiss)
lm1 <- lm(Fertility ~ . , data = swiss)
AIC(lm1)
stopifnot(all.equal(AIC(lm1),
                    AIC(logLik(lm1))))
## a version of BIC or Schwarz' BC :
AIC(lm1, k = log(nrow(swiss)))
```

alias   *Find Aliases (Dependencies) in a Model*

Description

Find aliases (linearly dependent terms) in a linear model specified by a formula.

Usage

```
alias(object, ...)

## S3 method for class 'formula':
alias(object, data, ...)

## S3 method for class 'lm':
alias(object, complete = TRUE, partial = FALSE,
      partial.pattern = FALSE, ...)
```

Arguments

object	A fitted model object, for example from lm or aov, or a formula for alias.formula.
data	Optionally, a data frame to search for the objects in the formula.
complete	Should information on complete aliasing be included?
partial	Should information on partial aliasing be included?
partial.pattern	Should partial aliasing be presented in a schematic way? If this is done, the results are presented in a more compact way, usually giving the deciles of the coefficients.
...	further arguments passed to or from other methods.

Details

Although the main method is for class "lm", alias is most useful for experimental designs and so is used with fits from aov. Complete aliasing refers to effects in linear models that cannot be estimated independently of the terms which occur earlier in the model and so have their coefficients omitted from the fit. Partial aliasing refers to effects that can be estimated less precisely because of correlations induced by the design.

Value

A list (of class `"listof"`) containing components

Model   Description of the model; usually the formula.

Complete A matrix with columns corresponding to effects that are linearly dependent on the rows; may be of class `"mtable"` which has its own print method.

Partial   The correlations of the estimable effects, with a zero diagonal.

Note

The aliasing pattern may depend on the contrasts in use: Helmert contrasts are probably most useful.

The defaults are different from those in S.

Author(s)

The design was inspired by the S function of the same name described in Chambers *et al.* (1992).

References

Chambers, J. M., Freeny, A and Heiberger, R. M. (1992) *Analysis of variance; designed experiments.* Chapter 5 of *Statistical Models in S* eds J. M. Chambers and T. J. Hastie, Wadsworth & Brooks/Cole.

Examples

```
had.VR <- "package:MASS" %in% search()
## The next line is for fractions() which gives neater
## results
if(!had.VR) res <- require(MASS)
## From Venables and Ripley (2002) p.165.
N <- c(0,1,0,1,1,1,0,0,0,1,1,0,1,1,0,0,1,0,1,0,1,1,0,0)
P <- c(1,1,0,0,0,1,0,1,1,1,0,0,0,1,0,1,1,0,0,1,0,1,1,0)
K <- c(1,0,0,1,0,1,1,0,0,1,0,1,0,1,1,0,0,0,1,1,1,0,1,0)
yield <- c(49.5,62.8,46.8,57.0,59.8,58.5,55.5,56.0,
           62.8,55.8,69.5,55.0,62.0,48.8,45.5,44.2,
           52.0,51.5,49.8,48.8,57.2,59.0,53.2,56.0)
npk <- data.frame(block=gl(6,4), N=factor(N), P=factor(P),
                  K=factor(K), yield=yield)
```

```
op <- options(contrasts=c("contr.helmert", "contr.poly"))
npk.aov <- aov(yield ~ block + N*P*K, npk)
alias(npk.aov)
if(!had.VR && res) detach(package:MASS)
options(op) # reset
```

anova *Anova Tables*

Description

Compute analysis of variance (or deviance) tables for one or more fitted model objects.

Usage

```
anova(object, ...)
```

Arguments

object   an object containing the results returned by a model fitting function (e.g., lm or glm).

...   additional objects of the same type.

Value

This (generic) function returns an object of class anova. These objects represent analysis-of-variance and analysis-of-deviance tables. When given a single argument it produces a table which tests whether the model terms are significant.

When given a sequence of objects, anova tests the models against one another in the order specified.

The print method for anova objects prints tables in a "pretty" form.

Warning

The comparison between two or more models will only be valid if they are fitted to the same dataset. This may be a problem if there are missing values and R's default of na.action = na.omit is used.

References

Chambers, J. M. and Hastie, T. J. (1992) *Statistical Models in S*, Wadsworth & Brooks/Cole.

See Also

coefficients, effects, fitted.values, residuals, summary, drop1, add1.

base — anova.glm

anova.glm   *Analysis of Deviance for Generalized Linear Model Fits*

Description

Compute an analysis of deviance table for one or more generalized linear model fits.

Usage

```
## S3 method for class 'glm':
anova(object, ..., dispersion = NULL, test = NULL)
```

Arguments

object, ...   objects of class glm, typically the result of a call to glm, or a list of objects for the "glmlist" method.

dispersion   the dispersion parameter for the fitting family. By default it is obtained from glm.obj.

test   a character string, (partially) matching one of "Chisq", "F" or "Cp". See stat.anova.

Details

Specifying a single object gives a sequential analysis of deviance table for that fit. That is, the reductions in the residual deviance as each term of the formula is added in turn are given in as the rows of a table, plus the residual deviances themselves.

If more than one object is specified, the table has a row for the residual degrees of freedom and deviance for each model. For all but the first model, the change in degrees of freedom and deviance is also given. (This only makes statistical sense if the models are nested.) It is conventional to list the models from smallest to largest, but this is up to the user.

The table will optionally contain test statistics (and P values) comparing the reduction in deviance for the row to the residuals. For models with known dispersion (e.g., binomial and Poisson fits) the chi-squared test is most appropriate, and for those with dispersion estimated by moments (e.g., gaussian, quasibinomial and quasipoisson fits) the F test is most appropriate. Mallows' C_p statistic is the residual deviance plus twice the estimate of σ^2 times the residual degrees of freedom, which is closely related to AIC (and a multiple of it if the dispersion is known).

Value

An object of class "anova" inheriting from class "data.frame".

Warning

The comparison between two or more models by anova or anova.glmlist will only be valid if they are fitted to the same dataset. This may be a problem if there are missing values and R's default of na.action = na.omit is used, and anova.glmlist will detect this with an error.

References

Hastie, T. J. and Pregibon, D. (1992) *Generalized linear models.* Chapter 6 of *Statistical Models in S* eds J. M. Chambers and T. J. Hastie, Wadsworth & Brooks/Cole.

See Also

glm, anova.

drop1 for so-called 'type II' anova where each term is dropped one at a time respecting their hierarchy.

Examples

```
# Continuing the Example from '?glm':
anova(glm.D93)
anova(glm.D93, test = "Cp")
anova(glm.D93, test = "Chisq")
```

base — anova.lm

anova.lm   *ANOVA for Linear Model Fits*

Description

Compute an analysis of variance table for one or more linear model fits.

Usage

```
## S3 method for class 'lm':
anova(object, ...)

anova.lmlist(object, ..., scale = 0, test = "F")
```

Arguments

object, ...   objects of class lm, usually, a result of a call to lm.

test   a character string specifying the test statistic to be used. Can be one of "F", "Chisq" or "Cp", with partial matching allowed, or NULL for no test.

scale   numeric. An estimate of the noise variance σ^2. If zero this will be estimated from the largest model considered.

Details

Specifying a single object gives a sequential analysis of variance table for that fit. That is, the reductions in the residual sum of squares as each term of the formula is added in turn are given in as the rows of a table, plus the residual sum of squares.

The table will contain F statistics (and P values) comparing the mean square for the row to the residual mean square.

If more than one object is specified, the table has a row for the residual degrees of freedom and sum of squares for each model. For all but the first model, the change in degrees of freedom and sum of squares is also given. (This only makes statistical sense if the models are nested.) It is conventional to list the models from smallest to largest, but this is up to the user.

Optionally the table can include test statistics. Normally the F statistic is most appropriate, which compares the mean square for a row to the residual sum of squares for the largest model considered. If scale is specified chi-squared tests can be used. Mallows' C_p statistic is the residual sum of squares plus twice the estimate of σ^2 times the residual degrees of freedom.

Value

An object of class `"anova"` inheriting from class `"data.frame"`.

Warning

The comparison between two or more models will only be valid if they are fitted to the same dataset. This may be a problem if there are missing values and R's default of `na.action = na.omit` is used, and `anova.lmlist` will detect this with an error.

Note

Versions of R prior to 1.2.0 based F tests on pairwise comparisons, and this behaviour can still be obtained by a direct call to `anovalist.lm`.

References

Chambers, J. M. (1992) *Linear models.* Chapter 4 of *Statistical Models in S* eds J. M. Chambers and T. J. Hastie, Wadsworth & Brooks/Cole.

See Also

The model fitting function `lm`, `anova`.

`drop1` for so-called 'type II' anova where each term is dropped one at a time respecting their hierarchy.

Examples

```
## sequential table
data(LifeCycleSavings)
fit <- lm(sr ~ ., data = LifeCycleSavings)
anova(fit)
## same effect via separate models
fit0 <- lm(sr ~ 1, data = LifeCycleSavings)
fit1 <- update(fit0, . ~ . + pop15)
fit2 <- update(fit1, . ~ . + pop75)
fit3 <- update(fit2, . ~ . + dpi)
fit4 <- update(fit3, . ~ . + ddpi)
anova(fit0, fit1, fit2, fit3, fit4, test="F")
anova(fit4, fit2, fit0, test="F") # unconventional order
```

aov   *Fit an Analysis of Variance Model*

Description

Fit an analysis of variance model by a call to `lm` for each stratum.

Usage

```
aov(formula, data = NULL, projections = FALSE, qr = TRUE,
    contrasts = NULL, ...)
```

Arguments

formula	A formula specifying the model.
data	A data frame in which the variables specified in the formula will be found. If missing, the variables are searched for in the standard way.
projections	Logical flag: should the projections be returned?
qr	Logical flag: should the QR decomposition be returned?
contrasts	A list of contrasts to be used for some of the factors in the formula. These are not used for any `Error` term, and supplying contrasts for factors only in the `Error` term will give a warning.
...	Arguments to be passed to `lm`, such as `subset` or `na.action`.

Details

This provides a wrapper to `lm` for fitting linear models to balanced or unbalanced experimental designs.

The main difference from `lm` is in the way `print`, `summary` and so on handle the fit: this is expressed in the traditional language of the analysis of variance rather than of linear models.

If the formula contains a single `Error` term, this is used to specify error strata, and appropriate models are fitted within each error stratum.

The formula can specify multiple responses.

Weights can be specified by a `weights` argument, but should not be used with an `Error` term, and are incompletely supported (e.g., not by `model.tables`).

Value

An object of class c("aov", "lm") or for multiple responses of class c("maov", "aov", "mlm", "lm") or for multiple error strata of class "aovlist". There are print and summary methods available for these.

Author(s)

The design was inspired by the S function of the same name described in Chambers *et al.* (1992).

References

Chambers, J. M., Freeny, A and Heiberger, R. M. (1992) *Analysis of variance; designed experiments.* Chapter 5 of *Statistical Models in S* eds J. M. Chambers and T. J. Hastie, Wadsworth & Brooks/Cole.

See Also

lm, summary.aov, alias, proj, model.tables, TukeyHSD

Examples

```
## From Venables and Ripley (2002) p.165.
N <- c(0,1,0,1,1,1,0,0,0,1,1,0,1,1,0,0,1,0,1,0,1,1,0,0)
P <- c(1,1,0,0,0,1,0,1,1,1,0,0,0,1,0,1,1,0,0,1,0,1,1,0)
K <- c(1,0,0,1,0,1,1,0,0,1,0,1,0,1,1,0,0,1,1,1,0,1,0)
yield <- c(49.5,62.8,46.8,57.0,59.8,58.5,55.5,56.0,
           62.8,55.8,69.5,55.0,62.0,48.8,45.5,44.2,
           52.0,51.5,49.8,48.8,57.2,59.0,53.2,56.0)
npk <- data.frame(block=gl(6,4), N=factor(N), P=factor(P),
                  K=factor(K), yield=yield)

( npk.aov <- aov(yield ~ block + N*P*K, npk) )
summary(npk.aov)
coefficients(npk.aov)

## as a test, not particularly sensible statistically
op <- options(contrasts = c("contr.helmert",
                            "contr.treatment"))
npk.aovE <- aov(yield ~  N*P*K + Error(block), npk)
npk.aovE
summary(npk.aovE)
options(op) # reset to previous
```

AsIs   Inhibit Interpretation/Conversion of Objects

Description

Change the class of an object to indicate that it should be treated "as is".

Usage

```
I(x)
```

Arguments

x   an object

Details

Function I has two main uses.

- In function `data.frame`. Protecting an object by enclosing it in I() in a call to data.frame inhibits the conversion of character vectors to factors. I can also be used to protect objects which are to be added to a data frame, or converted to a data frame via as.data.frame.

  It achieves this by prepending the class "AsIs" to the object's classes. Class "AsIs" has a few of its own methods, including for [, as.data.frame, print and format.

- In function `formula`. There it is used to inhibit the interpretation of operators such as "+", "-", "*" and "^" as formula operators, so they are used as arithmetical operators. This is interpreted as a symbol by terms.formula.

Value

A copy of the object with class "AsIs" prepended to the class(es).

References

Chambers, J. M. (1992) *Linear models.* Chapter 4 of *Statistical Models in S* eds J. M. Chambers and T. J. Hastie, Wadsworth & Brooks/Cole.

See Also

data.frame, formula

C *Sets Contrasts for a Factor*

Description

Sets the "contrasts" attribute for the factor.

Usage

```
C(object, contr, how.many, ...)
```

Arguments

object	a factor or ordered factor
contr	which contrasts to use. Can be a matrix with one row for each level of the factor or a suitable function like contr.poly or a character string giving the name of the function
how.many	the number of contrasts to set, by default one less than nlevels(object).
...	additional arguments for the function contr.

Details

For compatibility with S, contr can be treatment, helmert, sum or poly (without quotes) as shorthand for contr.treatment and so on.

Value

The factor object with the "contrasts" attribute set.

References

Chambers, J. M. and Hastie, T. J. (1992) *Statistical models.* Chapter 2 of *Statistical Models in S* eds J. M. Chambers and T. J. Hastie, Wadsworth & Brooks/Cole.

See Also

contrasts, contr.sum, etc.

Examples

```
## reset contrasts to defaults
options(contrasts=c("contr.treatment", "contr.poly"))
data(warpbreaks)
attach(warpbreaks)
tens <- C(tension, poly, 1)
attributes(tens)
detach()
## tension SHOULD be an ordered factor, but as it is not we
## can use
aov(breaks ~ wool + tens + tension, data=warpbreaks)

## show the use of ... The default contrast is
## contr.treatment here
summary(lm(breaks ~ wool + C(tension, base=2),
        data=warpbreaks))

data(esoph) # following on from help(esoph)
model3 <-
  glm(cbind(ncases, ncontrols) ~ agegp + C(tobgp, , 1) +
      C(alcgp, , 1), data = esoph, family = binomial())
summary(model3)
```

case/variable.names *Case and Variable Names of Fitted Models*

Description

Simple utilities returning (non-missing) case names, and (non-eliminated) variable names.

Usage

```
case.names(object, ...)
## S3 method for class 'lm':
case.names(object, full = FALSE, ...)

variable.names(object, ...)
## S3 method for class 'lm':
variable.names(object, full = FALSE, ...)
```

Arguments

object	an R object, typically a fitted model.
full	logical; if TRUE, all names (including zero weights, ...) are returned.
...	further arguments passed to or from other methods.

Value

A character vector.

See Also

lm

Examples

```
x <- 1:20
y <-  x + (x/4 - 2)^3 + rnorm(20, s=3)
names(y) <- paste("O",x,sep=".")
ww <- rep(1,20); ww[13] <- 0
summary(lmxy <- lm(y ~ x + I(x^2)+I(x^3) + I((x-10)^2),
                  weights = ww), cor = TRUE)
variable.names(lmxy)
```

```
variable.names(lmxy, full= TRUE) # includes the last
case.names(lmxy)
case.names(lmxy, full = TRUE) # includes the 0-weight case
```

coef *Extract Model Coefficients*

Description

`coef` is a generic function which extracts model coefficients from objects returned by modeling functions. `coefficients` is an *alias* for it.

Usage

```
coef(object, ...)
coefficients(object, ...)
```

Arguments

object
: an object for which the extraction of model coefficients is meaningful.

...
: other arguments.

Details

All object classes which are returned by model fitting functions should provide a `coef` method. (Note that the method is `coef` and not `coefficients`.)

Value

Coefficients extracted from the model object `object`.

References

Chambers, J. M. and Hastie, T. J. (1992) *Statistical Models in S.* Wadsworth & Brooks/Cole.

See Also

`fitted.values` and `residuals` for related methods; `glm`, `lm` for model fitting.

Examples

```
x <- 1:5; coef(lm(c(1:3,7,6) ~ x))
```

confint   *Confidence Intervals for Model Parameters*

Description

Computes confidence intervals for one or more parameters in a fitted model. Base has a method for objects inheriting from class "lm".

Usage

```
confint(object, parm, level = 0.95, ...)
```

Arguments

object   a fitted model object.

parm   a specification of which parameters are to be given confidence intervals, either a vector of numbers or a vector of names. If missing, all parameters are considered.

level   the confidence level required.

...   additional argument(s) for methods

Details

confint is a generic function with no default method. For objects of class "lm" the direct formulae based on t values are used.

Package **MASS** contains methods for "glm" and "nls" objects.

Value

A matrix (or vector) with columns giving lower and upper confidence limits for each parameter. These will be labelled as (1-level)/2 and 1 - (1-level)/2 in % (by default 2.5% and 97.5%).

See Also

confint.nls

Examples

```
data(mtcars)
fit <- lm(100/mpg ~ disp + hp + wt + am, data=mtcars)
confint(fit)
confint(fit, "wt")
```

constrOptim *Linearly constrained optimisation*

Description

Minimise a function subject to linear inequality constraints using an adaptive barrier algorithm.

Usage

```
constrOptim(theta, f, grad, ui, ci, mu = 1e-04,
    control = list(),
    method = if(is.null(grad)) "Nelder-Mead" else "BFGS",
    outer.iterations = 100, outer.eps = 1e-05, ...)
```

Arguments

theta	Starting value: must be in the feasible region.
f	Function to minimise.
grad	Gradient of f.
ui	Constraints (see below).
ci	Constraints (see below).
mu	(Small) tuning parameter.
control	Passed to optim.
method	Passed to optim.
outer.iterations	Iterations of the barrier algorithm.
outer.eps	Criterion for relative convergence of the barrier algorithm.
...	Other arguments passed to optim

Details

The feasible region is defined by ui %*% theta - ci >= 0. The starting value must be in the interior of the feasible region, but the minimum may be on the boundary.

A logarithmic barrier is added to enforce the constraints and then optim is called. The barrier function is chosen so that the objective function should decrease at each outer iteration. Minima in the interior of the feasible region are typically found quite quickly, but a substantial number of outer iterations may be needed for a minimum on the boundary.

The tuning parameter mu multiplies the barrier term. Its precise value is often relatively unimportant. As mu increases the augmented objective function becomes closer to the original objective function but also less smooth near the boundary of the feasible region.

Any optim method that permits infinite values for the objective function may be used (currently all but "L-BFGS-B"). The gradient function must be supplied except with method="Nelder-Mead".

As with optim, the default is to minimise and maximisation can be performed by setting control$fnscale to a negative value.

Value

As for optim, but with two extra components: barrier.value giving the value of the barrier function at the optimum and outer.iterations gives the number of outer iterations (calls to optim)

References

K. Lange *Numerical Analysis for Statisticians.* Springer 2001, p185ff

See Also

optim, especially method="L-BGFS-B" which does box-constrained optimisation.

Examples

```
## from optim
fr <- function(x) {    ## Rosenbrock Banana function
    x1 <- x[1]
    x2 <- x[2]
    100 * (x2 - x1 * x1)^2 + (1 - x1)^2
}
grr <- function(x) { ## Gradient of 'fr'
    x1 <- x[1]
    x2 <- x[2]
    c(-400 * x1 * (x2 - x1 * x1) - 2 * (1 - x1),
      200 *      (x2 - x1 * x1))
}

optim(c(-1.2,1), fr, grr)
```

```
# Box-constraint, optimum on the boundary
constrOptim(c(-1.2,0.9), fr, grr,
            ui=rbind(c(-1,0),c(0,-1)), ci=c(-1,-1))
# x<=0.9, y-x>0.1
constrOptim(c(.5,0), fr, grr,
            ui=rbind(c(-1,0),c(1,-1)), ci=c(-0.9,0.1))

## Solves linear and quadratic programming problems but
## needs a feasible starting value
#
# from example(solve.QP) in 'quadprog'
# no derivative
fQP <- function(b) {-sum(c(0,5,0)*b)+0.5*sum(b*b)}
Amat      <- matrix(c(-4,-3,0,2,1,0,0,-2,1),3,3)
bvec      <- c(-8,2,0)
constrOptim(c(2,-1,-1), fQP, NULL, ui=t(Amat),ci=bvec)
# derivative
gQP <- function(b) {-c(0,5,0)+b}
constrOptim(c(2,-1,-1), fQP, gQP, ui=t(Amat), ci=bvec)

## Now with maximisation instead of minimisation
hQP <- function(b) {sum(c(0,5,0)*b)-0.5*sum(b*b)}
constrOptim(c(2,-1,-1), hQP, NULL, ui=t(Amat), ci=bvec,
            control=list(fnscale=-1))
```

contrast *Contrast Matrices*

Description

Return a matrix of contrasts.

Usage

```
contr.helmert(n, contrasts = TRUE)
contr.poly(n, scores = 1:n, contrasts = TRUE)
contr.sum(n, contrasts = TRUE)
contr.treatment(n, base = 1, contrasts = TRUE)
```

Arguments

n
: a vector of levels for a factor, or the number of levels.

contrasts
: a logical indicating whether contrasts should be computed.

scores
: the set of values over which orthogonal polynomials are to be computed.

base
: an integer specifying which group is considered the baseline group. Ignored if contrasts is FALSE.

Details

These functions are used for creating contrast matrices for use in fitting analysis of variance and regression models. The columns of the resulting matrices contain contrasts which can be used for coding a factor with n levels. The returned value contains the computed contrasts. If the argument contrasts is FALSE then a square indicator matrix is returned.

cont.helmert returns Helmert contrasts, which contrast the second level with the first, the third with the average of the first two, and so on. contr.poly returns contrasts based on orthogonal polynomials. contr.sum uses "sum to zero contrasts".

contr.treatment contrasts each level with the baseline level (specified by base): the baseline level is omitted. Note that this does not produce "contrasts" as defined in the standard theory for linear models as they are not orthogonal to the constant.

Value

A matrix with n rows and k columns, with k=n-1 if contrasts is TRUE and k=n if contrasts is FALSE.

References

Chambers, J. M. and Hastie, T. J. (1992) *Statistical models.* Chapter 2 of *Statistical Models in S* eds J. M. Chambers and T. J. Hastie, Wadsworth & Brooks/Cole.

See Also

contrasts, C, and aov, glm, lm.

Examples

```
(cH <- contr.helmert(4))
apply(cH, 2,sum) # column sums are 0!
crossprod(cH) # diagonal -- columns are orthogonal
# just the 4 x 4 identity matrix
contr.helmert(4, contrasts = FALSE)

(cT <- contr.treatment(5))
all(crossprod(cT) == diag(4)) # TRUE: even orthonormal

(cP <- contr.poly(3)) # Linear and Quadratic
zapsmall(crossprod(cP), dig=15) # orthonormal up to fuzz
```

contrasts   *Get and Set Contrast Matrices*

Description

Set and view the contrasts associated with a factor.

Usage

```
contrasts(x, contrasts = TRUE)
contrasts(x, how.many) <- value
```

Arguments

x	a factor.
contrasts	logical. See Details.
how.many	How many contrasts should be made. Defaults to one less than the number of levels of x. This need not be the same as the number of columns of ctr.
value	either a matrix whose columns give coefficients for contrasts in the levels of x, or the (quoted) name of a function which computes such matrices.

Details

If contrasts are not set for a factor the default functions from options("contrasts") are used.

The argument contrasts is ignored if x has a matrix contrasts attribute set. Otherwise if contrasts = TRUE it is passed to a contrasts function such as contr.treatment and if contrasts = FALSE an identity matrix is returned.

References

Chambers, J. M. and Hastie, T. J. (1992) *Statistical models.* Chapter 2 of *Statistical Models in S* eds J. M. Chambers and T. J. Hastie, Wadsworth & Brooks/Cole.

See Also

C, contr.helmert, contr.poly, contr.sum, contr.treatment; glm, aov, lm.

Examples

```
example(factor)
fff <- ff[, drop=TRUE]   # reduce to 5 levels.
contrasts(fff) # treatment contrasts by default
contrasts(C(fff, sum))
contrasts(fff, contrasts = FALSE) # the 5x5 identity matrix

# set sum contrasts
contrasts(fff) <- contr.sum(5); contrasts(fff)
# set 2 contrasts
contrasts(fff, 2) <- contr.sum(5); contrasts(fff)
# supply 2 contrasts, compute 2 more to make full set of 4.
contrasts(fff) <- contr.sum(5)[,1:2]; contrasts(fff)
```

deviance *Model Deviance*

Description

Returns the deviance of a fitted model object.

Usage

```
deviance(object, ...)
```

Arguments

object
: an object for which the deviance is desired.

...
: additional optional argument.

Details

This is a generic function which can be used to extract deviances for fitted models. Consult the individual modeling functions for details on how to use this function.

Value

The value of the deviance extracted from the object `object`.

References

Chambers, J. M. and Hastie, T. J. (1992) *Statistical Models in S.* Wadsworth & Brooks/Cole.

See Also

`df.residual`, `extractAIC`, `glm`, `lm`.

df.residual — *Residual Degrees-of-Freedom*

Description

Returns the residual degrees-of-freedom extracted from a fitted model object.

Usage

```
df.residual(object, ...)
```

Arguments

object   an object for which the degrees-of-freedom are desired.
...   additional optional arguments.

Details

This is a generic function which can be used to extract residual degrees-of-freedom for fitted models. Consult the individual modeling functions for details on how to use this function.

The default method just extracts the `df.residual` component.

Value

The value of the residual degrees-of-freedom extracted from the object x.

See Also

`deviance`, `glm`, `lm`.

base — dummy.coef

dummy.coef *Extract Coefficients in Original Coding*

Description

This extracts coefficients in terms of the original levels of the coefficients rather than the coded variables.

Usage

```
dummy.coef(object, ...)

## S3 method for class 'lm':
dummy.coef(object, use.na = FALSE, ...)

## S3 method for class 'aovlist':
dummy.coef(object, use.na = FALSE, ...)
```

Arguments

object
: a linear model fit.

use.na
: logical flag for coefficients in a singular model. If use.na is true, undetermined coefficients will be missing; if false they will get one possible value.

...
: arguments passed to or from other methods.

Details

A fitted linear model has coefficients for the contrasts of the factor terms, usually one less in number than the number of levels. This function re-expresses the coefficients in the original coding; as the coefficients will have been fitted in the reduced basis, any implied constraints (e.g., zero sum for contr.helmert or contr.sum will be respected. There will be little point in using dummy.coef for contr.treatment contrasts, as the missing coefficients are by definition zero.

The method used has some limitations, and will give incomplete results for terms such as poly(x, 2)). However, it is adequate for its main purpose, aov models.

Value

A list giving for each term the values of the coefficients. For a multi-stratum aov model, such a list for each stratum.

Warning

This function is intended for human inspection of the output: it should not be used for calculations. Use coded variables for all calculations.

The results differ from S for singular values, where S can be incorrect.

See Also

`aov`, `model.tables`

Examples

```
options(contrasts=c("contr.helmert", "contr.poly"))
## From Venables and Ripley (2002) p.165.
N <- c(0,1,0,1,1,1,0,0,0,1,1,0,1,1,0,0,1,0,1,0,1,1,0,0)
P <- c(1,1,0,0,0,1,0,1,1,1,0,0,0,1,0,1,1,0,0,1,0,1,1,0)
K <- c(1,0,0,1,0,1,1,0,0,1,0,1,0,1,1,0,0,0,1,1,1,0,1,0)
yield <- c(49.5,62.8,46.8,57.0,59.8,58.5,55.5,56.0,
           62.8,55.8,69.5,55.0,62.0,48.8,45.5,44.2,
           52.0,51.5,49.8,48.8,57.2,59.0,53.2,56.0)

npk <- data.frame(block=gl(6,4), N=factor(N), P=factor(P),
                  K=factor(K), yield=yield)
npk.aov <- aov(yield ~ block + N*P*K, npk)
dummy.coef(npk.aov)


npk.aovE <- aov(yield ~ N*P*K + Error(block), npk)
dummy.coef(npk.aovE)
```

eff.aovlist   *Compute Efficiencies of Multistratum Analysis of Variance*

Description

Computes the efficiencies of fixed-effect terms in an analysis of variance model with multiple strata.

Usage

```
eff.aovlist(aovlist)
```

Arguments

aovlist   The result of a call to aov with an Error term.

Details

Fixed-effect terms in an analysis of variance model with multiple strata may be estimable in more than one stratum, in which case there is less than complete information in each. The efficiency is the fraction of the maximum possible precision (inverse variance) obtainable by estimating in just that stratum.

This is used to pick strata in which to estimate terms in model.tables.aovlist and elsewhere.

Value

A matrix giving for each non-pure-error stratum (row) the efficiencies for each fixed-effect term in the model.

See Also

aov, model.tables.aovlist, se.contrast.aovlist

Examples

```
## for balanced designs all efficiencies are zero or one.
## so as a statistically meaningless test:
options(contrasts=c("contr.helmert", "contr.poly"))
## From Venables and Ripley (2002) p.165.
N <- c(0,1,0,1,1,1,0,0,0,1,1,0,1,1,0,0,1,0,1,0,1,1,0,0)
P <- c(1,1,0,0,0,1,0,1,1,1,0,0,0,1,0,1,1,0,0,1,0,1,1,0)
```

```
K <- c(1,0,0,1,0,1,1,0,0,1,0,1,0,1,1,0,0,0,1,1,1,0,1,0)
yield <- c(49.5,62.8,46.8,57.0,59.8,58.5,55.5,56.0,
           62.8,55.8,69.5,55.0,62.0,48.8,45.5,44.2,
           52.0,51.5,49.8,48.8,57.2,59.0,53.2,56.0)

npk <- data.frame(block=gl(6,4), N=factor(N), P=factor(P),
                  K=factor(K), yield=yield)
npk.aovE <- aov(yield ~ N*P*K + Error(block), npk)
eff.aovlist(npk.aovE)
```

effects *Effects from Fitted Model*

Description

Returns (orthogonal) effects from a fitted model, usually a linear model. This is a generic function, but currently only has a methods for objects inheriting from classes "lm" and "glm".

Usage

```
effects(object, ...)

## S3 method for class 'lm':
effects(object, set.sign=FALSE, ...)
```

Arguments

object
: an R object; typically, the result of a model fitting function such as lm.

set.sign
: logical. If TRUE, the sign of the effects corresponding to coefficients in the model will be set to agree with the signs of the corresponding coefficients, otherwise the sign is arbitrary.

...
: arguments passed to or from other methods.

Details

For a linear model fitted by lm or aov, the effects are the uncorrelated single-degree-of-freedom values obtained by projecting the data onto the successive orthogonal subspaces generated by the QR decomposition during the fitting process. The first r (the rank of the model) are associated with coefficients and the remainder span the space of residuals (but are not associated with particular residuals).

Empty models do not have effects.

Value

A (named) numeric vector of the same length as residuals, or a matrix if there were multiple responses in the fitted model, in either case of class "coef".

The first r rows are labelled by the corresponding coefficients, and the remaining rows are unlabelled. Note that in rank-deficient models the "corresponding" coefficients will be in a different order if pivoting occurred.

References

Chambers, J. M. and Hastie, T. J. (1992) *Statistical Models in S*. Wadsworth & Brooks/Cole.

See Also

coef

Examples

```
y <- c(1:3,7,5)
x <- c(1:3,6:7)
( ee <- effects(lm(y ~ x)) )
# just the first is different
c(round(ee - effects(lm(y+10 ~ I(x-3.8))),3))
```

base — expand.grid

expand.grid *Create a Data Frame from All Combinations of Factors*

Description

Create a data frame from all combinations of the supplied vectors or factors. See the description of the return value for precise details of the way this is done.

Usage

```
expand.grid(...)
```

Arguments

...   Vectors, factors or a list containing these.

Value

A data frame containing one row for each combination of the supplied factors. The first factors vary fastest. The columns are labelled by the factors if these are supplied as named arguments or named components of a list.

References

Chambers, J. M. and Hastie, T. J. (1992) *Statistical Models in S.* Wadsworth & Brooks/Cole.

Examples

```
expand.grid(height = seq(60, 80, 5),
            weight = seq(100, 300, 50),
            sex = c("Male","Female"))
```

expand.model.frame — *Add new variables to a model frame*

Description

Evaluates new variables as if they had been part of the formula of the specified model. This ensures that the same `na.action` and `subset` arguments are applied and allows, for example, `x` to be recovered for a model using `sin(x)` as a predictor.

Usage

```
expand.model.frame(model, extras,
                envir=environment(formula(model)),
                na.expand = FALSE)
```

Arguments

`model` a fitted model

`extras` one-sided formula or vector of character strings describing new variables to be added

`envir` an environment to evaluate things in

`na.expand` logical; see below

Details

If `na.expand=FALSE` then `NA` values in the extra variables will be passed to the `na.action` function used in `model`. This may result in a shorter data frame (with `na.omit`) or an error (with `na.fail`). If `na.expand=TRUE` the returned data frame will have precisely the same rows as `model.frame(model)`, but the columns corresponding to the extra variables may contain `NA`.

Value

A data frame.

See Also

`model.frame`,`predict`

Examples

```
data(trees)
model <- lm(log(Volume) ~ log(Girth) + log(Height),
            data=trees)
expand.model.frame(model, ~ Girth) # prints data.frame like

dd <- data.frame(x=1:5, y=rnorm(5), z=c(1,2,NA,4,5))
model <- glm(y ~ x, data=dd, subset=1:4, na.action=na.omit)
expand.model.frame(model, "z", na.expand=FALSE) # = default
expand.model.frame(model, "z", na.expand=TRUE)
```

extractAIC *Extract AIC from a Fitted Model*

Description

Computes the (generalized) Akaike An Information Criterion for a fitted parametric model.

Usage

```
extractAIC(fit, scale,    k = 2, ...)
```

Arguments

`fit`   fitted model, usually the result of a fitter like `lm`.

`scale`   optional numeric specifying the scale parameter of the model, see `scale` in `step`.

`k`   numeric specifying the "weight" of the *equivalent degrees of freedom* ($\equiv$`edf`) part in the AIC formula.

`...`   further arguments (currently unused in base R).

Details

This is a generic function, with methods in base R for `"aov"`, `"coxph"`, `"glm"`, `"lm"`, `"negbin"` and `"survreg"` classes.

The criterion used is

$$AIC = -2\log L + k \times \text{edf},$$

where L is the likelihood and `edf` the equivalent degrees of freedom (i.e., the number of parameters for usual parametric models) of `fit`.

For linear models with unknown scale (i.e., for `lm` and `aov`), $-2\log L$ is computed from the *deviance* and uses a different additive constant to AIC.

`k = 2` corresponds to the traditional AIC, using `k = log(n)` provides the BIC (Bayes IC) instead.

For further information, particularly about `scale`, see `step`.

Value

A numeric vector of length 2, giving

edf	the "equivalent degrees of freedom" of the fitted model fit.
AIC	the (generalized) Akaike Information Criterion for fit.

Note

These functions are used in add1, drop1 and step and that may be their main use.

Author(s)

B. D. Ripley

References

Venables, W. N. and Ripley, B. D. (2002) *Modern Applied Statistics with S.* New York: Springer (4th ed).

See Also

AIC, deviance, add1, step

Examples

```
example(glm)
extractAIC(glm.D93) #   5   15.129
```

factor.scope *Compute Allowed Changes in Adding to or Dropping from a Formula*

Description

`add.scope` and `drop.scope` compute those terms that can be individually added to or dropped from a model while respecting the hierarchy of terms.

Usage

```
add.scope(terms1, terms2)
drop.scope(terms1, terms2)
factor.scope(factor, scope)
```

Arguments

terms1
: the terms or formula for the base model.

terms2
: the terms or formula for the upper (`add.scope`) or lower (`drop.scope`) scope. If missing for `drop.scope` it is taken to be the null formula, so all terms (except any intercept) are candidates to be dropped.

factor
: the `"factor"` attribute of the terms of the base object.

scope
: a list with one or both components `drop` and `add` giving the `"factor"` attribute of the lower and upper scopes respectively.

Details

`factor.scope` is not intended to be called directly by users.

Value

For `add.scope` and `drop.scope` a character vector of terms labels. For `factor.scope`, a list with components `drop` and `add`, character vectors of terms labels.

See Also

add1, drop1, aov, lm

Examples

```
add.scope( ~ a + b + c + a:b,   ~ (a + b + c)^3)
# [1] "a:c" "b:c"
drop.scope( ~ a + b + c + a:b)
# [1] "c"   "a:b"
```

| family | Family Objects for Models |

Description

Family objects provide a convenient way to specify the details of the models used by functions such as glm. See the documentation for glm for the details on how such model fitting takes place.

Usage

```
family(object, ...)

binomial(link = "logit")
gaussian(link ="identity")
Gamma(link = "inverse")
inverse.gaussian(link = "1/mu^2")
poisson(link = "log")
quasi(link = "identity", variance = "constant")
quasibinomial(link = "logit")
quasipoisson(link = "log")
```

Arguments

link   a specification for the model link function. The gaussian family accepts the links "identity", "log" and "inverse"; the binomial family the links "logit", "probit", "log" and "cloglog" (complementary log-log); the Gamma family the links "inverse", "identity" and "log"; the poisson family the links "log", "identity", and "sqrt" and the inverse.gaussian family the links "1/mu^2", "inverse", "inverse" and "log".

The quasi family allows the links "logit", "probit", "cloglog", "identity", "inverse", "log", "1/mu^2" and "sqrt". The function power can also be used to create a power link function for the quasi family.

variance   for all families, other than quasi, the variance function is determined by the family. The quasi family will accept the specifications "constant", "mu(1-

	mu)", "mu", "mu^2" and "mu^3" for the variance function.
object	the function `family` accesses the `family` objects which are stored within objects created by modelling functions (e.g., `glm`).
...	further arguments passed to methods.

Details

The `quasibinomial` and `quasipoisson` families differ from the `binomial` and `poisson` families only in that the dispersion parameter is not fixed at one, so they can "model" over-dispersion. For the binomial case see McCullagh and Nelder (1989, pp. 124–8). Although they show that there is (under some restrictions) a model with variance proportional to mean as in the quasi-binomial model, note that `glm` does not compute maximum-likelihood estimates in that model. The behaviour of S is closer to the quasi- variants.

Author(s)

The design was inspired by S functions of the same names described in Hastie & Pregibon (1992).

References

McCullagh P. and Nelder, J. A. (1989) *Generalized Linear Models.* London: Chapman and Hall.

Dobson, A. J. (1983) *An Introduction to Statistical Modelling.* London: Chapman and Hall.

Cox, D. R. and Snell, E. J. (1981). *Applied Statistics; Principles and Examples.* London: Chapman and Hall.

Hastie, T. J. and Pregibon, D. (1992) *Generalized linear models.* Chapter 6 of *Statistical Models in S* eds J. M. Chambers and T. J. Hastie, Wadsworth & Brooks/Cole.

See Also

`glm`, `power`.

Examples

```
nf <- gaussian() # Normal family
nf
str(nf) # internal STRucture

gf <- Gamma()
gf
str(gf)
gf$linkinv
gf$variance(-3:4) # == (.)^2

## quasipoisson. compare with example(glm)
counts <- c(18,17,15,20,10,20,25,13,12)
outcome <- gl(3,1,9)
treatment <- gl(3,3)
d.AD <- data.frame(treatment, outcome, counts)
glm.qD93 <- glm(counts ~ outcome + treatment,
                family=quasipoisson())
glm.qD93
anova(glm.qD93, test="F")
summary(glm.qD93)
## for Poisson results use
anova(glm.qD93, dispersion = 1, test="Chisq")
summary(glm.qD93, dispersion = 1)

## tests of quasi
x <- rnorm(100)
y <- rpois(100, exp(1+x))
glm(y ~x, family=quasi(var="mu", link="log"))
# which is the same as
glm(y ~x, family=poisson)
glm(y ~x, family=quasi(var="mu^2", link="log"))
# should fail
glm(y ~x, family=quasi(var="mu^3", link="log"))
y <- rbinom(100, 1, plogis(x))
# needs to set a starting value for the next fit
glm(y ~x, family=quasi(var="mu(1-mu)", link="logit"),
    start=c(0,1))
```

fitted   *Extract Model Fitted Values*

Description

`fitted` is a generic function which extracts fitted values from objects returned by modeling functions. `fitted.values` is an alias for it.

All object classes which are returned by model fitting functions should provide a `fitted` method. (Note that the generic is `fitted` and not `fitted.values`.)

Methods can make use of `napredict` methods to compensate for the omission of missing values. The default, `lm` and `glm` methods do.

Usage

```
fitted(object, ...)
fitted.values(object, ...)
```

Arguments

object	an object for which the extraction of model fitted values is meaningful.
...	other arguments.

Value

Fitted values extracted from the object x.

References

Chambers, J. M. and Hastie, T. J. (1992) *Statistical Models in S.* Wadsworth & Brooks/Cole.

See Also

`coefficients`, `glm`, `lm`, `residuals`.

formula   *Model Formulae*

Description

The generic function `formula` and its specific methods provide a way of extracting formulae which have been included in other objects.

`as.formula` is almost identical, additionally preserving attributes when `object` already inherits from `"formula"`. The default value of the `env` argument is used only when the formula would otherwise lack an environment.

Usage

```
y ~ model
formula(x, ...)
as.formula(object, env = parent.frame())
```

Arguments

`x, object`   an object

`...`   further arguments passed to or from other methods.

`env`   the environment to associate with the result.

Details

The models fit by, e.g., the `lm` and `glm` functions are specified in a compact symbolic form. The ~ operator is basic in the formation of such models. An expression of the form `y ~ model` is interpreted as a specification that the response `y` is modelled by a linear predictor specified symbolically by `model`. Such a model consists of a series of terms separated by + operators. The terms themselves consist of variable and factor names separated by : operators. Such a term is interpreted as the interaction of all the variables and factors appearing in the term.

In addition to + and :, a number of other operators are useful in model formulae. The * operator denotes factor crossing: `a*b` interpreted as `a+b+a:b`. The ^ operator indicates crossing to the specified degree. For example `(a+b+c)^2` is identical to `(a+b+c)*(a+b+c)` which in turn expands to a formula containing the main effects for a, b and c together with their second-order interactions. The %in% operator indicates that the terms on its left are nested within those on the right. For example `a+b%in%a` expands to the formula `a+a:b`. The - operator removes the specified terms, so that (a+b+c)^2 - a:b is identical to a + b + c + b:c + a:c. It can also used to remove the intercept term: y~x - 1 is a line through the origin. A model with no intercept can be also specified as y~x + 0 or 0 + y~x.

While formulae usually involve just variable and factor names, they can also involve arithmetic expressions. The formula log(y) ~ a + log(x) is quite legal. When such arithmetic expressions involve operators which are also used symbolically in model formulae, there can be confusion between arithmetic and symbolic operator use.

To avoid this confusion, the function I() can be used to bracket those portions of a model formula where the operators are used in their arithmetic sense. For example, in the formula y ~ a + I(b+c), the term b+c is to be interpreted as the sum of b and c.

As from R 1.8.0 variable names can be quoted by backticks 'like this' in formulae, although there is no guarantee that all code using formulae will accept such non-syntactic names.

Value

All the functions above produce an object of class "formula" which contains a symbolic model formula.

Environments

A formula object has an associated environment, and this environment (rather than the parent environment) is used by model.frame to evaluate variables that are not found in the supplied data argument.

Formulas created with the ~ operator use the environment in which they were created. Formulas created with as.formula will use the env argument for their environment. Pre-existing formulas extracted with as.formula will only have their environment changed if env is explicitly given.

References

Chambers, J. M. and Hastie, T. J. (1992) *Statistical models.* Chapter 2 of *Statistical Models in S* eds J. M. Chambers and T. J. Hastie, Wadsworth & Brooks/Cole.

See Also

I.

For formula manipulation: `terms`, and `all.vars`; for typical use: `lm`, `glm`, and `coplot`.

Examples

```
class(fo <- y ~ x1*x2) # "formula"
fo
typeof(fo) # R internal : "language"
terms(fo)

environment(fo)
environment(as.formula("y ~ x"))
environment(as.formula("y ~ x",env=new.env()))

## Create a formula for a model with a large number of
## variables:
xnam <- paste("x", 1:25, sep="")
(fmla <-
  as.formula(paste("y ~ ", paste(xnam, collapse= "+"))))
```

glm *Fitting Generalized Linear Models*

Description

glm is used to fit generalized linear models, specified by giving a symbolic description of the linear predictor and a description of the error distribution.

Usage

```
glm(formula, family = gaussian, data, weights = NULL,
    subset = NULL, na.action, start = NULL, etastart = NULL,
    mustart = NULL, offset = NULL,
    control = glm.control(...), model = TRUE,
    method = "glm.fit", x = FALSE, y = TRUE,
    contrasts = NULL, ...)

glm.fit(x, y, weights = rep(1, nobs),
    start = NULL, etastart = NULL, mustart = NULL,
    offset = rep(0, nobs), family = gaussian(),
    control = glm.control(), intercept = TRUE)

## S3 method for class 'glm':
weights(object, type = c("prior", "working"), ...)
```

Arguments

formula	a symbolic description of the model to be fit. The details of model specification are given below.
family	a description of the error distribution and link function to be used in the model. This can be a character string naming a family function, a family function or the result of a call to a family function. (See family for details of family functions.)
data	an optional data frame containing the variables in the model. By default the variables are taken from environment(formula), typically the environment from which glm is called.
weights	an optional vector of weights to be used in the fitting process.

subset	an optional vector specifying a subset of observations to be used in the fitting process.
na.action	a function which indicates what should happen when the data contain NAs. The default is set by the na.action setting of options, and is na.fail if that is unset. The "factory-fresh" default is na.omit.
start	starting values for the parameters in the linear predictor.
etastart	starting values for the linear predictor.
mustart	starting values for the vector of means.
offset	this can be used to specify an *a priori* known component to be included in the linear predictor during fitting.
control	a list of parameters for controlling the fitting process. See the documentation for glm.control for details.
model	a logical value indicating whether *model frame* should be included as a component of the returned value.
method	the method to be used in fitting the model. The default method "glm.fit" uses iteratively reweighted least squares (IWLS). The only current alternative is "model.frame" which returns the model frame and does no fitting.
x, y	For glm: logical values indicating whether the response vector and model matrix used in the fitting process should be returned as components of the returned value. For glm.fit: x is a design matrix of dimension n * p, and y is a vector of observations of length n.
contrasts	an optional list. See the contrasts.arg of model.matrix.default.
object	an object inheriting from class "glm".
type	character, partial matching allowed. Type of weights to extract from the fitted model object.
intercept	logical. Should an intercept be included?
...	further arguments passed to or from other methods.

Details

A typical predictor has the form response ~ terms where response is the (numeric) response vector and terms is a series of terms which specifies a linear predictor for response. For binomial models the response can also be specified as a factor (when the first level denotes failure and all others success) or as a two-column matrix with the columns giving the numbers of successes and failures. A terms specification of the form first + second indicates all the terms in first together with all the terms in second with duplicates removed.

A specification of the form first:second indicates the the set of terms obtained by taking the interactions of all terms in first with all terms in second. The specification first*second indicates the *cross* of first and second. This is the same as first + second + first:second.

glm.fit and glm.fit.null are the workhorse functions: the former calls the latter for a null model (with no intercept).

If more than one of etastart, start and mustart is specified, the first in the list will be used.

Value

glm returns an object of class inheriting from "glm" which inherits from the class "lm". See later in this section.

The function summary (i.e., summary.glm) can be used to obtain or print a summary of the results and the function anova (i.e., anova.glm) to produce an analysis of variance table.

The generic accessor functions coefficients, effects, fitted.values and residuals can be used to extract various useful features of the value returned by glm.

weights extracts a vector of weights, one for each case in the fit (after subsetting and na.action).

An object of class "glm" is a list containing at least the following components:

coefficients	a named vector of coefficients
residuals	the *working* residuals, that is the residuals in the final iteration of the IWLS fit.
fitted.values	the fitted mean values, obtained by transforming the linear predictors by the inverse of the link function.
rank	the numeric rank of the fitted linear model.

family	the family object used.
linear.predictors	the linear fit on link scale.
deviance	up to a constant, minus twice the maximized log-likelihood. Where sensible, the constant is chosen so that a saturated model has deviance zero.
aic	Akaike's *An Information Criterion*, minus twice the maximized log-likelihood plus twice the number of coefficients (so assuming that the dispersion is known.
null.deviance	The deviance for the null model, comparable with deviance. The null model will include the offset, and an intercept if there is one in the model
iter	the number of iterations of IWLS used.
weights	the *working* weights, that is the weights in the final iteration of the IWLS fit.
prior.weights	the case weights initially supplied.
df.residual	the residual degrees of freedom.
df.null	the residual degrees of freedom for the null model.
y	the y vector used. (It is a vector even for a binomial model.)
converged	logical. Was the IWLS algorithm judged to have converged?
boundary	logical. Is the fitted value on the boundary of the attainable values?
call	the matched call.
formula	the formula supplied.
terms	the terms object used.
data	the data argument.
offset	the offset vector used.
control	the value of the control argument used.
method	the name of the fitter function used, in R always "glm.fit".
contrasts	(where relevant) the contrasts used.
xlevels	(where relevant) a record of the levels of the factors used in fitting.

In addition, non-empty fits will have components qr, R and effects relating to the final weighted linear fit.

Objects of class "glm" are normally of class c("glm", "lm"), that is inherit from class "lm", and well-designed methods for class "lm" will be applied to the weighted linear model at the final iteration of IWLS. However, care is needed, as extractor functions for class "glm" such as residuals and weights do **not** just pick out the component of the fit with the same name.

If a binomial glm model is specified by giving a two-column response, the weights returned by prior.weights are the total numbers of cases (factored by the supplied case weights) and the component y of the result is the proportion of successes.

Author(s)

The original R implementation of glm was written by Simon Davies working for Ross Ihaka at the University of Auckland, but has since been extensively re-written by members of the R Core team.

The design was inspired by the S function of the same name described in Hastie & Pregibon (1992).

References

Dobson, A. J. (1990) *An Introduction to Generalized Linear Models.* London: Chapman and Hall.

Hastie, T. J. and Pregibon, D. (1992) *Generalized linear models.* Chapter 6 of *Statistical Models in S* eds J. M. Chambers and T. J. Hastie, Wadsworth & Brooks/Cole.

McCullagh P. and Nelder, J. A. (1989) *Generalized Linear Models.* London: Chapman and Hall.

Venables, W. N. and Ripley, B. D. (2002) *Modern Applied Statistics with S.* New York: Springer.

See Also

anova.glm, summary.glm, etc. for glm methods, and the generic functions anova, summary, effects, fitted.values, and residuals. Further, lm for non-generalized *linear* models.

esoph, infert and predict.glm have examples of fitting binomial glms.

Examples

```
## Dobson (1990) Page 93: Randomized Controlled Trial :
counts <- c(18,17,15,20,10,20,25,13,12)
outcome <- gl(3,1,9)
treatment <- gl(3,3)
print(d.AD <- data.frame(treatment, outcome, counts))
glm.D93 <- glm(counts ~ outcome + treatment,
               family=poisson())
anova(glm.D93)
summary(glm.D93)

## an example with offsets from Venables & Ripley (2002,
## p.189)

## Need the anorexia data from a recent version of the
## package 'MASS':
library(MASS)
data(anorexia)

anorex.1 <- glm(Postwt ~ Prewt + Treat + offset(Prewt),
           family = gaussian, data = anorexia)
summary(anorex.1)

# A Gamma example, from McCullagh & Nelder (1989, pp.300-2)
clotting <- data.frame(
    u = c(5,10,15,20,30,40,60,80,100),
    lot1 = c(118,58,42,35,27,25,21,19,18),
    lot2 = c(69,35,26,21,18,16,13,12,12))
summary(glm(lot1 ~ log(u), data=clotting, family=Gamma))
summary(glm(lot2 ~ log(u), data=clotting, family=Gamma))
```

glm.control   *Auxiliary for Controlling GLM Fitting*

Description

Auxiliary function as user interface for `glm` fitting. Typically only used when calling `glm` or `glm.fit`.

Usage

```
glm.control(epsilon=1e-8, maxit=25, trace=FALSE)
```

Arguments

- `epsilon`   positive convergence tolerance *epsilon*; the iterations converge when $|dev - devold|/(|dev| + 0.1) < epsilon$.
- `maxit`   integer giving the maximal number of IWLS iterations.
- `trace`   logical indicating if output should be produced for each iteration.

Details

If `epsilon` is small, it is also used as the tolerance for the least squares solution.

When `trace` is true, calls to `cat` produce the output for each IWLS iteration. Hence, `options(digits = *)` can be used to increase the precision, see the example.

Value

A list with the arguments as components.

References

Hastie, T. J. and Pregibon, D. (1992) *Generalized linear models.* Chapter 6 of *Statistical Models in S* eds J. M. Chambers and T. J. Hastie, Wadsworth & Brooks/Cole.

See Also

`glm.fit`, the fitting procedure used by `glm`.

Examples

```
### A variation on example(glm) :
## Annette Dobson's example ...
counts <- c(18,17,15,20,10,20,25,13,12)
outcome <- gl(3,1,9)
treatment <- gl(3,3)
oo <- options(digits = 12) # to see more when tracing :
glm.D93X <-
  glm(counts ~ outcome + treatment, family=poisson(),
      trace = TRUE, epsilon = 1e-14)
options(oo)
# the last two are closer to 0 than in ?glm's  glm.D93
coef(glm.D93X)
# put less so than in R < 1.8.0 when the default was 1e-4
```

glm.summaries *Accessing Generalized Linear Model Fits*

Description

These functions are all methods for class `glm` or `summary.glm` objects.

Usage

```
## S3 method for class 'glm':
family(object, ...)

## S3 method for class 'glm':
residuals(object,
          type = c("deviance", "pearson", "working",
                   "response", "partial"), ...)
```

Arguments

object	an object of class `glm`, typically the result of a call to `glm`.
type	the type of residuals which should be returned. The alternatives are: `"deviance"` (default), `"pearson"`, `"working"`, `"response"`, and `"partial"`.
...	further arguments passed to or from other methods.

Details

The references define the types of residuals: Davison & Snell is a good reference for the usages of each.

The partial residuals are a matrix of working residuals, with each column formed by omitting a term from the model.

References

Davison, A. C. and Snell, E. J. (1991) *Residuals and diagnostics.* In: Statistical Theory and Modelling. In Honour of Sir David Cox, FRS, eds. Hinkley, D. V., Reid, N. and Snell, E. J., Chapman & Hall.

Hastie, T. J. and Pregibon, D. (1992) *Generalized linear models.* Chapter 6 of *Statistical Models in S* eds J. M. Chambers and T. J. Hastie, Wadsworth & Brooks/Cole.

McCullagh P. and Nelder, J. A. (1989) *Generalized Linear Models*. London: Chapman and Hall.

See Also

`glm` for computing `glm.obj`, `anova.glm`; the corresponding *generic* functions, `summary.glm`, `coef`, `deviance`, `df.residual`, `effects`, `fitted`, `residuals`.

influence.measures   *Regression Deletion Diagnostics*

Description

This suite of functions can be used to compute some of the regression (leave-one-out deletion) diagnostics for linear and generalized linear models discussed in Belsley, Kuh and Welsch (1980), Cook and Weisberg (1982), etc.

Usage

```
influence.measures(model)

rstandard(model, ...)
## S3 method for class 'lm':
rstandard(model, infl = lm.influence(model, do.coef=FALSE),
    sd = sqrt(deviance(model)/df.residual(model)), ...)
## S3 method for class 'glm':
rstandard(model, infl = lm.influence(model, do.coef=FALSE),
          ...)

rstudent(model, ...)
## S3 method for class 'lm':
rstudent(model, infl = lm.influence(model, do.coef=FALSE),
         res = infl$wt.res, ...)
## S3 method for class 'glm':
rstudent(model, infl = influence(model, do.coef=FALSE),
         ...)

dffits(model, infl = , res = )

dfbeta(model, ...)
## S3 method for class 'lm':
dfbeta(model, infl = lm.influence(model, do.coef=TRUE),
       ...)

dfbetas(model, ...)
## S3 method for class 'lm':
dfbetas(model, infl = lm.influence(model, do.coef=TRUE),
        ...)
```

```
covratio(model, infl = lm.influence(model, do.coef=FALSE),
         res = weighted.residuals(model))

cooks.distance(model, ...)
## S3 method for class 'lm':
cooks.distance(model,
  infl = lm.influence(model, do.coef=FALSE),
  res = weighted.residuals(model),
  sd = sqrt(deviance(model)/df.residual(model)),
  hat = infl$hat, ...)
## S3 method for class 'glm':
cooks.distance(model,
  infl = influence(model, do.coef=FALSE),
  res = infl$pear.res,
  dispersion = summary(model)$dispersion,
  hat = infl$hat, ...)

hatvalues(model, ...)
## S3 method for class 'lm':
hatvalues(model,
   infl = lm.influence(model, do.coef=FALSE), ...)

hat(x, intercept = TRUE)
```

Arguments

model	an R object, typically returned by lm or glm.
infl	influence structure as returned by lm.influence or influence (the latter only for the glm method of rstudent and cooks.distance).
res	(possibly weighted) residuals, with proper default.
sd	standard deviation to use, see default.
dispersion	dispersion (for glm objects) to use, see default.
hat	hat values H_{ii}, see default.
x	the X or design matrix.
intercept	should an intercept column be pre-prended to x?
...	further arguments passed to or from other methods.

Details

The primary high-level function is influence.measures which produces a class "infl" object tabular display showing the DFBETAS for each model variable, DFFITS, covariance ratios, Cook's distances and the diagonal elements of the hat matrix. Cases which are influential with respect to any of these measures are marked with an asterisk.

The functions `dfbetas`, `dffits`, `covratio` and `cooks.distance` provide direct access to the corresponding diagnostic quantities. Functions `rstandard` and `rstudent` give the standardized and Studentized residuals respectively. (These re-normalize the residuals to have unit variance, using an overall and leave-one-out measure of the error variance respectively.)

Values for generalized linear models are approximations, as described in Williams (1987) (except that Cook's distances are scaled as F rather than as chi-square values).

The optional `infl`, `res` and `sd` arguments are there to encourage the use of these direct access functions, in situations where, e.g., the underlying basic influence measures (from `lm.influence` or the generic `influence`) are already available.

Note that cases with `weights` == 0 are *dropped* from all these functions, but that if a linear model has been fitted with `na.action = na.exclude`, suitable values are filled in for the cases excluded during fitting.

The function `hat()` exists mainly for S (version 2) compatibility; we recommend using `hatvalues()` instead.

Note

For `hatvalues`, `dfbeta`, and `dfbetas`, the method for linear models also works for generalized linear models.

Author(s)

Several R core team members and John Fox, originally in his 'car' package.

References

Belsley, D. A., Kuh, E. and Welsch, R. E. (1980) *Regression Diagnostics.* New York: Wiley.

Cook, R. D. and Weisberg, S. (1982) *Residuals and Influence in Regression.* London: Chapman and Hall.

Williams, D. A. (1987) Generalized linear model diagnostics using the deviance and single case deletions. *Applied Statistics* **36**, 181–191.

Fox, J. (1997) *Applied Regression, Linear Models, and Related Methods.* Sage.

Fox, J. (2002) *An R and S-Plus Companion to Applied Regression.* Sage Publ.

See Also

influence (containing lm.influence).

Examples

```
## Analysis of the life-cycle savings data given in
## Belsley, Kuh and Welsch.
data(LifeCycleSavings)
lm.SR <- lm(sr ~ pop15 + pop75 + dpi + ddpi,
            data = LifeCycleSavings)

inflm.SR <- influence.measures(lm.SR)
# which observations 'are' influential
which(apply(inflm.SR$is.inf, 1, any))
summary(inflm.SR) # only these
inflm.SR          # all
# recommended by some
plot(rstudent(lm.SR) ~ hatvalues(lm.SR))

## The 'infl' argument is not needed, but avoids
## recomputation:
rs <- rstandard(lm.SR)
iflSR <- influence(lm.SR)
identical(rs, rstandard(lm.SR, infl = iflSR))
## to "see" the larger values:
1000 * round(dfbetas(lm.SR, infl = iflSR), 3)

## Huber's data [Atkinson 1985]
xh <- c(-4:0, 10)
yh <- c(2.48, .73, -.04, -1.44, -1.32, 0)
summary(lmH <- lm(yh ~ xh))
(im <- influence.measures(lmH))
plot(xh, yh,
  main = "Huber's data: L.S. line and influential obs.")
abline(lmH);
points(xh[im$is.inf], yh[im$is.inf], pch=20, col=2)
```

is.empty.model   *Check if a Model is Empty*

Description

R model notation allows models with no intercept and no predictors. These require special handling internally. is.empty.model() checks whether an object describes an empty model.

Usage

```
is.empty.model(x)
```

Arguments

x   A terms object or an object with a terms method.

Value

TRUE if the model is empty

See Also

lm, glm

Examples

```
y <- rnorm(20)
is.empty.model(y ~ 0)
is.empty.model(y ~ -1)
is.empty.model(lm(y ~ 0))
```

labels — *Find Labels from Object*

Description

Find a suitable set of labels from an object for use in printing or plotting, for example. A generic function.

Usage

```
labels(object, ...)
```

Arguments

object   Any R object: the function is generic.

...   further arguments passed to or from other methods.

Value

A character vector or list of such vectors. For a vector the results is the names or seq(along=x), for a data frame or array it is the dimnames (with NULL expanded to seq(len=d[i])), for a terms object it is the term labels and for an lm object it is the term labels for estimable terms.

References

Chambers, J. M. and Hastie, T. J. (1992) *Statistical Models in S.* Wadsworth & Brooks/Cole.

lm   *Fitting Linear Models*

Description

lm is used to fit linear models. It can be used to carry out regression, single stratum analysis of variance and analysis of covariance (although aov may provide a more convenient interface for these).

Usage

```
lm(formula, data, subset, weights, na.action,
   method = "qr", model = TRUE, x = FALSE, y = FALSE,
   qr = TRUE, singular.ok = TRUE, contrasts = NULL,
   offset = NULL, ...)
```

Arguments

formula
: a symbolic description of the model to be fit. The details of model specification are given below.

data
: an optional data frame containing the variables in the model. By default the variables are taken from environment(formula), typically the environment from which lm is called.

subset
: an optional vector specifying a subset of observations to be used in the fitting process.

weights
: an optional vector of weights to be used in the fitting process. If specified, weighted least squares is used with weights weights (that is, minimizing sum(w*e^2)); otherwise ordinary least squares is used.

na.action
: a function which indicates what should happen when the data contain NAs. The default is set by the na.action setting of options, and is na.fail if that is unset. The "factory-fresh" default is na.omit.

method
: the method to be used; for fitting, currently only method="qr" is supported; method="model.frame" returns the model frame (the same as with model = TRUE, see below).

model, x, y, qr
: logicals. If TRUE the corresponding components of the fit (the model frame, the model matrix, the response, the QR decomposition) are returned.

singular.ok	logical. If FALSE (the default in S but not in R) a singular fit is an error.
contrasts	an optional list. See the contrasts.arg of model.matrix.default.
offset	this can be used to specify an *a priori* known component to be included in the linear predictor during fitting. An offset term can be included in the formula instead or as well, and if both are specified their sum is used.
...	additional arguments to be passed to the low level regression fitting functions (see below).

Details

Models for lm are specified symbolically. A typical model has the form response ~ terms where response is the (numeric) response vector and terms is a series of terms which specifies a linear predictor for response. A terms specification of the form first + second indicates all the terms in first together with all the terms in second with duplicates removed. A specification of the form first:second indicates the set of terms obtained by taking the interactions of all terms in first with all terms in second. The specification first*second indicates the *cross* of first and second. This is the same as first + second + first:second. If response is a matrix, a linear model is fitted to each column of the matrix. See model.matrix for some further details.

lm calls the lower level functions lm.fit, etc, see below, for the actual numerical computations. For programming only, you may consider doing likewise.

Value

lm returns an object of class "lm" or for multiple responses of class c("mlm", "lm").

The functions summary and anova are used to obtain and print a summary and analysis of variance table of the results. The generic accessor functions coefficients, effects, fitted.values and residuals extract various useful features of the value returned by lm.

An object of class "lm" is a list containing at least the following components:

coefficients	a named vector of coefficients
residuals	the residuals, that is response minus fitted values.

`fitted.values`	the fitted mean values.
`rank`	the numeric rank of the fitted linear model.
`weights`	(only for weighted fits) the specified weights.
`df.residual`	the residual degrees of freedom.
`call`	the matched call.
`terms`	the `terms` object used.
`contrasts`	(only where relevant) the contrasts used.
`xlevels`	(only where relevant) a record of the levels of the factors used in fitting.
`y`	if requested, the response used.
`x`	if requested, the model matrix used.
`model`	if requested (the default), the model frame used.

In addition, non-null fits will have components `assign`, `effects` and (unless not requested) `qr` relating to the linear fit, for use by extractor functions such as `summary` and `effects`.

Note

Offsets specified by `offset` will not be included in predictions by `predict.lm`, whereas those specified by an offset term in the formula will be.

Author(s)

The design was inspired by the S function of the same name described in Chambers (1992). The implementation of model formula by Ross Ihaka was based on Wilkinson & Rogers (1973).

References

Chambers, J. M. (1992) *Linear models.* Chapter 4 of *Statistical Models in S* eds J. M. Chambers and T. J. Hastie, Wadsworth & Brooks/Cole.

Wilkinson, G. N. and Rogers, C. E. (1973) Symbolic descriptions of factorial models for analysis of variance. *Applied Statistics*, **22**, 392–9.

See Also

`summary.lm` for summaries and `anova.lm` for the ANOVA table; `aov` for a different interface.

The generic functions `coefficients`, `effects`, `residuals`, `fitted.values`.

`predict.lm` (via `predict`) for prediction, including confidence and prediction intervals.

`lm.influence` for regression diagnostics, and `glm` for **generalized linear models**.

The underlying low level functions, `lm.fit` for plain, and `lm.wfit` for weighted regression fitting.

Examples

```
## Annette Dobson (1990) "An Introduction to Generalized
## Linear Models". Page 9: Plant Weight Data.
ctl <- c(4.17,5.58,5.18,6.11,4.50,4.61,5.17,4.53,5.33,5.14)
trt <- c(4.81,4.17,4.41,3.59,5.87,3.83,6.03,4.89,4.32,4.69)
group <- gl(2,10,20, labels=c("Ctl","Trt"))
weight <- c(ctl, trt)
anova(lm.D9 <- lm(weight ~ group))
# omitting intercept
summary(lm.D90 <- lm(weight ~ group - 1))
# residuals almost identical
summary(resid(lm.D9) - resid(lm.D90))

opar <- par(mfrow = c(2,2), oma = c(0, 0, 1.1, 0))
plot(lm.D9, las = 1)      # Residuals, Fitted, ...
par(opar)

## model frame :
stopifnot(
  identical(lm(weight ~ group, method = "model.frame"),
            model.frame(lm.D9))
)
```

lm.fit   *Fitter Functions for Linear Models*

Description

These are the basic computing engines called by lm used to fit linear models. These should usually *not* be used directly unless by experienced users.

Usage

```
lm.fit (x, y,    offset = NULL, method = "qr", tol = 1e-7,
        singular.ok = TRUE, ...)

lm.wfit(x, y, w, offset = NULL, method = "qr", tol = 1e-7,
        singular.ok = TRUE, ...)
```

Arguments

x	design matrix of dimension n * p.
y	vector of observations of length n.
w	vector of weights (length n) to be used in the fitting process for the wfit functions. Weighted least squares is used with weights w, i.e., sum(w * e^2) is minimized.
offset	numeric of length n). This can be used to specify an *a priori* known component to be included in the linear predictor during fitting.
method	currently, only method="qr" is supported.
tol	tolerance for the qr decomposition. Default is 1e-7.
singular.ok	logical. If FALSE, a singular model is an error.
...	currently disregarded.

Details

The functions lm.{w}fit.null are called by lm.fit or lm.wfit respectively, when x has zero columns.

Value

a list with components

coefficients	p vector
residuals	n vector
fitted.values	n vector
effects	(not null fits)n vector of orthogonal single-df effects. The first rank of them correspond to non-aliased coefficients, and are named accordingly.
weights	n vector — *only* for the *wfit* functions.
rank	integer, giving the rank
df.residual	degrees of freedom of residuals
qr	(not null fits) the QR decomposition, see qr.

See Also

lm which you should use for linear least squares regression, unless you know better.

Examples

```
set.seed(129)
n <- 7 ; p <- 2
X <- matrix(rnorm(n * p), n,p) # no intercept!
y <- rnorm(n)
w <- rnorm(n)^2

str(lmw <- lm.wfit(x=X, y=y, w=w))

str(lm. <- lm.fit (x=X, y=y))
```

base — lm.influence

lm.influence *Regression Diagnostics*

Description

This function provides the basic quantities which are used in forming a wide variety of diagnostics for checking the quality of regression fits.

Usage

```
influence(model, ...)
## S3 method for class 'lm':
influence(model, do.coef = TRUE, ...)
## S3 method for class 'glm':
influence(model, do.coef = TRUE, ...)

lm.influence(model, do.coef = TRUE)
```

Arguments

model	an object as returned by `lm`.
do.coef	logical indicating if the changed `coefficients` (see below) are desired. These need $O(n^2 p)$ computing time.
...	further arguments passed to or from other methods.

Details

The `influence.measures()` and other functions listed in **See Also** provide a more user oriented way of computing a variety of regression diagnostics. These all build on `lm.influence`.

An attempt is made to ensure that computed hat values that are probably one are treated as one, and the corresponding rows in `sigma` and `coefficients` are NaN. (Dropping such a case would normally result in a variable being dropped, so it is not possible to give simple drop-one diagnostics.)

Value

A list containing the following components of the same length or number of rows n, which is the number of non-zero weights. Cases omitted in the fit are omitted unless a `na.action` method was used (such as `na.exclude`) which restores them.

hat	a vector containing the diagonal of the "hat" matrix.
coefficients	(unless do.coef is false) a matrix whose i-th row contains the change in the estimated coefficients which results when the i-th case is dropped from the regression. Note that aliased coefficients are not included in the matrix.
sigma	a vector whose i-th element contains the estimate of the residual standard deviation obtained when the i-th case is dropped from the regression.
wt.res	a vector of *weighted* (or for class glm rather *deviance*) residuals.

Note

The coefficients returned by the R version of lm.influence differ from those computed by S. Rather than returning the coefficients which result from dropping each case, we return the changes in the coefficients. This is more directly useful in many diagnostic measures.

Since these need $O(n^2 p)$ computing time, they can be omitted by do.coef = FALSE.

Note that cases with weights == 0 are *dropped* (contrary to the situation in S).

If a model has been fitted with na.action=na.exclude (see na.exclude), cases excluded in the fit *are* considered here.

References

See the list in the documentation for influence.measures.

Chambers, J. M. (1992) *Linear models.* Chapter 4 of *Statistical Models in S* eds J. M. Chambers and T. J. Hastie, Wadsworth & Brooks/Cole.

See Also

summary.lm for summary and related methods;
influence.measures,
hat for the hat matrix diagonals,
dfbetas, dffits, covratio, cooks.distance, lm.

Examples

```
## Analysis of the life-cycle savings data given in
## Belsley, Kuh and Welsch.
```

```
data(LifeCycleSavings)
summary(lm.SR <- lm(sr ~ pop15 + pop75 + dpi + ddpi,
                    data = LifeCycleSavings),
        corr = TRUE)
str(lmI <- lm.influence(lm.SR))

## For more "user level" examples, use
## example(influence.measures)
```

lm.summaries   *Accessing Linear Model Fits*

Description

All these functions are **methods** for class "lm" objects.

Usage

```
## S3 method for class 'lm':
family(object, ...)

## S3 method for class 'lm':
formula(x, ...)

## S3 method for class 'lm':
residuals(object, type = c("working", "response",
   "deviance","pearson", "partial"), ...)

weights(object, ...)
```

Arguments

object, x
: an object inheriting from class lm, usually the result of a call to lm or aov.

...
: further arguments passed to or from other methods.

type
: the type of residuals which should be returned.

Details

The generic accessor functions coef, effects, fitted and residuals can be used to extract various useful features of the value returned by lm.

The working and response residuals are "observed - fitted". The deviance and pearson residuals are weighted residuals, scaled by the square root of the weights used in fitting. The partial residuals are a matrix with each column formed by omitting a term from the model. In all these, zero weight cases are never omitted (as opposed to the standardized rstudent residuals).

References

Chambers, J. M. (1992) *Linear models.* Chapter 4 of *Statistical Models in S* eds J. M. Chambers and T. J. Hastie, Wadsworth & Brooks/Cole.

See Also

The model fitting function lm, anova.lm.

coef, deviance, df.residual, effects, fitted, glm for **generalized** linear models, influence (etc on that page) for regression diagnostics, weighted.residuals, residuals, residuals.glm, summary.lm.

Examples

```
## Continuing the lm(.) example:
coef(lm.D90) # the bare coefficients

## The 2 basic regression diagnostic plots [plot.lm(.) is
## preferred]
plot(resid(lm.D90), fitted(lm.D90)) # Tukey-Anscombe's
abline(h=0, lty=2, col = 'gray')

qqnorm(residuals(lm.D90))
```

logLik   *Extract Log-Likelihood*

Description

This function is generic; method functions can be written to handle specific classes of objects. Classes which already have methods for this function include: `glm`, `lm`, `nls` in package **nls** and `gls`, `lme` and others in package **nlme**.

Usage

```
logLik(object, ...)

## S3 method for class 'logLik':
as.data.frame(x, row.names = NULL, optional = FALSE)
```

Arguments

`object`   any object from which a log-likelihood value, or a contribution to a log-likelihood value, can be extracted.

`...`   some methods for this generic function require additional arguments.

`x`   an object of class `logLik`.

`row.names, optional`   arguments to the `as.data.frame` method; see its documentation.

Value

Returns an object, say r, of class `logLik` which is a number with attributes, `attr(r, "df")` (degrees of freedom) giving the number of parameters in the model. There's a simple `print` method for `logLik` objects.

The details depend on the method function used; see the appropriate documentation.

Author(s)

Jose Pinheiro and Douglas Bates

See Also

logLik.lm, logLik.glm, logLik.gls, logLik.lme, etc.

Examples

```
## see the method function documentation
x <- 1:5
lmx <- lm(x ~ 1)
logLik(lmx) # using print.logLik() method
str(logLik(lmx))
```

logLik.glm   *Extract Log-Likelihood from an glm Object*

Description

Returns the log-likelihood value of the generalized linear model represented by object evaluated at the estimated coefficients.

Usage

```
## S3 method for class 'glm':
logLik(object, ...)
```

Arguments

object   an object inheriting from class "glm".

...   further arguments to be passed to or from methods.

Details

As a **family** does not have to specify how to calculate the log-likelihood, this is based on the family's function to compute the AIC. For **gaussian**, **Gamma** and **inverse.gaussian** families it assumed that the dispersion of the GLM is estimated and has been included in the AIC, and for all other families it is assumed that the dispersion is known.

Not that this procedure is not completely accurate for the gamma and inverse gaussian families, as the estimate of dispersion used is not the MLE.

Value

the log-likelihood of the linear model represented by object evaluated at the estimated coefficients.

See Also

glm, logLik.lm

logLik.lm   *Extract Log-Likelihood from an lm Object*

Description

If `REML = FALSE`, returns the log-likelihood value of the linear model represented by `object` evaluated at the estimated coefficients; else, the restricted log-likelihood evaluated at the estimated coefficients is returned.

Usage

```
## S3 method for class 'lm':
logLik(object, REML = FALSE, ...)
```

Arguments

object
: an object inheriting from class `"lm"`.

REML
: an optional logical value. If `TRUE` the restricted log-likelihood is returned, else, if `FALSE`, the log-likelihood is returned. Defaults to `FALSE`.

...
: further arguments to be passed to or from methods.

Value

an object of class `logLik`, the (restricted) log-likelihood of the linear model represented by `object` evaluated at the estimated coefficients. Note that error variance σ^2 is estimated in `lm()` and hence counted as well.

Author(s)

Jose Pinheiro and Douglas Bates

References

Harville, D.A. (1974). Bayesian inference for variance components using only error contrasts. *Biometrika*, **61**, 383–385.

See Also

lm

Examples

```
data(attitude)
(fm1 <- lm(rating ~ ., data = attitude))
logLik(fm1)
logLik(fm1, REML = TRUE)

res <- try(data(Orthodont, package="nlme"))
if(!inherits(res, "try-error")) {
  fm1 <- lm(distance ~ Sex * age, Orthodont)
  print(logLik(fm1))
  print(logLik(fm1, REML = TRUE))
}
```

loglin *Fitting Log-Linear Models*

Description

loglin is used to fit log-linear models to multidimensional contingency tables by Iterative Proportional Fitting.

Usage

```
loglin(table, margin, start = rep(1, length(table)),
   fit = FALSE, eps = 0.1, iter = 20, param = FALSE,
   print = TRUE)
```

Arguments

table
: a contingency table to be fit, typically the output from table.

margin
: a list of vectors with the marginal totals to be fit.

    (Hierarchical) log-linear models can be specified in terms of these marginal totals which give the "maximal" factor subsets contained in the model. For example, in a three-factor model, list(c(1, 2), c(1, 3)) specifies a model which contains parameters for the grand mean, each factor, and the 1-2 and 1-3 interactions, respectively (but no 2-3 or 1-2-3 interaction), i.e., a model where factors 2 and 3 are independent conditional on factor 1 (sometimes represented as '[12][13]').

    The names of factors (i.e., names(dimnames(table))) may be used rather than numeric indices.

start
: a starting estimate for the fitted table. This optional argument is important for incomplete tables with structural zeros in table which should be preserved in the fit. In this case, the corresponding entries in start should be zero and the others can be taken as one.

fit
: a logical indicating whether the fitted values should be returned.

eps
: maximum deviation allowed between observed and fitted margins.

iter	maximum number of iterations.
param	a logical indicating whether the parameter values should be returned.
print	a logical. If TRUE, the number of iterations and the final deviation are printed.

Details

The Iterative Proportional Fitting algorithm as presented in Haberman (1972) is used for fitting the model. At most `iter` iterations are performed, convergence is taken to occur when the maximum deviation between observed and fitted margins is less than `eps`. All internal computations are done in double precision; there is no limit on the number of factors (the dimension of the table) in the model.

Assuming that there are no structural zeros, both the Likelihood Ratio Test and Pearson test statistics have an asymptotic chi-squared distribution with `df` degrees of freedom.

Package **MASS** contains `loglm`, a front-end to `loglin` which allows the log-linear model to be specified and fitted in a formula-based manner similar to that of other fitting functions such as `lm` or `glm`.

Value

A list with the following components.

lrt	the Likelihood Ratio Test statistic.
pearson	the Pearson test statistic (X-squared).
df	the degrees of freedom for the fitted model. There is no adjustment for structural zeros.
margin	list of the margins that were fit. Basically the same as the input `margin`, but with numbers replaced by names where possible.
fit	An array like `table` containing the fitted values. Only returned if `fit` is TRUE.
param	A list containing the estimated parameters of the model. The "standard" constraints of zero marginal sums (e.g., zero row and column sums for a two factor parameter) are employed. Only returned if `param` is TRUE.

Author(s)

Kurt Hornik

References

Becker, R. A., Chambers, J. M. and Wilks, A. R. (1988) *The New S Language.* Wadsworth & Brooks/Cole.

Haberman, S. J. (1972) Log-linear fit for contingency tables—Algorithm AS51. *Applied Statistics*, **21**, 218–225.

Agresti, A. (1990) *Categorical data analysis.* New York: Wiley.

See Also

`table`

Examples

```
data(HairEyeColor)
## Model of joint independence of sex from hair and eye
## color.
fm <- loglin(HairEyeColor, list(c(1, 2), c(1, 3), c(2, 3)))
fm
1 - pchisq(fm$lrt, fm$df)
## Model with no three-factor interactions fits well.
```

ls.diag   *Compute Diagnostics for 'lsfit' Regression Results*

Description

Computes basic statistics, including standard errors, t- and p-values for the regression coefficients.

Usage

```
ls.diag(ls.out)
```

Arguments

ls.out   Typically the result of `lsfit()`

Value

A `list` with the following numeric components.

std.dev	The standard deviation of the errors, an estimate of σ.
hat	diagonal entries h_{ii} of the hat matrix H
std.res	standardized residuals
stud.res	studentized residuals
cooks	Cook's distances
dfits	DFITS statistics
correlation	correlation matrix
std.err	standard errors of the regression coefficients
cov.scaled	Scaled covariance matrix of the coefficients
cov.unscaled	Unscaled covariance matrix of the coefficients

References

Belsley, D. A., Kuh, E. and Welsch, R. E. (1980) *Regression Diagnostics.* New York: Wiley.

See Also

`hat` for the hat matrix diagonals, `ls.print`, `lm.influence`, `summary`. `lm`, `anova`.

Examples

```
## Using the same data as the lm(.) example:
lsD9 <- lsfit(x = as.numeric(gl(2, 10, 20)), y = weight)
dlsD9 <- ls.diag(lsD9)
str(dlsD9, give.attr=FALSE)
# sum(h.ii) = p
abs(1 - sum(dlsD9$hat) / 2) < 10*.Machine$double.eps
plot(dlsD9$hat, dlsD9$stud.res, xlim=c(0,0.11))
abline(h = 0, lty = 2, col = "lightgray")
```

ls.print   *Print 'lsfit' Regression Results*

Description

Computes basic statistics, including standard errors, t- and p-values for the regression coefficients and prints them if print.it is TRUE.

Usage

```
ls.print(ls.out, digits = 4, print.it = TRUE)
```

Arguments

ls.out	Typically the result of lsfit()
digits	The number of significant digits used for printing
print.it	a logical indicating whether the result should also be printed

Value

A list with the components

summary	The ANOVA table of the regression
coef.table	matrix with regression coefficients, standard errors, t- and p-values

Note

Usually, you'd rather use summary(lm(...)) and anova(lm(...)) for obtaining similar output.

See Also

ls.diag, lsfit, also for examples; lm, lm.influence which usually are preferable.

lsfit   *Find the Least Squares Fit*

Description

The least squares estimate of β in the model

$$Y = X\beta + \epsilon$$

is found.

Usage

```
lsfit(x, y, wt=NULL, intercept=TRUE, tolerance=1e-07,
    yname=NULL)
```

Arguments

x
: a matrix whose rows correspond to cases and whose columns correspond to variables.

y
: the responses, possibly a matrix if you want to fit multiple left hand sides.

wt
: an optional vector of weights for performing weighted least squares.

intercept
: whether or not an intercept term should be used.

tolerance
: the tolerance to be used in the matrix decomposition.

yname
: names to be used for the response variables.

Details

If weights are specified then a weighted least squares is performed with the weight given to the jth case specified by the jth entry in wt.

If any observation has a missing value in any field, that observation is removed before the analysis is carried out. This can be quite inefficient if there is a lot of missing data.

The implementation is via a modification of the LINPACK subroutines which allow for multiple left-hand sides.

Value

A list with the following named components:

coef
: the least squares estimates of the coefficients in the model (β as stated above).

residuals
: residuals from the fit.

intercept
: indicates whether an intercept was fitted.

qr
: the QR decomposition of the design matrix.

References

Becker, R. A., Chambers, J. M. and Wilks, A. R. (1988) *The New S Language.* Wadsworth & Brooks/Cole.

See Also

lm which usually is preferable; ls.print, ls.diag.

Examples

```
## Using the same data as the lm(.) example:
lsD9 <- lsfit(x = unclass(gl(2,10)), y = weight)
ls.print(lsD9)
```

make.link   *Create a Link for GLM families*

Description

This function is used with the `family` functions in `glm()`. Given a link, it returns a link function, an inverse link function, the derivative $d\mu/d\eta$ and a function for domain checking.

Usage

```
make.link(link)
```

Arguments

link   character or numeric; one of "logit", "probit", "cloglog", "identity", "log", "sqrt", "1/mu^2", "inverse", or number, say λ resulting in power link $= \mu^\lambda$.

Value

A list with components

linkfun   Link function `function(mu)`
linkinv   Inverse link function `function(eta)`
mu.eta   Derivative `function(eta)` $d\mu/d\eta$
valideta   `function(eta){ TRUE if all of eta is in the domain of linkinv }`.

See Also

`glm`, `family`.

Examples

```
str(make.link("logit"))

l2 <- make.link(2)
l2$linkfun(0:3) # 0 1 4 9
l2$mu.eta(eta= 1:2) # = 1/(2*sqrt(eta))
```

makepredictcall   *Utility Function for Safe Prediction*

Description

A utility to help `model.frame.default` create the right matrices when predicting from models with terms like `poly` or `ns`.

Usage

```
makepredictcall(var, call)
```

Arguments

var   A variable.

call   The term in the formula, as a call.

Details

This is a generic function with methods for `poly`, `bs` and `ns`: the default method handles `scale`. If `model.frame.default` encounters such a term when creating a model frame, it modifies the `predvars` attribute of the terms supplied to replace the term with one that will work for predicting new data. For example `makepredictcall.ns` adds arguments for the knots and intercept.

To make use of this, have your model-fitting function return the `terms` attribute of the model frame, or copy the `predvars` attribute of the `terms` attribute of the model frame to your `terms` object.

To extend this, make sure the term creates variables with a class, and write a suitable method for that class.

Value

A replacement for `call` for the `predvars` attribute of the terms.

See Also

`model.frame`, `poly`, `scale`, `bs`, `ns`, `cars`

Examples

```
## using poly: this did not work in R < 1.5.0
data(women)
fm <- lm(weight ~ poly(height, 2), data = women)
plot(women, xlab = "Height (in)", ylab = "Weight (lb)")
ht <- seq(57, 73, len = 200)
lines(ht, predict(fm, data.frame(height=ht)))

## see also example(cars)
## see bs and ns for spline examples.
```

manova	*Multivariate Analysis of Variance*

Description

A class for the multivariate analysis of variance.

Usage

```
manova(...)
```

Arguments

...   Arguments to be passed to aov.

Details

Class "manova" differs from class "aov" in selecting a different summary method. Function manova calls aov and then add class "manova" to the result object for each stratum.

Value

See aov and the comments in Details here.

Note

manova does not support multistratum analysis of variance, so the formula should not include an Error term.

References

Krzanowski, W. J. (1988) *Principles of Multivariate Analysis. A User's Perspective.* Oxford.

Hand, D. J. and Taylor, C. C. (1987) *Multivariate Analysis of Variance and Repeated Measures.* Chapman and Hall.

See Also

aov, summary.manova, the latter containing examples.

model.extract   *Extract Components from a Model Frame*

Description

Returns the response, offset, subset, weights or other special components of a model frame passed as optional arguments to `model.frame`.

Usage

```
model.extract(frame, component)
model.offset(x)
model.response(data, type = "any")
model.weights(x)
```

Arguments

frame, x, data
: A model frame.

component
: literal character string or name. The name of a component to extract, such as `"weights"`, `"subset"`.

type
: One of `"any"`, `"numeric"`, `"double"`. Using the either of latter two coerces the result to have storage mode `"double"`.

Details

`model.extract` is provided for compatibility with S, which does not have the more specific functions.

`model.offset` and `model.response` are equivalent to `model.frame(, "offset")` and `model.frame(, "response")` respectively.

`model.weights` is slightly different from `model.frame(, "weights")` in not naming the vector it returns.

Value

The specified component of the model frame, usually a vector.

See Also

`model.frame`, `offset`

Examples

```
data(esoph)
a <-
  model.frame(cbind(ncases,ncontrols) ~ agegp+tobgp+alcgp,
              data=esoph)
model.extract(a, "response")
stopifnot(
  model.extract(a, "response") == model.response(a)
)

a <-
  model.frame(ncases/(ncases+ncontrols) ~ agegp+tobgp+alcgp,
              data = esoph, weights = ncases+ncontrols)
model.response(a)
model.extract(a, "weights")

a <- model.frame(cbind(ncases,ncontrols) ~ agegp,
                 something = tobgp, data = esoph)
names(a)
stopifnot(model.extract(a, "something") == esoph$tobgp)
```

model.frame   *Extracting the "Environment" of a Model Formula*

Description

model.frame (a generic function) and its methods return a data.frame with the variables needed to use formula and any ... arguments.

Usage

```
model.frame(formula, ...)

## Default S3 method:
model.frame(formula, data = NULL,
            subset = NULL, na.action = na.fail,
            drop.unused.levels = FALSE, xlev = NULL, ...)

## S3 method for class 'aovlist':
model.frame(formula, data = NULL, ...)

## S3 method for class 'glm':
model.frame(formula, data, na.action, ...)

## S3 method for class 'lm':
model.frame(formula, data, na.action, ...)
```

Arguments

formula	a model formula
data	data.frame, list, environment or object coercible to data.frame containing the variables in formula.
subset	a specification of the rows to be used: defaults to all rows. This can be any valid indexing vector (see [.data.frame for the rows of data or if that is not supplied, a data frame made up of the variables used in formula).
na.action	how NAs are treated. The default is first, any na.action attribute of data, second a na.action setting of options, and third na.fail if that is unset. The "factory-fresh" default is na.omit.

`drop.unused.levels`
: should factors have unused levels dropped? Defaults to FALSE.

`xlev`
: a named list of character vectors giving the full set of levels to be assumed for each factor.

`...`
: further arguments such as `subset`, `offset` and `weights`. NULL arguments are treated as missing.

Details

Variables in the formula, `subset` and in ... are looked for first in `data` and then in the environment of `formula`: see the help for `formula()` for further details.

First all the variables needed are collected into a data frame. Then `subset` expression is evaluated, and it is used as a row index to the data frame. Then the `na.action` function is applied to the data frame (and may well add attributes). The levels of any factors in the data frame are adjusted according to the `drop.unused.levels` and `xlev` arguments.

Value

A `data.frame` containing the variables used in `formula` plus those specified ....

References

Chambers, J. M. (1992) *Data for models.* Chapter 3 of *Statistical Models in S* eds J. M. Chambers and T. J. Hastie, Wadsworth & Brooks/Cole.

See Also

`model.matrix` for the "design matrix", `formula` for formulas and `expand.model.frame` for model.frame manipulation.

Examples

```
data(cars)
data.class(model.frame(dist ~ speed, data = cars))
```

model.matrix   *Construct Design Matrices*

Description

model.matrix creates a design matrix.

Usage

```
model.matrix(object, ...)

## Default S3 method:
model.matrix(object, data = environment(object),
             contrasts.arg = NULL, xlev = NULL, ...)
```

Arguments

object	an object of an appropriate class. For the default method, a model formula or terms object.
data	a data frame created with model.frame.
contrasts.arg	A list, whose entries are contrasts suitable for input to the contrasts replacement function and whose names are the names of columns of data containing factors.
xlev	to be used as argument of model.frame if data has no "terms" attribute.
...	further arguments passed to or from other methods.

Details

model.matrix creates a design matrix from the description given in terms(formula), using the data in data which must contain columns with the same names as would be created by a call to model.frame(formula) or, more precisely, by evaluating attr(terms(formula), "variables"). There may be other columns and the order is not important. If contrasts is specified it overrides the default factor coding for that variable.

In interactions, the variable whose levels vary fastest is the first one to appear in the formula (and not in the term), so in ~ a + b + b:a the interaction will have a varying fastest.

By convention, if the response variable also appears on the right-hand side of the formula it is dropped (with a warning), although interactions involving the term are retained.

Value

The design matrix for a regression model with the specified formula and data.

References

Chambers, J. M. (1992) *Data for models.* Chapter 3 of *Statistical Models in S* eds J. M. Chambers and T. J. Hastie, Wadsworth & Brooks/Cole.

See Also

`model.frame, model.extract, terms`

Examples

```
data(trees)
ff <- log(Volume) ~ log(Height) + log(Girth)
str(m <- model.frame(ff, trees))
mat <- model.matrix(ff, m)

# balanced 2-way
dd <- data.frame(a = gl(3,4), b = gl(4,1,12))
options("contrasts")
model.matrix(~ a + b, dd)
model.matrix(~ a + b, dd, contrasts = list(a="contr.sum"))
model.matrix(~ a + b, dd, contrasts = list(a="contr.sum",
             b="contr.poly"))
m.orth <- model.matrix(~a+b, dd,
  contrasts = list(a="contr.helmert"))
crossprod(m.orth) # m.orth is  ALMOST orthogonal
```

model.tables   *Compute Tables of Results from an Aov Model Fit*

Description

Computes summary tables for model fits, especially complex `aov` fits.

Usage

```
model.tables(x, ...)

## S3 method for class 'aov':
model.tables(x, type = "effects", se = FALSE, cterms, ...)

## S3 method for class 'aovlist':
model.tables(x, type = "effects", se = FALSE, ...)
```

Arguments

x	a model object, usually produced by `aov`
type	type of table: currently only `"effects"` and `"means"` are implemented.
se	should standard errors be computed?
cterms	A character vector giving the names of the terms for which tables should be computed. The default is all tables.
...	further arguments passed to or from other methods.

Details

For `type = "effects"` give tables of the coefficients for each term, optionally with standard errors.

For `type = "means"` give tables of the mean response for each combinations of levels of the factors in a term.

Value

An object of class `"tables.aov"`, as list which may contain components

tables	A list of tables for each requested term.
n	The replication information for each term.
se	Standard error information.

Warning

The implementation is incomplete, and only the simpler cases have been tested thoroughly.

Weighted aov fits are not supported.

See Also

aov, proj, replications, TukeyHSD, se.contrast

Examples

```
## From Venables and Ripley (2002) p.165.
N <- c(0,1,0,1,1,1,0,0,0,1,1,0,1,1,0,0,1,0,1,0,1,1,0,0)
P <- c(1,1,0,0,0,1,0,1,1,1,0,0,0,1,0,1,1,0,0,1,0,1,1,0)
K <- c(1,0,0,1,0,1,1,0,0,1,0,1,0,1,1,0,0,1,1,1,0,1,0)
yield <- c(49.5,62.8,46.8,57.0,59.8,58.5,55.5,56.0,
           62.8,55.8,69.5,55.0,62.0,48.8,45.5,44.2,
           52.0,51.5,49.8,48.8,57.2,59.0,53.2,56.0)

npk <- data.frame(block=gl(6,4), N=factor(N), P=factor(P),
                  K=factor(K), yield=yield)
npk.aov <- aov(yield ~ block + N*P*K, npk)
model.tables(npk.aov, "means", se=TRUE)

## as a test, not particularly sensible statistically
options(contrasts=c("contr.helmert", "contr.treatment"))
npk.aovE <- aov(yield ~  N*P*K + Error(block), npk)
model.tables(npk.aovE,   se=TRUE)
model.tables(npk.aovE, "means")
```

naprint *Adjust for Missing Values*

Description

Use missing value information to report the effects of an `na.action`.

Usage

```
naprint(x, ...)
```

Arguments

x
: An object produced by an `na.action` function.

...
: further arguments passed to or from other methods.

Details

This is a generic function, and the exact information differs by method. `naprint.omit` reports the number of rows omitted: `naprint.default` reports an empty string.

Value

A character string providing information on missing values, for example the number.

naresid   *Adjust for Missing Values*

Description
Use missing value information to adjust residuals and predictions.

Usage
```
naresid(omit, x, ...)
napredict(omit, x, ...)
```

Arguments

omit
: An object produced by an na.action function.

x
: A vector, data frame, or matrix to be adjusted based upon the missing value information.

...
: further arguments passed to or from other methods.

Details
These are utility functions used to allow predict and resid methods for modelling functions to compensate for the removal of NAs in the fitting process. These are used by the default "lm" and "glm" methods, and by further methods in packages **MASS**, **rpart** and **survival**.

The default methods do nothing. The default method for the na.exclude action is to pad the object with NAs in the correct positions to have the same number of rows as the original data frame.

Currently naresid and napredict are identical, but future methods need not be. naresid is used for residuals, and napredict for fitted values and predictions.

Value
These return a similar object to x.

Note
Packages **rpart** and **survival5** used to contain versions of these functions that had an na.omit action equivalent to that now used for na.exclude.

nlm   *Non-Linear Minimization*

Description

This function carries out a minimization of the function f using a Newton-type algorithm. See the references for details.

Usage

```
nlm(f, p, hessian = FALSE, typsize = rep(1, length(p)),
    fscale = 1, print.level = 0, ndigit = 12, gradtol = 1e-6,
    stepmax = max(1000 * sqrt(sum((p/typsize)^2)), 1000),
    steptol = 1e-6, iterlim = 100, check.analyticals = TRUE,
    ...)
```

Arguments

f	the function to be minimized. If the function value has an attribute called gradient or both gradient and hessian attributes, these will be used in the calculation of updated parameter values. Otherwise, numerical derivatives are used. deriv returns a function with suitable gradient attribute. This should be a function of a vector of the length of p followed by any other arguments specified in dots.
p	starting parameter values for the minimization.
hessian	if TRUE, the hessian of f at the minimum is returned.
typsize	an estimate of the size of each parameter at the minimum.
fscale	an estimate of the size of f at the minimum.
print.level	this argument determines the level of printing which is done during the minimization process. The default value of 0 means that no printing occurs, a value of 1 means that initial and final details are printed and a value of 2 means that full tracing information is printed.
ndigit	the number of significant digits in the function f.
gradtol	a positive scalar giving the tolerance at which the scaled gradient is considered close enough to zero to terminate the algorithm. The scaled gradient is a

measure of the relative change in f in each direction p[i] divided by the relative change in p[i].

stepmax
a positive scalar which gives the maximum allowable scaled step length. stepmax is used to prevent steps which would cause the optimization function to overflow, to prevent the algorithm from leaving the area of interest in parameter space, or to detect divergence in the algorithm. stepmax would be chosen small enough to prevent the first two of these occurrences, but should be larger than any anticipated reasonable step.

steptol
A positive scalar providing the minimum allowable relative step length.

iterlim
a positive integer specifying the maximum number of iterations to be performed before the program is terminated.

check.analyticals
a logical scalar specifying whether the analytic gradients and Hessians, if they are supplied, should be checked against numerical derivatives at the initial parameter values. This can help detect incorrectly formulated gradients or Hessians.

...
additional arguments to f.

Details

If a gradient or hessian is supplied but evaluates to the wrong mode or length, it will be ignored if check.analyticals = TRUE (the default) with a warning. The hessian is not even checked unless the gradient is present and passes the sanity checks.

From the three methods available in the original source, we always use method "1" which is line search.

Value

A list containing the following components:

minimum
the value of the estimated minimum of f.

estimate
the point at which the minimum value of f is obtained.

gradient
the gradient at the estimated minimum of f.

hessian
the hessian at the estimated minimum of f (if requested).

`code`   an integer indicating why the optimization process terminated.

  1: relative gradient is close to zero, current iterate is probably solution.

  2: successive iterates within tolerance, current iterate is probably solution.

  3: last global step failed to locate a point lower than `estimate`. Either `estimate` is an approximate local minimum of the function or `steptol` is too small.

  4: iteration limit exceeded.

  5: maximum step size `stepmax` exceeded five consecutive times. Either the function is unbounded below, becomes asymptotic to a finite value from above in some direction or `stepmax` is too small.

`iterations`   the number of iterations performed.

References

Dennis, J. E. and Schnabel, R. B. (1983) *Numerical Methods for Unconstrained Optimization and Nonlinear Equations.* Prentice-Hall, Englewood Cliffs, NJ.

Schnabel, R. B., Koontz, J. E. and Weiss, B. E. (1985) A modular system of algorithms for unconstrained minimization. *ACM Trans. Math. Software*, **11**, 419–440.

See Also

`optim`. `optimize` for one-dimensional minimization and `uniroot` for root finding. `deriv` to calculate analytical derivatives.

For nonlinear regression, `nls` (in package **nls**), may be of better use.

Examples

```
f <- function(x) sum((x-1:length(x))^2)
nlm(f, c(10,10))
nlm(f, c(10,10), print.level = 2)
str(nlm(f, c(5), hessian = TRUE))

f <- function(x, a) sum((x-a)^2)
nlm(f, c(10,10), a=c(3,5))
f <- function(x, a)
```

```
{
    res <- sum((x-a)^2)
    attr(res, "gradient") <- 2*(x-a)
    res
}
nlm(f, c(10,10), a=c(3,5))

## more examples, including the use of derivatives.
demo(nlm)
```

offset — Include an Offset in a Model Formula

Description

An offset is a term to be added to a linear predictor, such as in a generalised linear model, with known coefficient 1 rather than an estimated coefficient.

Usage

```
offset(object)
```

Arguments

object   An offset to be included in a model frame

Value

The input value.

See Also

`model.offset`, `model.frame`.

For examples see `glm`, `Insurance`.

optim   *General-purpose Optimization*

Description

General-purpose optimization based on Nelder–Mead, quasi-Newton and conjugate-gradient algorithms. It includes an option for box-constrained optimization and simulated annealing.

Usage

```
optim(par, fn, gr = NULL,
  method = c("Nelder-Mead","BFGS","CG","L-BFGS-B","SANN"),
  lower = -Inf, upper = Inf,
  control = list(), hessian = FALSE, ...)
```

Arguments

par	Initial values for the parameters to be optimized over.
fn	A function to be minimized (or maximized), with first argument the vector of parameters over which minimization is to take place. It should return a scalar result.
gr	A function to return the gradient for the `"BFGS"`, `"CG"` and `"L-BFGS-B"` methods. If it is `NULL`, a finite-difference approximation will be used.
	For the `"SANN"` method it specifies a function to generate a new candidate point. If it is `NULL` a default Gaussian Markov kernel is used.
method	The method to be used. See **Details**.
lower, upper	Bounds on the variables for the `"L-BFGS-B"` method.
control	A list of control parameters. See **Details**.
hessian	Logical. Should a numerically differentiated Hessian matrix be returned?
...	Further arguments to be passed to `fn` and `gr`.

Details

By default this function performs minimization, but it will maximize if `control$fnscale` is negative.

The default method is an implementation of that of Nelder and Mead (1965), that uses only function values and is robust but relatively slow. It will work reasonably well for non-differentiable functions.

Method "BFGS" is a quasi-Newton method (also known as a variable metric algorithm), specifically that published simultaneously in 1970 by Broyden, Fletcher, Goldfarb and Shanno. This uses function values and gradients to build up a picture of the surface to be optimized.

Method "CG" is a conjugate gradients method based on that by Fletcher and Reeves (1964) (but with the option of Polak–Ribiere or Beale–Sorenson updates). Conjugate gradient methods will generally be more fragile that the BFGS method, but as they do not store a matrix they may be successful in much larger optimization problems.

Method "L-BFGS-B" is that of Byrd *et. al.* (1994) which allows *box constraints*, that is each variable can be given a lower and/or upper bound. The initial value must satisfy the constraints. This uses a limited-memory modification of the BFGS quasi-Newton method. If non-trivial bounds are supplied, this method will be selected, with a warning.

Nocedal and Wright (1999) is a comprehensive reference for the previous three methods.

Method "SANN" is by default a variant of simulated annealing given in Belisle (1992). Simulated-annealing belongs to the class of stochastic global optimization methods. It uses only function values but is relatively slow. It will also work for non-differentiable functions. This implementation uses the Metropolis function for the acceptance probability. By default the next candidate point is generated from a Gaussian Markov kernel with scale proportional to the actual temperature. If a function to generate a new candidate point is given, method "SANN" can also be used to solve combinatorial optimization problems. Temperatures are decreased according to the logarithmic cooling schedule as given in Belisle (1992, p. 890). Note that the "SANN" method depends critically on the settings of the control parameters. It is not a general-purpose method but can be very useful in getting to a good value on a very rough surface.

Function `fn` can return `NA` or `Inf` if the function cannot be evaluated at the supplied value, but the initial value must have a computable finite value of `fn`. (Except for method "L-BFGS-B" where the values should always be finite.)

`optim` can be used recursively, and for a single parameter as well as many.

The `control` argument is a list that can supply any of the following components:

`trace` Non-negative integer. If positive, tracing information on the progress of the optimization is produced. Higher values may produce more tracing information: for method "L-BFGS-B" there are six levels of tracing. (To understand exactly what these do see the source code: higher levels give more detail.)

`fnscale` An overall scaling to be applied to the value of `fn` and `gr` during optimization. If negative, turns the problem into a maximization problem. Optimization is performed on `fn(par)/fnscale`.

`parscale` A vector of scaling values for the parameters. Optimization is performed on `par/parscale` and these should be comparable in the sense that a unit change in any element produces about a unit change in the scaled value.

`ndeps` A vector of step sizes for the finite-difference approximation to the gradient, on `par/parscale` scale. Defaults to 1e-3.

`maxit` The maximum number of iterations. Defaults to 100 for the derivative-based methods, and 500 for "Nelder-Mead". For "SANN" `maxit` gives the total number of function evaluations. There is no other stopping criterion. Defaults to 10000.

`abstol` The absolute convergence tolerance. Only useful for non-negative functions, as a tolerance for reaching zero.

`reltol` Relative convergence tolerance. The algorithm stops if it is unable to reduce the value by a factor of `reltol * (abs(val) + reltol)` at a step. Defaults to `sqrt(.Machine$double.eps)`, typically about 1e-8.

`alpha, beta, gamma` Scaling parameters for the "Nelder-Mead" method. `alpha` is the reflection factor (default 1.0), `beta` the contraction factor (0.5) and `gamma` the expansion factor (2.0).

`REPORT` The frequency of reports for the "BFGS" and "L-BFGS-B" methods if `control$trace` is positive. Defaults to every 10 iterations.

`type` for the conjugate-gradients method. Takes value 1 for the Fletcher–Reeves update, 2 for Polak–Ribiere and 3 for Beale–Sorenson.

`lmm` is an integer giving the number of BFGS updates retained in the "L-BFGS-B" method, It defaults to 5.

`factr` controls the convergence of the "L-BFGS-B" method. Convergence occurs when the reduction in the objective is within this factor of the machine tolerance. Default is 1e7, that is a tolerance of about 1e-8.

pgtol helps to control the convergence of the "L-BFGS-B" method. It is a tolerance on the projected gradient in the current search direction. This defaults to zero, when the check is suppressed.

temp controls the "SANN" method. It is the starting temperature for the cooling schedule. Defaults to 10.

tmax is the number of function evaluations at each temperature for the "SANN" method. Defaults to 10.

Value

A list with components:

par   The best set of parameters found.

value   The value of fn corresponding to par.

counts   A two-element integer vector giving the number of calls to fn and gr respectively. This excludes those calls needed to compute the Hessian, if requested, and any calls to fn to compute a finite-difference approximation to the gradient.

convergence   An integer code. 0 indicates successful convergence. Error codes are

    1 indicates that the iteration limit maxit had been reached.

    10 indicates degeneracy of the Nelder–Mead simplex.

    51 indicates a warning from the "L-BFGS-B" method; see component message for further details.

    52 indicates an error from the "L-BFGS-B" method; see component message for further details.

message   A character string giving any additional information returned by the optimizer, or NULL.

hessian   Only if argument hessian is true. A symmetric matrix giving an estimate of the Hessian at the solution found. Note that this is the Hessian of the unconstrained problem even if the box constraints are active.

Note

optim will work with one-dimensional pars, but the default method does not work well (and will warn). Use optimize instead.

The code for methods "Nelder-Mead", "BFGS" and "CG" was based originally on Pascal code in Nash (1990) that was translated by p2c and then hand-optimized. Dr Nash has agreed that the code can be made freely available.

The code for method "L-BFGS-B" is based on Fortran code by Zhu, Byrd, Lu-Chen and Nocedal obtained from Netlib (file 'opt/lbfgs_bcm.shar': another version is in 'toms/778').

The code for method "SANN" was contributed by A. Trapletti.

References

Belisle, C. J. P. (1992) Convergence theorems for a class of simulated annealing algorithms on R^d. *J Applied Probability*, **29**, 885–895.

Byrd, R. H., Lu, P., Nocedal, J. and Zhu, C. (1995) A limited memory algorithm for bound constrained optimization. *SIAM J. Scientific Computing*, **16**, 1190–1208.

Fletcher, R. and Reeves, C. M. (1964) Function minimization by conjugate gradients. *Computer Journal* **7**, 148–154.

Nash, J. C. (1990) *Compact Numerical Methods for Computers. Linear Algebra and Function Minimisation.* Adam Hilger.

Nelder, J. A. and Mead, R. (1965) A simplex algorithm for function minimization. *Computer Journal* **7**, 308–313.

Nocedal, J. and Wright, S. J. (1999) *Numerical Optimization.* Springer.

See Also

`nlm`, `optimize`, `constrOptim`

Examples

```
fr <- function(x) {    ## Rosenbrock Banana function
    x1 <- x[1]
    x2 <- x[2]
    100 * (x2 - x1 * x1)^2 + (1 - x1)^2
}
grr <- function(x) { ## Gradient of 'fr'
    x1 <- x[1]
    x2 <- x[2]
    c(-400 * x1 * (x2 - x1 * x1) - 2 * (1 - x1),
       200 *      (x2 - x1 * x1))
}
```

base — optim

```
optim(c(-1.2,1), fr)
optim(c(-1.2,1), fr, grr, method = "BFGS")
optim(c(-1.2,1), fr, NULL, method = "BFGS", hessian = TRUE)
optim(c(-1.2,1), fr, grr, method = "CG")
optim(c(-1.2,1), fr, grr, method = "CG",
      control=list(type=2))
optim(c(-1.2,1), fr, grr, method = "L-BFGS-B")

flb <- function(x) {
  p <- length(x);
  sum(c(1, rep(4, p-1)) * (x - c(1, x[-p])^2)^2)
}
## 25-dimensional box constrained
# par[24] is not at boundary
optim(rep(3, 25), flb, NULL, "L-BFGS-B",
  lower=rep(2, 25), upper=rep(4, 25))

## "wild" function , global minimum at about -15.81515
fw <- function (x)
    10*sin(0.3*x)*sin(1.3*x^2) + 0.00001*x^4 + 0.2*x+80
plot(fw, -50, 50, n=1000,
     main = "optim() minimising 'wild function'")

res <- optim(50, fw, method="SANN",
  control=list(maxit=20000, temp=20, parscale=20))
res
## Now improve locally
(r2 <- optim(res$par, fw, method="BFGS"))
points(r2$par, r2$val, pch = 8, col = "red", cex = 2)

## Combinatorial optimization: Traveling salesman problem
library(mva) # normally loaded
library(ts)  # for embed, normally loaded

data(eurodist)
eurodistmat <- as.matrix(eurodist)

# Target function
distance <- function(sq) {
    sq2 <- embed(sq, 2)
    return(sum(eurodistmat[cbind(sq2[,2],sq2[,1])]))
}
```

```
# Generate new candidate sequence
genseq <- function(sq) {
    idx <- seq(2, NROW(eurodistmat)-1, by=1)
    changepoints <- sample(idx, size=2, replace=FALSE)
    tmp <- sq[changepoints[1]]
    sq[changepoints[1]] <- sq[changepoints[2]]
    sq[changepoints[2]] <- tmp
    return(sq)
}

sq <- c(1,2:NROW(eurodistmat),1)  # Initial sequence
distance(sq)

set.seed(2222) # chosen to get a good soln quickly
res <- optim(sq, distance, genseq, method="SANN",
  control = list(maxit=6000, temp=2000, trace=TRUE))
res  # Near optimum distance around 12842

loc <- cmdscale(eurodist)
rx <- range(x <- loc[,1])
ry <- range(y <- -loc[,2])
tspinit <- loc[sq,]
tspres <- loc[res$par,]
s <- seq(NROW(tspres)-1)

plot(x, y, type="n", asp=1, xlab="", ylab="",
  main="initial solution of traveling salesman problem")
arrows(tspinit[s,1], -tspinit[s,2], tspinit[s+1,1],
  -tspinit[s+1,2], angle=10, col="green")
text(x, y, names(eurodist), cex=0.8)

plot(x, y, type="n", asp=1, xlab="", ylab="",
    main="optim() 'solving' traveling salesman problem")
arrows(tspres[s,1], -tspres[s,2], tspres[s+1,1],
  -tspres[s+1,2], angle=10, col="red")
text(x, y, names(eurodist), cex=0.8)
```

optimize   One Dimensional Optimization

Description

The function `optimize` searches the interval from `lower` to `upper` for a minimum or maximum of the function `f` with respect to its first argument.

`optimise` is an alias for `optimize`.

Usage

```
optimize(f = , interval = , lower = min(interval),
        upper = max(interval), maximum = FALSE,
        tol = .Machine$double.eps^0.25, ...)
optimise(f = , interval = , lower = min(interval),
        upper = max(interval), maximum = FALSE,
        tol = .Machine$double.eps^0.25, ...)
```

Arguments

f	the function to be optimized. The function is either minimized or maximized over its first argument depending on the value of `maximum`.
interval	a vector containing the end-points of the interval to be searched for the minimum.
lower	the lower end point of the interval to be searched.
upper	the upper end point of the interval to be searched.
maximum	logical. Should we maximize or minimize (the default)?
tol	the desired accuracy.
...	additional arguments to `f`.

Details

The method used is a combination of golden section search and successive parabolic interpolation. Convergence is never much slower than that for a Fibonacci search. If `f` has a continuous second derivative which is positive at the minimum (which is not at `lower` or `upper`), then convergence is superlinear, and usually of the order of about 1.324.

The function f is never evaluated at two points closer together than $\epsilon|x_0| + (tol/3)$, where ϵ is approximately sqrt(.Machine$double.eps) and x_0 is the final abscissa optimize()$minimum.

If f is a unimodal function and the computed values of f are always unimodal when separated by at least $\epsilon |x| + (tol/3)$, then x_0 approximates the abscissa of the global minimum of f on the interval lower, upper with an error less than $\epsilon|x_0| + tol$.

If f is not unimodal, then optimize() may approximate a local, but perhaps non-global, minimum to the same accuracy.

The first evaluation of f is always at $x_1 = a + (1-\phi)(b-a)$ where (a,b) = (lower, upper) and $\phi = (\sqrt{5}-1)/2 = 0.61803..$ is the golden section ratio. Almost always, the second evaluation is at $x_2 = a + \phi(b - a)$. Note that a local minimum inside $[x_1, x_2]$ will be found as solution, even when f is constant in there, see the last example.

It uses a C translation of Fortran code (from Netlib) based on the Algol 60 procedure localmin given in the reference.

Value

A list with components minimum (or maximum) and objective which give the location of the minimum (or maximum) and the value of the function at that point.

References

Brent, R. (1973) *Algorithms for Minimization without Derivatives.* Englewood Cliffs N.J.: Prentice-Hall.

See Also

nlm, uniroot.

Examples

```
f <- function (x,a) (x-a)^2
xmin <- optimize(f, c(0, 1), tol = 0.0001, a = 1/3)
xmin

## See where the function is evaluated:
optimize(function(x) x^2*(print(x)-1), l=0, u=10)

## "wrong" solution with unlucky interval and piecewise
## constant f():
```

```
f  <- function(x)
   ifelse(x > -1, ifelse(x < 4, exp(-1/abs(x - 1)), 10), 10)
fp <- function(x) { print(x); f(x) }

plot(f, -2,5, ylim = 0:1, col = 2)
optimize(fp, c(-4, 20)) # doesn't see the minimum
optimize(fp, c(-7, 20)) # ok
```

power — Create a Power Link Object

Description

Creates a link object based on the link function $\eta = \mu^\lambda$.

Usage

```
power(lambda = 1)
```

Arguments

lambda   a real number.

Details

If `lambda` is non-negative, it is taken as zero, and the log link is obtained. The default `lambda = 1` gives the identity link.

Value

A list with components `linkfun`, `linkinv`, `mu.eta`, and `valideta`. See `make.link` for information on their meaning.

References

Chambers, J. M. and Hastie, T. J. (1992) *Statistical Models in S*. Wadsworth & Brooks/Cole.

See Also

`make.link`, `family`

To raise a number to a power, see `Arithmetic`.

To calculate the power of a test, see various functions in the **ctest** package, e.g., `power.t.test`.

Examples

```
power()
quasi(link=power(1/3))[c("linkfun", "linkinv")]
```

predict.glm   *Predict Method for GLM Fits*

Description

Obtains predictions and optionally estimates standard errors of those predictions from a fitted generalized linear model object.

Usage

```
## S3 method for class 'glm':
predict(object, newdata = NULL,
    type = c("link", "response", "terms"),
    se.fit = FALSE, dispersion = NULL, terms = NULL,
    na.action = na.pass, ...)
```

Arguments

object
: a fitted object of class inheriting from "glm".

newdata
: optionally, a new data frame from which to make the predictions. If omitted, the fitted linear predictors are used.

type
: the type of prediction required. The default is on the scale of the linear predictors; the alternative "response" is on the scale of the response variable. Thus for a default binomial model the default predictions are of log-odds (probabilities on logit scale) and type = "response" gives the predicted probabilities. The "terms" option returns a matrix giving the fitted values of each term in the model formula on the linear predictor scale.

  The value of this argument can be abbreviated.

se.fit
: logical switch indicating if standard errors are required.

dispersion
: the dispersion of the GLM fit to be assumed in computing the standard errors. If omitted, that returned by summary applied to the object is used.

terms
: with type="terms" by default all terms are returned. A character vector specifies which terms are to be returned

na.action   function determining what should be done with missing values in newdata. The default is to predict NA.

...   further arguments passed to or from other methods.

Value

If se = FALSE, a vector or matrix of predictions. If se = TRUE, a list with components

fit   Predictions

se.fit   Estimated standard errors

residual.scale
                 A scalar giving the square root of the dispersion used in computing the standard errors.

See Also

glm, SafePrediction

Examples

```
## example from Venables and Ripley (2002, pp. 190-2.)
ldose <- rep(0:5, 2)
numdead <- c(1, 4, 9, 13, 18, 20, 0, 2, 6, 10, 12, 16)
sex <- factor(rep(c("M", "F"), c(6, 6)))
SF <- cbind(numdead, numalive=20-numdead)
budworm.lg <- glm(SF ~ sex*ldose, family=binomial)
summary(budworm.lg)

plot(c(1,32), c(0,1), type = "n", xlab = "dose",
     ylab = "prob", log = "x")
text(2^ldose, numdead/20, as.character(sex))
ld <- seq(0, 5, 0.1)
lines(2^ld, predict(budworm.lg, data.frame(ldose=ld,
    sex=factor(rep("M", length(ld)), levels=levels(sex))),
    type = "response"))
lines(2^ld, predict(budworm.lg, data.frame(ldose=ld,
    sex=factor(rep("F", length(ld)), levels=levels(sex))),
    type = "response"))
```

predict.lm *Predict method for Linear Model Fits*

Description

Predicted values based on linear model object

Usage

```
## S3 method for class 'lm':
predict(object, newdata, se.fit = FALSE, scale = NULL,
  df = Inf,
  interval = c("none", "confidence", "prediction"),
  level = 0.95, type = c("response", "terms"),
  terms = NULL, na.action = na.pass, ...)
```

Arguments

object	Object of class inheriting from "lm"
newdata	Data frame in which to predict
se.fit	A switch indicating if standard errors are required.
scale	Scale parameter for std.err. calculation
df	Degrees of freedom for scale
interval	Type of interval calculation
level	Tolerance/confidence level
type	Type of prediction (response or model term)
terms	If type="terms", which terms (default is all terms)
na.action	function determining what should be done with missing values in newdata. The default is to predict NA.
...	further arguments passed to or from other methods.

Details

predict.lm produces predicted values, obtained by evaluating the regression function in the frame newdata (which defaults to model.frame(object). If the logical se.fit is TRUE, standard errors of the predictions are calculated. If the numeric argument scale is set (with optional df), it is used as the residual standard deviation in the computation of the standard errors, otherwise this is extracted from the model fit. Setting `intervals` specifies computation of confidence or prediction (tolerance) intervals at the specified `level`.

If the fit is rank-deficient, some of the columns of the design matrix will have been dropped. Prediction from such a fit only makes sense if `newdata` is contained in the same subspace as the original data. That cannot be checked accurately, so a warning is issued.

Value

`predict.lm` produces a vector of predictions or a matrix of predictions and bounds with column names `fit`, `lwr`, and `upr` if `interval` is set. If `se.fit` is TRUE, a list with the following components is returned:

`fit`	vector or matrix as above
`se.fit`	standard error of predictions
`residual.scale`	
	residual standard deviations
`df`	degrees of freedom for residual

Note

Offsets specified by `offset` in the fit by `lm` will not be included in predictions, whereas those specified by an offset term in the formula will be.

See Also

The model fitting function `lm`, `predict`, `SafePrediction`

Examples

```
## Predictions
x <- rnorm(15)
y <- x + rnorm(15)
predict(lm(y ~ x))
new <- data.frame(x = seq(-3, 3, 0.5))
predict(lm(y ~ x), new, se.fit = TRUE)
pred.w.plim <-
   predict(lm(y ~ x), new, interval="prediction")
pred.w.clim <-
   predict(lm(y ~ x), new, interval="confidence")
matplot(new$x,cbind(pred.w.clim, pred.w.plim[,-1]),
        lty=c(1,2,3,3), type="l", ylab="predicted y")
```

profile *Generic Function for Profiling Models*

Description

Investigates behavior of objective function near the solution represented by `fitted`.

See documentation on method functions for further details.

Usage

```
profile(fitted, ...)
```

Arguments

`fitted`   the original fitted model object.

`...`   additional parameters. See documentation on individual methods.

Value

A list with an element for each parameter being profiled. See the individual methods for further details.

See Also

`profile.nls` in package **nls**, `profile.glm` in package **MASS**, ...

For profiling code, see `Rprof`.

proj — *Projections of Models*

Description

proj returns a matrix or list of matrices giving the projections of the data onto the terms of a linear model. It is most frequently used for aov models.

Usage

```
proj(object, ...)

## S3 method for class 'aov':
proj(object, onedf = FALSE, unweighted.scale = FALSE, ...)

## S3 method for class 'aovlist':
proj(object, onedf = FALSE, unweighted.scale = FALSE, ...)

## Default S3 method:
proj(object, onedf = TRUE, ...)

## S3 method for class 'lm':
proj(object, onedf = FALSE, unweighted.scale = FALSE, ...)
```

Arguments

object	An object of class "lm" or a class inheriting from it, or an object with a similar structure including in particular components qr and effects.
onedf	A logical flag. If TRUE, a projection is returned for all the columns of the model matrix. If FALSE, the single-column projections are collapsed by terms of the model (as represented in the analysis of variance table).
unweighted.scale	
	If the fit producing object used weights, this determines if the projections correspond to weighted or unweighted observations.
...	Swallow and ignore any other arguments.

Details

A projection is given for each stratum of the object, so for `aov` models with an `Error` term the result is a list of projections.

Value

A projection matrix or (for multi-stratum objects) a list of projection matrices.

Each projection is a matrix with a row for each observations and either a column for each term (`onedf = FALSE`) or for each coefficient (`onedf = TRUE`). Projection matrices from the default method have orthogonal columns representing the projection of the response onto the column space of the Q matrix from the QR decomposition. The fitted values are the sum of the projections, and the sum of squares for each column is the reduction in sum of squares from fitting that column (after those to the left of it).

The methods for `lm` and `aov` models add a column to the projection matrix giving the residuals (the projection of the data onto the orthogonal complement of the model space).

Strictly, when `onedf = FALSE` the result is not a projection, but the columns represent sums of projections onto the columns of the model matrix corresponding to that term. In this case the matrix does not depend on the coding used.

Author(s)

The design was inspired by the S function of the same name described in Chambers *et al.* (1992).

References

Chambers, J. M., Freeny, A and Heiberger, R. M. (1992) *Analysis of variance; designed experiments.* Chapter 5 of *Statistical Models in S* eds J. M. Chambers and T. J. Hastie, Wadsworth & Brooks/Cole.

See Also

`aov, lm, model.tables`

Examples

```
N <- c(0,1,0,1,1,1,0,0,0,1,1,0,1,1,0,0,1,0,1,0,1,1,0,0)
P <- c(1,1,0,0,0,1,0,1,1,1,0,0,0,1,0,1,1,0,0,1,0,1,1,0)
```

```
K <- c(1,0,0,1,0,1,1,0,0,1,0,1,0,1,1,0,0,0,1,1,1,0,1,0)
yield <- c(49.5,62.8,46.8,57.0,59.8,58.5,55.5,56.0,
           62.8,55.8,69.5,55.0,62.0,48.8,45.5,44.2,
           52.0,51.5,49.8,48.8,57.2,59.0,53.2,56.0)

npk <- data.frame(block=gl(6,4), N=factor(N), P=factor(P),
                  K=factor(K), yield=yield)
npk.aov <- aov(yield ~ block + N*P*K, npk)
proj(npk.aov)

## as a test, not particularly sensible
options(contrasts=c("contr.helmert", "contr.treatment"))
npk.aovE <- aov(yield ~  N*P*K + Error(block), npk)
proj(npk.aovE)
```

relevel — Reorder Levels of Factor

Description

The levels of a factor are re-ordered so that the level specified by `ref` is first and the others are moved down. This is useful for `contr.treatment` contrasts which take the first level as the reference.

Usage

```
relevel(x, ref, ...)
```

Arguments

`x`	An unordered factor.
`ref`	The reference level.
`...`	Additional arguments for future methods.

Value

A factor of the same length as `x`.

See Also

`factor`, `contr.treatment`

Examples

```
data(warpbreaks)
warpbreaks$tension <- relevel(warpbreaks$tension, ref="M")
summary(lm(breaks ~ wool + tension, data=warpbreaks))
```

replications — *Number of Replications of Terms*

Description

Returns a vector or a list of the number of replicates for each term in the formula.

Usage

```
replications(formula, data=NULL, na.action)
```

Arguments

formula   a formula or a terms object or a data frame.

data   a data frame used to find the objects in `formula`.

na.action function for handling missing values. Defaults to a na.action attribute of `data`, then a setting of the option `na.action`, or `na.fail` if that is not set.

Details

If `formula` is a data frame and `data` is missing, `formula` is used for data with the formula ~ ..

Value

A vector or list with one entry for each term in the formula giving the number(s) of replications for each level. If all levels are balanced (have the same number of replications) the result is a vector, otherwise it is a list with a component for each terms, as a vector, matrix or array as required.

A test for balance is `!is.list(replications(formula,data))`.

Author(s)

The design was inspired by the S function of the same name described in Chambers *et al.* (1992).

References

Chambers, J. M., Freeny, A and Heiberger, R. M. (1992) *Analysis of variance; designed experiments.* Chapter 5 of *Statistical Models in S* eds J. M. Chambers and T. J. Hastie, Wadsworth & Brooks/Cole.

See Also

model.tables

Examples

```
## From Venables and Ripley (2002) p.165.
N <- c(0,1,0,1,1,1,0,0,0,1,1,0,1,1,0,0,1,0,1,0,1,1,0,0)
P <- c(1,1,0,0,0,1,0,1,1,1,0,0,0,1,0,1,1,0,0,1,0,1,1,0)
K <- c(1,0,0,1,0,1,1,0,0,1,0,1,0,1,1,0,0,1,1,1,0,1,0)
yield <- c(49.5,62.8,46.8,57.0,59.8,58.5,55.5,56.0,
           62.8,55.8,69.5,55.0,62.0,48.8,45.5,44.2,
           52.0,51.5,49.8,48.8,57.2,59.0,53.2,56.0)

npk <- data.frame(block=gl(6,4), N=factor(N), P=factor(P),
                  K=factor(K), yield=yield)
replications(~ . - yield, npk)
```

residuals — *Extract Model Residuals*

Description

residuals is a generic function which extracts model residuals from objects returned by modeling functions.

The abbreviated form **resid** is an alias for **residuals**. It is intended to encourage users to access object components through an accessor function rather than by directly referencing an object slot.

All object classes which are returned by model fitting functions should provide a **residuals** method. (Note that the method is 'residuals' and not 'resid'.)

Methods can make use of **naresid** methods to compensate for the omission of missing values. The default method does.

Usage

```
residuals(object, ...)
resid(object, ...)
```

Arguments

object   an object for which the extraction of model residuals is meaningful.

...   other arguments.

Value

Residuals extracted from the object **object**.

References

Chambers, J. M. and Hastie, T. J. (1992) *Statistical Models in S.* Wadsworth & Brooks/Cole.

See Also

coefficients, fitted.values, glm, lm.

se.contrast   *Standard Errors for Contrasts in Model Terms*

Description

Returns the standard errors for one or more contrasts in an `aov` object.

Usage

```
se.contrast(object, ...)
## S3 method for class 'aov':
se.contrast(object, contrast.obj,
            coef = contr.helmert(ncol(contrast))[, 1],
            data = NULL, ...)
```

Arguments

`object`	A suitable fit, usually from `aov`.
`contrast.obj`	The contrasts for which standard errors are requested. This can be specified via a list or via a matrix. A single contrast can be specified by a list of logical vectors giving the cells to be contrasted. Multiple contrasts should be specified by a matrix, each column of which is a numerical contrast vector (summing to zero).
`coef`	used when `contrast.obj` is a list; it should be a vector of the same length as the list with zero sum. The default value is the first Helmert contrast, which contrasts the first and second cell means specified by the list.
`data`	The data frame used to evaluate `contrast.obj`.
`...`	further arguments passed to or from other methods.

Details

Contrasts are usually used to test if certain means are significantly different; it can be easier to use `se.contrast` than compute them directly from the coefficients.

In multistratum models, the contrasts can appear in more than one stratum; the contrast and standard error are computed in the lowest stratum and adjusted for efficiencies and comparisons between strata.

Suitable matrices for use with `coef` can be found by calling `contrasts` and indexing the columns by a factor.

Value

A vector giving the standard errors for each contrast.

See Also

contrasts, model.tables

Examples

```
## From Venables and Ripley (2002) p.165.
N <- c(0,1,0,1,1,1,0,0,0,1,1,0,1,1,0,0,1,0,1,0,1,1,0,0)
P <- c(1,1,0,0,0,1,0,1,1,1,0,0,0,1,0,1,1,0,0,1,0,1,1,0)
K <- c(1,0,0,1,0,1,1,0,0,1,0,1,0,1,1,0,0,0,1,1,1,0,1,0)
yield <- c(49.5,62.8,46.8,57.0,59.8,58.5,55.5,56.0,
           62.8,55.8,69.5,55.0,62.0,48.8,45.5,44.2,
           52.0,51.5,49.8,48.8,57.2,59.0,53.2,56.0)

npk <- data.frame(block = gl(6,4), N = factor(N),
  P = factor(P), K = factor(K), yield = yield)
options(contrasts=c("contr.treatment", "contr.poly"))
npk.aov1 <- aov(yield ~ block + N + K, npk)
se.contrast(npk.aov1, list(N=="0", N=="1"), data=npk)
# or via a matrix
cont <- matrix(c(-1,1), 2, 1, dimnames=list(NULL, "N"))
se.contrast(npk.aov1, cont[N, , drop=FALSE]/12, data=npk)

## test a multi-stratum model
npk.aov2 <- aov(yield ~ N + K + Error(block/(N + K)), npk)
se.contrast(npk.aov2, list(N == "0", N == "1"))
```

stat.anova   *GLM Anova Statistics*

Description

This is a utility function, used in `lm` and `glm` methods for `anova(..., test != NULL)` and should not be used by the average user.

Usage

```
stat.anova(table, test = c("Chisq", "F", "Cp"), scale,
   df.scale, n)
```

Arguments

`table`	numeric matrix as results from `anova.glm(..., test=NULL)`.
`test`	a character string, matching one of `"Chisq"`, `"F"` or `"Cp"`.
`scale`	a weighted residual sum of squares.
`df.scale`	degrees of freedom corresponding to scale.
`n`	number of observations.

Value

A matrix which is the original `table`, augmented by a column of test statistics, depending on the `test` argument.

References

Hastie, T. J. and Pregibon, D. (1992) *Generalized linear models.* Chapter 6 of *Statistical Models in S* eds J. M. Chambers and T. J. Hastie, Wadsworth & Brooks/Cole.

See Also

`anova.lm`, `anova.glm`.

Examples

```
## Continued from '?glm':
print(ag <- anova(glm.D93))
stat.anova(ag$table, test = "Cp",
  scale = sum(resid(glm.D93, "pearson")^2)/4, df = 4,
  n = 9)
```

step — Choose a model by AIC in a Stepwise Algorithm

Description

Select a formula-based model by AIC.

Usage

```
step(object, scope, scale = 0,
    direction = c("both", "backward", "forward"),
    trace = 1, keep = NULL, steps = 1000, k = 2, ...)
```

Arguments

object an object representing a model of an appropriate class (mainly "lm" and "glm"). This is used as the initial model in the stepwise search.

scope defines the range of models examined in the stepwise search. This should be either a single formula, or a list containing components upper and lower, both formulae. See the details for how to specify the formulae and how they are used.

scale used in the definition of the AIC statistic for selecting the models, currently only for lm, aov and glm models.

direction the mode of stepwise search, can be one of "both", "backward", or "forward", with a default of "both". If the scope argument is missing the default for direction is "backward".

trace if positive, information is printed during the running of step. Larger values may give more detailed information.

keep a filter function whose input is a fitted model object and the associated AIC statistic, and whose output is arbitrary. Typically keep will select a subset of the components of the object and return them. The default is not to keep anything.

steps the maximum number of steps to be considered. The default is 1000 (essentially as many as required). It is typically used to stop the process early.

k	the multiple of the number of degrees of freedom used for the penalty. Only k = 2 gives the genuine AIC: k = log(n) is sometimes referred to as BIC or SBC.
...	any additional arguments to extractAIC.

Details

step uses add1 and drop1 repeatedly; it will work for any method for which they work, and that is determined by having a valid method for extractAIC. When the additive constant can be chosen so that AIC is equal to Mallows' C_p, this is done and the tables are labelled appropriately.

The set of models searched is determined by the scope argument. The right-hand-side of its lower component is always included in the model, and right-hand-side of the model is included in the upper component. If scope is a single formula, it specifies the upper component, and the lower model is empty. If scope is missing, the initial model is used as the upper model.

Models specified by scope can be templates to update object as used by update.formula.

There is a potential problem in using glm fits with a variable scale, as in that case the deviance is not simply related to the maximized log-likelihood. The function extractAIC.glm makes the appropriate adjustment for a gaussian family, but may need to be amended for other cases. (The binomial and poisson families have fixed scale by default and do not correspond to a particular maximum-likelihood problem for variable scale.)

Value

the stepwise-selected model is returned, with up to two additional components. There is an "anova" component corresponding to the steps taken in the search, as well as a "keep" component if the keep= argument was supplied in the call. The "Resid. Dev" column of the analysis of deviance table refers to a constant minus twice the maximized log likelihood: it will be a deviance only in cases where a saturated model is well-defined (thus excluding lm, aov and survreg fits, for example).

Warning

The model fitting must apply the models to the same dataset. This may be a problem if there are missing values and R's default of na.action = na.omit is used. We suggest you remove the missing values first.

Note

This function differs considerably from the function in S, which uses a number of approximations and does not compute the correct AIC.

This is a minimal implementation. Use `stepAIC` for a wider range of object classes.

Author(s)

B. D. Ripley: `step` is a slightly simplified version of `stepAIC` in package **MASS** (Venables & Ripley, 2002 and earlier editions).

The idea of a `step` function follows that described in Hastie & Pregibon (1992); but the implementation in R is more general.

References

Hastie, T. J. and Pregibon, D. (1992) *Generalized linear models.* Chapter 6 of *Statistical Models in S* eds J. M. Chambers and T. J. Hastie, Wadsworth & Brooks/Cole.

Venables, W. N. and Ripley, B. D. (2002) *Modern Applied Statistics with S.* New York: Springer (4th ed).

See Also

`stepAIC`, `add1`, `drop1`

Examples

```
example(lm)
step(lm.D9)

data(swiss)
summary(lm1 <- lm(Fertility ~ ., data = swiss))
slm1 <- step(lm1)
summary(slm1)
slm1$anova
```

summary.aov   *Summarize an Analysis of Variance Model*

Description

Summarize an analysis of variance model.

Usage

```
## S3 method for class 'aov':
summary(object, intercept = FALSE, split,
        expand.split = TRUE, keep.zero.df = TRUE, ...)

## S3 method for class 'aovlist':
summary(object, ...)
```

Arguments

`object`	An object of class `"aov"` or `"aovlist"`.
`intercept`	logical: should intercept terms be included?
`split`	an optional named list, with names corresponding to terms in the model. Each component is itself a list with integer components giving contrasts whose contributions are to be summed.
`expand.split`	logical: should the split apply also to interactions involving the factor?
`keep.zero.df`	logical: should terms with no degrees of freedom be included?
`...`	Arguments to be passed to or from other methods, for `summary.aovlist` including those for `summary.aov`.

Value

An object of class `c("summary.aov", "listof")` or `"summary.aovlist"` respectively.

Note

The use of `expand.split = TRUE` is little tested: it is always possible to set it to `FALSE` and specify exactly all the splits required.

See Also

aov, summary, model.tables, TukeyHSD

Examples

```
## From Venables and Ripley (2002) p.165.
N <- c(0,1,0,1,1,1,0,0,0,1,1,0,1,1,0,0,1,0,1,0,1,1,0,0)
P <- c(1,1,0,0,0,1,0,1,1,1,0,0,0,1,0,1,1,0,0,1,0,1,1,0)
K <- c(1,0,0,1,0,1,1,0,0,1,0,1,0,1,1,0,0,0,1,1,1,0,1,0)
yield <- c(49.5,62.8,46.8,57.0,59.8,58.5,55.5,56.0,
           62.8,55.8,69.5,55.0,62.0,48.8,45.5,44.2,
           52.0,51.5,49.8,48.8,57.2,59.0,53.2,56.0)
npk <- data.frame(block=gl(6,4), N=factor(N), P=factor(P),
                  K=factor(K), yield=yield)

( npk.aov <- aov(yield ~ block + N*P*K, npk) )
summary(npk.aov)
coefficients(npk.aov)

# Cochran and Cox (1957, p.164)
# 3x3 factorial with ordered factors, each is average
# of 12.
CC <- data.frame(
    y = c(449, 413, 326, 409, 358, 291, 341, 278, 312)/12,
    P = ordered(gl(3, 3)), N = ordered(gl(3, 1, 9))
)
CC.aov <- aov(y ~ N * P, data = CC , weights = rep(12, 9))
summary(CC.aov)

# Split both main effects into linear and quadratic parts.
summary(CC.aov, split = list(N = list(L = 1, Q = 2),
  P = list(L = 1, Q = 2)))

# Split only the interaction
summary(CC.aov, split = list("N:P" = list(L.L = 1,
  Q = 2:4)))

# split on just one var
summary(CC.aov, split = list(P = list(lin = 1, quad = 2)))
summary(CC.aov, split = list(P = list(lin = 1, quad = 2)),
        expand.split=FALSE)
```

summary.glm *Summarizing Generalized Linear Model Fits*

Description

These functions are all methods for class glm or summary.glm objects.

Usage

```
## S3 method for class 'glm':
summary(object, dispersion = NULL, correlation = FALSE,
        symbolic.cor = FALSE, ...)

## S3 method for class 'summary.glm':
print(x, digits = max(3, getOption("digits") - 3),
      symbolic.cor = x$symbolic.cor,
      signif.stars = getOption("show.signif.stars"), ...)
```

Arguments

object	an object of class "glm", usually, a result of a call to glm.
x	an object of class "summary.glm", usually, a result of a call to summary.glm.
dispersion	the dispersion parameter for the fitting family. By default it is obtained from object.
correlation	logical; if TRUE, the correlation matrix of the estimated parameters is returned and printed.
digits	the number of significant digits to use when printing.
symbolic.cor	logical. If TRUE, print the correlations in a symbolic form (see symnum) rather than as numbers.
signif.stars	logical. If TRUE, "significance stars" are printed for each coefficient.
...	further arguments passed to or from other methods.

Details

print.summary.glm tries to be smart about formatting the coefficients, standard errors, etc. and additionally gives "significance stars" if signif.stars is TRUE.

Aliased coefficients are omitted in the returned object but (as from R 1.8.0) restored by the `print` method.

Correlations are printed to two decimal places (or symbolically): to see the actual correlations print `summary(object)$correlation` directly.

Value

`summary.glm` returns an object of class `"summary.glm"`, a list with components

`call`	the component from `object`.
`family`	the component from `object`.
`deviance`	the component from `object`.
`contrasts`	the component from `object`.
`df.residual`	the component from `object`.
`null.deviance`	the component from `object`.
`df.null`	the component from `object`.
`deviance.resid`	the deviance residuals: see `residuals.glm`.
`coefficients`	the matrix of coefficients, standard errors, z-values and p-values. Aliased coefficients are omitted.
`aliased`	named logical vector showing if the original coefficients are aliased.
`dispersion`	either the supplied argument or the estimated dispersion if the latter in `NULL`
`df`	a 3-vector of the rank of the model and the number of residual degrees of freedom, plus number of non-aliased coefficients.
`cov.unscaled`	the unscaled (`dispersion = 1`) estimated covariance matrix of the estimated coefficients.
`cov.scaled`	ditto, scaled by `dispersion`.
`correlation`	(only if `correlation` is true.) The estimated correlations of the estimated coefficients.
`symbolic.cor`	(only if `correlation` is true.) The value of the argument `symbolic.cor`.

See Also

`glm`, `summary`.

Examples

```
# Continuing the Example from '?glm':
summary(glm.D93)
```

summary.lm *Summarizing Linear Model Fits*

Description

summary method for class "lm".

Usage

```
## S3 method for class 'lm':
summary(object, correlation = FALSE, symbolic.cor = FALSE,
    ...)

## S3 method for class 'summary.lm':
print(x, digits = max(3, getOption("digits") - 3),
      symbolic.cor = x$symbolic.cor,
      signif.stars = getOption("show.signif.stars"), ...)
```

Arguments

object	an object of class "lm", usually, a result of a call to lm.
x	an object of class "summary.lm", usually, a result of a call to summary.lm.
correlation	logical; if TRUE, the correlation matrix of the estimated parameters is returned and printed.
digits	the number of significant digits to use when printing.
symbolic.cor	logical. If TRUE, print the correlations in a symbolic form (see symnum) rather than as numbers.
signif.stars	logical. If TRUE, "significance stars" are printed for each coefficient.
...	further arguments passed to or from other methods.

Details

print.summary.lm tries to be smart about formatting the coefficients, standard errors, etc. and additionally gives "significance stars" if signif.stars is TRUE.

Correlations are printed to two decimal places (or symbolically): to see the actual correlations print summary(object)$correlation directly.

Value

The function summary.lm computes and returns a list of summary statistics of the fitted linear model given in object, using the components (list elements) "call" and "terms" from its argument, plus

residuals
: the *weighted* residuals, the usual residuals rescaled by the square root of the weights specified in the call to lm.

coefficients
: a $p \times 4$ matrix with columns for the estimated coefficient, its standard error, t-statistic and corresponding (two-sided) p-value. Aliased coefficients are omitted.

aliased
: named logical vector showing if the original coefficients are aliased.

sigma
: the square root of the estimated variance of the random error

$$\hat{\sigma}^2 = \frac{1}{n-p} \sum_i R_i^2,$$

where R_i is the i-th residual, residuals[i].

df
: degrees of freedom, a 3-vector $(p, n-p, p*)$, the last being the number of non-aliased coefficients.

fstatistic
: (for models including non-intercept terms) a 3-vector with the value of the F-statistic with its numerator and denominator degrees of freedom.

r.squared
: R^2, the "fraction of variance explained by the model",

$$R^2 = 1 - \frac{\sum_i R_i^2}{\sum_i (y_i - y^*)^2},$$

where y^* is the mean of y_i if there is an intercept and zero otherwise.

adj.r.squared
: the above R^2 statistic "*adjusted*", penalizing for higher p.

cov.unscaled
: a $p \times p$ matrix of (unscaled) covariances of the $\hat{\beta}_j$, $j = 1, \ldots, p$.

correlation
: the correlation matrix corresponding to the above cov.unscaled, if correlation = TRUE is specified.

symbolic.cor
: (only if correlation is true.) The value of the argument symbolic.cor.

See Also

The model fitting function lm, summary.

Examples

```
## Continuing the lm(.) example:
coef(lm.D90) # the bare coefficients
# omitting intercept
sld90 <- summary(lm.D90 <- lm(weight ~ group -1))
sld90
coef(sld90) # much more
```

summary.manova   *Summary Method for Multivariate Analysis of Variance*

Description

A summary method for class "manova".

Usage

```
## S3 method for class 'manova':
summary(object,
   test = c("Pillai", "Wilks", "Hotelling-Lawley", "Roy"),
   intercept = FALSE, ...)
```

Arguments

object	An object of class "manova" or an aov object with multiple responses.
test	The name of the test statistic to be used. Partial matching is used so the name can be abbreviated.
intercept	logical. If TRUE, the intercept term is included in the table.
...	further arguments passed to or from other methods.

Details

The summary.manova method uses a multivariate test statistic for the summary table. Wilks' statistic is most popular in the literature, but the default Pillai-Bartlett statistic is recommended by Hand and Taylor (1987).

Value

A list with components

SS	A named list of sums of squares and product matrices.
Eigenvalues	A matrix of eigenvalues,
stats	A matrix of the statistics, approximate F value and degrees of freedom.

References

Krzanowski, W. J. (1988) *Principles of Multivariate Analysis. A User's Perspective.* Oxford.

Hand, D. J. and Taylor, C. C. (1987) *Multivariate Analysis of Variance and Repeated Measures.* Chapman and Hall.

See Also

manova, aov

Examples

```
## Example on producing plastic film from Krzanowski (1998,
## p. 381)
tear <- c(6.5, 6.2, 5.8, 6.5, 6.5, 6.9, 7.2, 6.9, 6.1, 6.3,
          6.7, 6.6, 7.2, 7.1, 6.8, 7.1, 7.0, 7.2, 7.5, 7.6)
gloss <- c(9.5, 9.9, 9.6, 9.6, 9.2, 9.1, 10.0, 9.9, 9.5,
           9.4, 9.1, 9.3, 8.3, 8.4, 8.5, 9.2, 8.8, 9.7,
           10.1, 9.2)
opacity <- c(4.4, 6.4, 3.0, 4.1, 0.8, 5.7, 2.0, 3.9, 1.9,
             5.7, 2.8, 4.1, 3.8, 1.6, 3.4, 8.4, 5.2, 6.9,
             2.7, 1.9)
Y <- cbind(tear, gloss, opacity)
rate <- factor(gl(2,10), labels=c("Low", "High"))
additive <- factor(gl(2, 5, len=20),
  labels=c("Low", "High"))

fit <- manova(Y ~ rate * additive)
summary.aov(fit)          # univariate ANOVA tables
summary(fit, test="Wilks") # ANOVA table of Wilks' lambda
```

terms   *Model Terms*

Description

The function `terms` is a generic function which can be used to extract *terms* objects from various kinds of R data objects.

Usage

```
terms(x, ...)
```

Arguments

x   object used to select a method to dispatch.
...   further arguments passed to or from other methods.

Details

There are methods for classes `"aovlist"`, and `"terms"` `"formula"` (see `terms.formula`): the default method just extracts the `terms` component of the object (if any).

Value

An object of class `c("terms", "formula")` which contains the *terms* representation of a symbolic model. See `terms.object` for its structure.

References

Chambers, J. M. and Hastie, T. J. (1992) *Statistical models.* Chapter 2 of *Statistical Models in S* eds J. M. Chambers and T. J. Hastie, Wadsworth & Brooks/Cole.

See Also

`terms.object`, `terms.formula`, `lm`, `glm`, `formula`.

base — terms.formula

terms.formula *Construct a terms Object from a Formula*

Description

This function takes a formula and some optional arguments and constructs a terms object. The terms object can then be used to construct a model.matrix.

Usage

```
## S3 method for class 'formula':
terms(x, specials = NULL, abb = NULL, data = NULL,
  neg.out = TRUE, keep.order = FALSE, simplify = FALSE,
  ...)
```

Arguments

x	a formula.
specials	which functions in the formula should be marked as special in the terms object.
abb	Not implemented in R.
data	a data frame from which the meaning of the special symbol . can be inferred. It is unused if there is no . in the formula.
neg.out	Not implemented in R.
keep.order	a logical value indicating whether the terms should keep their positions. If FALSE the terms are reordered so that main effects come first, followed by the interactions, all second-order, all third-order and so on. Effects of a given order are kept in the order specified.
simplify	should the formula be expanded and simplified, the pre-1.7.0 behaviour?
...	further arguments passed to or from other methods.

Details

Not all of the options work in the same way that they do in S and not all are implemented.

Value

A `terms.object` object is returned. The object itself is the re-ordered (unless `keep.order` = TRUE) formula. In all cases variables within an interaction term in the formula are re-ordered by the ordering of the "variables" attribute, which is the order in which the variables occur in the formula.

See Also

`terms`, `terms.object`

terms.object *Description of Terms Objects*

Description

An object of class `terms` holds information about a model. Usually the model was specified in terms of a `formula` and that formula was used to determine the terms object.

Value

The object itself is simply the formula supplied to the call of `terms.formula`. The object has a number of attributes and they are used to construct the model frame:

factors
: A matrix of variables by terms showing which variables appear in which terms. The entries are 0 if the variable does not occur in the term, 1 if it does occur and should be coded by contrasts, and 2 if it occurs and should be coded via dummy variables for all levels (as when an intercept or lower-order term is missing).

term.labels
: A character vector containing the labels for each of the terms in the model. Non-syntactic names will be quoted by backticks.

variables
: A call to `list` of the variables in the model.

intercept
: Either 0, indicating no intercept is to be fit, or 1 indicating that an intercept is to be fit.

order
: A vector of the same length as `term.labels` indicating the order of interaction for each term

response
: The index of the variable (in variables) of the response (the left hand side of the formula).

offset
: If the model contains `offset` terms there is an `offset` attribute indicating which variables are offsets

specials
: If the `specials` argument was given to `terms.formula` there is a `specials` attribute, a list of vectors indicating the terms that contain these special functions.

The object has class `c("terms", "formula")`.

Note

These objects are different from those found in S. In particular there is no `formula` attribute, instead the object is itself a formula. Thus, the mode of a terms object is different as well.

Examples of the `specials` argument can be seen in the `aov` and `coxph` functions.

See Also

`terms`, `formula`.

TukeyHSD — Compute Tukey Honest Significant Differences

Description

Create a set of confidence intervals on the differences between the means of the levels of a factor with the specified family-wise probability of coverage. The intervals are based on the Studentized range statistic, Tukey's 'Honest Significant Difference' method. There is a `plot` method.

Usage

```
TukeyHSD(x, which, ordered = FALSE, conf.level = 0.95, ...)
```

Arguments

x	A fitted model object, usually an `aov` fit.
which	A list of terms in the fitted model for which the intervals should be calculated. Defaults to all the terms.
ordered	A logical value indicating if the levels of the factor should be ordered according to increasing average in the sample before taking differences. If `ordered` is true then the calculated differences in the means will all be positive. The significant differences will be those for which the `lwr` end point is positive.
conf.level	A numeric value between zero and one giving the family-wise confidence level to use.
...	Optional additional arguments. None are used at present.

Details

When comparing the means for the levels of a factor in an analysis of variance, a simple comparison using t-tests will inflate the probability of declaring a significant difference when it is not in fact present. This is because the intervals are calculated with a given coverage probability for each interval but the interpretation of the coverage is usually with respect to the entire family of intervals.

John Tukey introduced intervals based on the range of the sample means rather than the individual differences. The intervals returned by this function are based on this Studentized range statistics.

Technically the intervals constructed in this way would only apply to balanced designs where there are the same number of observations made at each level of the factor. This function incorporates an adjustment for sample size that produces sensible intervals for mildly unbalanced designs.

Value

A list with one component for each term requested in `which`. Each component is a matrix with columns `diff` giving the difference in the observed means, `lwr` giving the lower end point of the interval, and `upr` giving the upper end point.

Author(s)

Douglas Bates

References

Miller, R. G. (1981) *Simultaneous Statistical Inference.* Springer.

Yandell, B. S. (1997) *Practical Data Analysis for Designed Experiments.* Chapman & Hall.

See Also

`aov`, `qtukey`, `model.tables`

Examples

```
data(warpbreaks)
summary(fm1 <- aov(breaks ~ wool + tension,
  data = warpbreaks))
TukeyHSD(fm1, "tension", ordered = TRUE)
plot(TukeyHSD(fm1, "tension"))
```

uniroot   *One Dimensional Root (Zero) Finding*

Description

The function uniroot searches the interval from lower to upper for a root (i.e., zero) of the function f with respect to its first argument.

Usage

```
uniroot(f, interval, lower = min(interval),
    upper = max(interval), tol = .Machine$double.eps^0.25,
    maxiter = 1000, ...)
```

Arguments

f	the function for which the root is sought.
interval	a vector containing the end-points of the interval to be searched for the root.
lower	the lower end point of the interval to be searched.
upper	the upper end point of the interval to be searched.
tol	the desired accuracy (convergence tolerance).
maxiter	the maximum number of iterations.
...	additional arguments to f.

Details

Either interval or both lower and upper must be specified. The function uses Fortran subroutine '"zeroin"' (from Netlib) based on algorithms given in the reference below.

If the algorithm does not converge in maxiter steps, a warning is printed and the current approximation is returned.

Value

A list with four components: root and f.root give the location of the root and the value of the function evaluated at that point. iter and estim.prec give the number of iterations used and an approximate estimated precision for root.

References

Brent, R. (1973) *Algorithms for Minimization without Derivatives.* Englewood Cliffs, NJ: Prentice-Hall.

See Also

polyroot for all complex roots of a polynomial; optimize, nlm.

Examples

```
f <- function (x,a) x - a
str(xmin <- uniroot(f, c(0, 1), tol = 0.0001, a = 1/3))
str(uniroot(function(x) x*(x^2-1) + .5,
     low = -2, up = 2, tol = 0.0001), dig = 10)
str(uniroot(function(x) x*(x^2-1) + .5,
     low = -2, up =2 , tol = 1e-10 ), dig = 10)

## Find the smallest value x for which exp(x) > 0
## (numerically):
r <- uniroot(function(x) 1e80*exp(x) -1e-300, , -1000, 0,
          tol=1e-20)
str(r, digits= 15) ## around -745.1332191

exp(r$r)         # = 0, but not for r$r * 0.999...
minexp <- r$r * (1 - .Machine$double.eps)
exp(minexp)      # typically denormalized
```

update — *Update and Re-fit a Model Call*

Description

update will update and (by default) re-fit a model. It does this by extracting the call stored in the object, updating the call and (by default) evaluating that call. Sometimes it is useful to call update with only one argument, for example if the data frame has been corrected.

Usage

```
update(object, ...)

## Default S3 method:
update(object, formula., ..., evaluate = TRUE)
```

Arguments

object	An existing fit from a model function such as lm, glm and many others.
formula.	Changes to the formula – see update.formula for details.
...	Additional arguments to the call, or arguments with changed values. Use name=NULL to remove the argument name.
evaluate	If true evaluate the new call else return the call.

Value

If evaluate = TRUE the fitted object, otherwise the updated call.

References

Chambers, J. M. (1992) *Linear models.* Chapter 4 of *Statistical Models in S* eds J. M. Chambers and T. J. Hastie, Wadsworth & Brooks/Cole.

See Also

update.formula

Examples

```
oldcon <- options(contrasts = c("contr.treatment",
                                "contr.poly"))
## Annette Dobson (1990) "An Introduction to Generalized
## Linear Models". Page 9: Plant Weight Data.
ctl <- c(4.17,5.58,5.18,6.11,4.50,4.61,5.17,4.53,5.33,5.14)
trt <- c(4.81,4.17,4.41,3.59,5.87,3.83,6.03,4.89,4.32,4.69)
group <- gl(2, 10, 20, labels = c("Ctl", "Trt"))
weight <- c(ctl, trt)
lm.D9 <- lm(weight ~ group)
lm.D9
summary(lm.D90 <- update(lm.D9, . ~ . - 1))
options(contrasts = c("contr.helmert", "contr.poly"))
update(lm.D9)
options(oldcon)
```

update.formula   *Model Updating*

Description

`update.formula` is used to update model formulae. This typically involves adding or dropping terms, but updates can be more general.

Usage

```
## S3 method for class 'formula':
update(old, new, ...)
```

Arguments

old
: a model formula to be updated.

new
: a formula giving a template which specifies how to update.

...
: further arguments passed to or from other methods.

Details

The function works by first identifying the *left-hand side* and *right-hand side* of the `old` formula. It then examines the `new` formula and substitutes the *lhs* of the `old` formula for any occurrence of "." on the left of `new`, and substitutes the *rhs* of the `old` formula for any occurrence of "." on the right of `new`.

Value

The updated formula is returned.

See Also

`terms`, `model.matrix`.

Examples

```
update(y ~ x,    ~ . + x2) # y ~ x + x2
update(y ~ x, log(.) ~ . ) # log(y) ~ x
```

vcov   *Calculate Variance-Covariance Matrix for a Fitted Model Object*

Description

Returns the variance-covariance matrix of the main parameters of a fitted model object.

Usage

```
vcov(object, ...)
```

Arguments

object   a fitted model object.

...   additional arguments for method functions. For the glm method this can be used to pass a dispersion parameter.

Details

This is a generic function. Functions with names beginning in vcov. will be methods for this function. Classes with methods for this function include: lm, glm, nls, lme, gls, coxph and survreg

Value

A matrix of the estimated covariances between the parameter estimates in the linear or non-linear predictor of the model.

weighted.residuals   *Compute Weighted Residuals*

Description

Computed weighted residuals from a linear model fit.

Usage

```
weighted.residuals(obj, drop0 = TRUE)
```

Arguments

obj	R object, typically of class lm or glm.
drop0	logical. If TRUE, drop all cases with weights == 0.

Details

Weighted residuals are the usual residuals R_i, multiplied by $\sqrt{w_i}$, where w_i are the weights as specified in lm's call.

Dropping cases with weights zero is compatible with influence and related functions.

Value

Numeric vector of length n', where n' is the number of of non-0 weights (drop0 = TRUE) or the number of observations, otherwise.

See Also

residuals, lm.influence, etc.

Examples

```
example("lm")
all.equal(weighted.residuals(lm.D9),
          residuals(lm.D9))
x <- 1:10
w <- 0:9
y <- rnorm(x)
weighted.residuals(lmxy <- lm(y ~ x, weights = w))
weighted.residuals(lmxy, drop0 = FALSE)
```

Chapter 5

Base package — dates, time and time-series

as.POSIX*   *Date-time Conversion Functions*

Description

Functions to manipulate objects of classes "POSIXlt" and "POSIXct" representing calendar dates and times (to the nearest second).

Usage

```
as.POSIXct(x, tz = "")
as.POSIXlt(x, tz = "")
```

Arguments

x   An object to be converted.

tz   A timezone specification to be used for the conversion, *if one is required.* System-specific, but "" is the current timezone, and "GMT" is UTC (Coordinated Universal Time, in French).

Details

The as.POSIX* functions convert an object to one of the two classes used to represent date/times (calendar dates plus time to the nearest second). They can convert a wide variety of objects, including objects of the other class and of classes "date" (from package [date:as.date]date or [date:as.date]survival), "chron" and "dates" (from package [chron]chron) to these classes. They can also convert character strings of the formats "2001-02-03" and "2001/02/03" optionally followed by white space and a time in the format "14:52" or "14:52:03". (Formats such as "01/02/03" are ambiguous but can be converted via a format specification by strptime.)

Logical NAs can be converted to either of the classes, but no other logical vectors can be.

Value

as.POSIXct and as.POSIXlt return an object of the appropriate class. If tz was specified, as.POSIXlt will give an appropriate "tzone" attribute.

Note

If you want to extract specific aspects of a time (such as the day of the week) just convert it to class "POSIXlt" and extract the relevant component(s) of the list, or if you want a character representation (such as a named day of the week) use format.POSIXlt or format.POSIXct.

If a timezone is needed and that specified is invalid on your system, what happens is system-specific but it will probably be ignored.

See Also

DateTimeClasses for details of the classes; strptime for conversion to and from character representations.

Examples

```
(z <- Sys.time())    # the current date, as class "POSIXct"
unclass(z)           # a large integer
floor(unclass(z)/86400) # number of days since 1970-01-01
(z <- as.POSIXlt(Sys.time())) # current date,
                              # as class "POSIXlt"
unlist(unclass(z)) # a list shown as a named vector

as.POSIXlt(Sys.time(), "GMT") # the current time in GMT
```

cut.POSIXt *Convert a Date-Time Object to a Factor*

Description

Method for cut applied to date-time objects.

Usage

```
## S3 method for class 'POSIXt':
cut(x, breaks, labels = NULL, start.on.monday = TRUE,
    right = FALSE, ...)
```

Arguments

x
: an object inheriting from class "POSIXt".

breaks
: a vector of cut points *or* number giving the number of intervals which x is to be cut into *or* an interval specification, one of "sec", "min", "hour", "day", "DSTday", "week", "month" or "year", optionally preceded by an integer and a space, or followed by "s".

labels
: labels for the levels of the resulting category. By default, labels are constructed from the left-hand end of the intervals (which are include for the default value of right). If labels = FALSE, simple integer codes are returned instead of a factor.

start.on.monday
: logical. If breaks = "weeks", should the week start on Mondays or Sundays?

right, ...
: arguments to be passed to or from other methods.

Value

A factor is returned, unless labels = FALSE which returns the integer level codes.

See Also

seq.POSIXt, cut

Examples

```
## random dates in a 10-week period
cut(ISOdate(2001, 1, 1) + 70*86400*runif(100), "weeks")
```

DateTimeClasses   *Date-Time Classes*

Description

Description of the classes "POSIXlt" and "POSIXct" representing calendar dates and times (to the nearest second).

Usage

```
## S3 method for class 'POSIXct':
print(x, ...)

## S3 method for class 'POSIXct':
summary(object, digits = 15, ...)

time + number
time - number
time1 lop time2
```

Arguments

x, object	An object to be printed or summarized from one of the date-time classes.
digits	Number of significant digits for the computations: should be high enough to represent the least important time unit exactly.
...	Further arguments to be passed from or to other methods.
time, time1, time2	date-time objects.
number	a numeric object.
lop	One of ==, !=, <, <=, > or >=.

Details

There are two basic classes of date/times. Class "POSIXct" represents the (signed) number of seconds since the beginning of 1970 as a numeric vector. Class "POSIXlt" is a named list of vectors representing

sec 0-61: seconds

min 0-59: minutes

hour 0–23: hours

mday 1–31: day of the month

mon 0–11: months after the first of the year.

year Years since 1900.

wday 0–6 day of the week, starting on Sunday.

yday 0–365: day of the year.

isdst Daylight savings time flag. Positive if in force, zero if not, negative if unknown.

The classes correspond to the ANSI C constructs of "calendar time" (the time_t data type) and "local time" (or broken-down time, the struct tm data type), from which they also inherit their names.

"POSIXct" is more convenient for including in data frames, and "POSIXlt" is closer to human-readable forms. A virtual class "POSIXt" inherits from both of the classes: it is used to allow operations such as subtraction to mix the two classes.

Logical comparisons and limited arithmetic are available for both classes. One can add or subtract a number of seconds or a difftime object from a date-time object, but not add two date-time objects. Subtraction of two date-time objects is equivalent to using difftime. Be aware that "POSIXlt" objects will be interpreted as being in the current timezone for these operations, unless a timezone has been specified.

"POSIXlt" objects will often have an attribute "tzone", a character vector of length 3 giving the timezone name from the TZ environment variable and the names of the base timezone and the alternate (daylight-saving) timezone. Sometimes this may just be of length one, giving the timezone name.

Unfortunately, the conversion is complicated by the operation of time zones and leap seconds (22 days have been 86401 seconds long so far: the times of the extra seconds are in the object .leap.seconds). The details of this are entrusted to the OS services where possible. This will usually cover the period 1970–2037, and on Unix machines back to 1902 (when time zones were in their infancy). Outside those ranges we use our own C code. This uses the offset from GMT in use in the timezone in 2000, and uses the alternate (daylight-saving) timezone only if isdst is positive.

It seems that some systems use leap seconds but most do not. This is detected and corrected for at build time, so all "POSIXct" times used by R do not include leap seconds. (Conceivably this could be wrong if the system has changed since build time, just possibly by changing locales.)

Using c on "POSIXlt" objects converts them to the current time zone.

Warning

Some Unix-like systems (especially GNU/Linux ones) do not have "TZ" set, yet have internal code that expects it (as does POSIX). We have tried to work around this, but if you get unexpected results try setting "TZ".

See Also

as.POSIXct and as.POSIXlt for conversion between the classes.

strptime for conversion to and from character representations.

Sys.time for clock time as a "POSIXct" object.

difftime for time intervals.

cut.POSIXt, seq.POSIXt, round.POSIXt and trunc.POSIXt for methods for these classes.

weekdays.POSIXt for convenience extraction functions.

Examples

```
(z <- Sys.time()) # the current date, as class "POSIXct"

Sys.time() - 3600 # an hour ago

as.POSIXlt(Sys.time(), "GMT") # the current time in GMT
format(.leap.seconds) # all 22 leapseconds in your timezone
```

diff *Lagged Differences*

Description

Returns suitably lagged and iterated differences.

Usage

```
diff(x, ...)

## Default S3 method:
diff(x, lag = 1, differences = 1, ...)

## S3 method for class 'POSIXt':
diff(x, lag = 1, differences = 1, ...)
```

Arguments

x	a numeric vector or matrix containing the values to be differenced.
lag	an integer indicating which lag to use.
differences	an integer indicating the order of the difference.
...	further arguments to be passed to or from methods.

Details

`diff` is a generic function with a default method and ones for classes "ts" and "POSIXt". NA's propagate.

Value

If x is a vector of length n and `differences=1`, then the computed result is equal to the successive differences `x[(1+lag):n] - x[1:(n-lag)]`.

If `difference` is larger than one this algorithm is applied recursively to x. Note that the returned value is a vector which is shorter than x.

If x is a matrix then the difference operations are carried out on each column separately.

References

Becker, R. A., Chambers, J. M. and Wilks, A. R. (1988) *The New S Language.* Wadsworth & Brooks/Cole.

See Also

diff.ts, diffinv.

Examples

```
diff(1:10, 2)
diff(1:10, 2, 2)
x <- cumsum(cumsum(1:10))
diff(x, lag = 2)
diff(x, differences = 2)

diff(.leap.seconds)
```

difftime   *Time Intervals*

Description

Create, print and round time intervals.

Usage

```
time1 - time2
difftime(time1, time2, tz = "",
  units = c("auto","secs","mins","hours","days","weeks"))
as.difftime(tim, format = "%X")

## S3 method for class 'difftime':
round(x, digits = 0)
```

Arguments

time1, time2	date-time objects.
tz	a timezone specification to be used for the conversion. System-specific, but "" is the current time zone, and "GMT" is UTC.
units	character. Units in which the results are desired. Can be abbreviated.
tim	character string specifying a time interval.
format	character specifying the format of tim.
x	an object inheriting from class "difftime".
digits	integer. Number of significant digits to retain.

Details

Function difftime takes a difference of two date/time objects (of either class) and returns an object of class "difftime" with an attribute indicating the units. There is a round method for objects of this class, as well as methods for the group-generic (see Ops) logical and arithmetic operations.

If units = "auto", a suitable set of units is chosen, the largest possible (excluding "weeks") in which all the absolute differences are greater than one.

Subtraction of two date-time objects gives an object of this class, by calling difftime with units="auto". Alternatively, as.difftime() works on character-coded time intervals.

Limited arithmetic is available on "difftime" objects: they can be added or subtracted, and multiplied or divided by a numeric vector. In addition, adding or subtracting a numeric vector implicitly converts the numeric vector to a "difftime" object with the same units as the "difftime" object.

See Also

DateTimeClasses.

Examples

```
(z <- Sys.time() - 3600)
Sys.time() - z                # just over 3600 seconds.

## time interval between releases of 1.2.2 and 1.2.3.
ISOdate(2001, 4, 26) - ISOdate(2001, 2, 26)

as.difftime(c("0:3:20", "11:23:15"))
# 3rd gives NA
as.difftime(c("3:20", "23:15", "2:"), format= "%H:%M")
```

hist.POSIXt   *Histogram of a Date-Time Object*

Description

Method for `hist` applied to date-time objects.

Usage

```
## S3 method for class 'POSIXt':
hist(x, breaks, ..., plot = TRUE, freq = FALSE,
     start.on.monday = TRUE, format)
```

Arguments

x	an object inheriting from class `"POSIXt"`.
breaks	a vector of cut points *or* number giving the number of intervals which x is to be cut into *or* an interval specification, one of `"secs"`, `"mins"`, `"hours"`, `"days"`, `"weeks"`, `"months"` or `"years"`.
...	graphical parameters, or arguments to `hist.default` such as `include.lowest`, `right` and `labels`.
plot	logical. If TRUE (default), a histogram is plotted, otherwise a list of breaks and counts is returned.
freq	logical; if TRUE, the histogram graphic is a representation of frequencies, i.e, the `counts` component of the result; if FALSE, *relative* frequencies ("probabilities") are plotted.
start.on.monday	logical. If `breaks = "weeks"`, should the week start on Mondays or Sundays?
format	for the x-axis labels. See `strptime`.

Value

An object of class `"histogram"`: see `hist`.

See Also

`seq.POSIXt`, `axis.POSIXct`, `hist`

Examples

```
hist(.leap.seconds, "years", freq = TRUE)
hist(.leap.seconds,
 seq(ISOdate(1970, 1, 10), ISOdate(2002, 1, 1), "5 years"))

## 100 random dates in a 10-week period
random.dates <- ISOdate(2001, 1, 1) + 70*86400*runif(100)
hist(random.dates, "weeks", format = "%d %b")
```

print.ts *Printing Time-Series Objects*

Description

Print method for time series objects.

Usage

```
## S3 method for class 'ts':
print(x, calendar, ...)
```

Arguments

x   a time series object.

calendar   enable/disable the display of information about month names, quarter names or year when printing. The default is TRUE for a frequency of 4 or 12, FALSE otherwise.

...   additional arguments to print.

Details

This is the print methods for objects inheriting from class "ts".

See Also

print, ts.

Examples

```
print(ts(1:10, freq = 7, start = c(12, 2)), calendar=TRUE)
```

| rep | *Replicate Elements of Vectors and Lists* |

Description

`rep` replicates the values in `x`. It is a generic function, and the default method is described here.

`rep.int` is a faster simplified version for the commonest case.

Usage

```
rep(x, times, ...)

## Default S3 method:
rep(x, times, length.out, each, ...)

rep.int(x, times)
```

Arguments

x	a vector (of any mode including a list) or a pairlist or a `POSIXct` or `POSIXlt` object.
times	non-negative integer. A vector giving the number of times to repeat each element if of length `length(x)`, or to repeat the whole vector if of length 1.
length.out	integer. (Optional.) The desired length of the output vector.
each	optional integer. Each element of `x` is repeated `each` times.
...	further arguments to be passed to or from other methods.

Details

If `times` consists of a single integer, the result consists of the values in `x` repeated this many times. If `times` is a vector of the same length as `x`, the result consists of `x[1]` repeated `times[1]` times, `x[2]` repeated `times[2]` times and so on.

`length.out` may be given in place of `times`, in which case `x` is repeated as many times as is necessary to create a vector of this length. If both `length.out` and `times` are specified, `times` determines the replication, and `length.out` can be used to truncate the output vector (or extend it by `NA`s).

Non-integer values of `times` will be truncated towards zero. If `times` is a computed quantity it is prudent to add a small fuzz.

Value

A vector of the same class as x.

Note

If the original vector has names, these are also replicated and so will almost always contain duplicates.

If `length.out` is used to extend the vector, the behaviour is different from that of S-PLUS, which recycles the existing vector.

Function `rep.int` is a simple case handled by internal code, and provided as a separate function purely for S compatibility.

References

Becker, R. A., Chambers, J. M. and Wilks, A. R. (1988) *The New S Language.* Wadsworth & Brooks/Cole.

See Also

`seq`, `sequence`.

Examples

```
rep(1:4, 2)
rep(1:4, each = 2)         # not the same.
rep(1:4, c(2,2,2,2))       # same as second.
rep(1:4, c(2,1,2,1))
rep(1:4, each = 2, len = 4)   # first 4 only.
rep(1:4, each = 2, len = 10)  # 8 integers plus two NAs

rep(1, 40*(1-.8)) # length 7 on most platforms
rep(1, 40*(1-.8)+1e-7) # better

## replicate a list
fred <- list(happy = 1:10, name = "squash")
rep(fred, 5)
```

```
# date-time objects
x <- .leap.seconds[1:3]
rep(x, 2)
rep(as.POSIXlt(x), rep(2, 3))
```

round.POSIXt   *Round / Truncate Data-Time Objects*

Description

Round or truncate date-time objects.

Usage

```
## S3 method for class 'POSIXt':
round(x, units=c("secs", "mins", "hours", "days"))
## S3 method for class 'POSIXt':
trunc(x, units=c("secs", "mins", "hours", "days"))
```

Arguments

x   an object inheriting from "POSIXt".

units   one of the units listed. Can be abbreviated.

Details

The time is rounded or truncated to the second, minute, hour or day. Timezones are only relevant to days, when midnight in the current timezone is used.

Value

An object of class "POSIX1t".

See Also

DateTimeClasses

Examples

```
round(.leap.seconds + 1000, "hour")
trunc.POSIXt(Sys.time(), "day")
```

seq.POSIXt   *Generate Regular Sequences of Dates*

Description

The method for seq for data-time classes.

Usage

```
## S3 method for class 'POSIXt':
seq(from, to, by, length.out=NULL, along.with=NULL, ...)
```

Arguments

from	starting date. Required
to	end date. Optional. If supplied must be after from.
by	increment of the sequence. Optional. See Details.
length.out	integer, optional. desired length of the sequence.
along.with	take the length from the length of this argument.
...	arguments passed to or from other methods.

Details

by can be specified in several ways.

- A number, taken to be in seconds.
- A object of class difftime
- A character string, containing one of "sec", "min", "hour", "day", "DSTday", "week", "month" or "year". This can optionally be preceded by an integer and a space, or followed by "s".

The difference between "day" and "DSTday" is that the former ignores changes to/from daylight savings time and the latter takes the same clock time each day. ("week" ignores DST, but "7 DSTdays") can be used as an alternative. "month" and "year" allow for DST as from R 1.5.0.)

Value

A vector of class "POSIXct".

See Also

`DateTimeClasses`

Examples

```
## first days of years
seq(ISOdate(1910,1,1), ISOdate(1999,1,1), "years")
## by month
seq(ISOdate(2000,1,1), by="month", length=12)
## quarters
seq(ISOdate(1990,1,1), ISOdate(2000,1,1), by="3 months")
## days vs DSTdays
seq(ISOdate(2000,3,20), by="day", length = 10)
seq(ISOdate(2000,3,20), by="DSTday", length = 10)
seq(ISOdate(2000,3,20), by="7 DSTdays", length = 4)
```

start *Encode the Terminal Times of Time Series*

Description

Extract and encode the times the first and last observations were taken. Provided only for compatibility with S version 2.

Usage

```
start(x, ...)
end(x, ...)
```

Arguments

x
: a univariate or multivariate time-series, or a vector or matrix.

...
: extra arguments for future methods.

Details

These are generic functions, which will use the tsp attribute of x if it exists. Their default methods decode the start time from the original time units, so that for a monthly series 1995.5 is represented as c(1995, 7). For a series of frequency f, time n+i/f is presented as c(n, i+1) (even for i = 0 and f = 1).

Warning

The representation used by start and end has no meaning unless the frequency is supplied.

See Also

ts, time, tsp.

strptime *Date-time Conversion Functions to and from Character*

Description

Functions to convert between character representations and objects of classes "POSIXlt" and "POSIXct" representing calendar dates and times.

Usage

```
## S3 method for class 'POSIXct':
format(x, format = "", tz = "", usetz = FALSE, ...)
## S3 method for class 'POSIXlt':
format(x, format = "", usetz = FALSE, ...)

## S3 method for class 'POSIXt':
as.character(x, ...)

strftime(x, format="", usetz = FALSE, ...)
strptime(x, format)

ISOdatetime(year, month, day, hour, min, sec, tz = "")
ISOdate(year, month, day, hour=12, min=0, sec=0, tz="GMT")
```

Arguments

x	An object to be converted.
tz	A timezone specification to be used for the conversion. System-specific, but "" is the current time zone, and "GMT" is UTC.
format	A character string. The default is "%Y-%m-%d %H:%M:%S" if any component has a time component which is not midnight, and "%Y-%m-%d" otherwise.
...	Further arguments to be passed from or to other methods.
usetz	logical. Should the timezone be appended to the output? This is used in printing time, and as a workaround for problems with using "%Z" on most GNU/Linux systems.

`year, month, day`
    numerical values to specify a day.

`hour, min, sec`
    numerical values for a time within a day.

Details

`strftime` is an alias for `format.POSIXlt`, and `format.POSIXct` first converts to class `"POSIXct"` by calling `as.POSIXct`. Note that only that conversion depends on the time zone.

The usual vector re-cycling rules are applied to `x` and `format` so the answer will be of length that of the longer of the vectors.

Locale-specific conversions to and from character strings are used where appropriate and available. This affects the names of the days and months, the AM/PM indicator (if used) and the separators in formats such as %x and %X.

The details of the formats are system-specific, but the following are defined by the POSIX standard for `strftime` and are likely to be widely available. Any character in the format string other than the % escapes is interpreted literally (and %% gives %).

%a Abbreviated weekday name.

%A Full weekday name.

%b Abbreviated month name.

%B Full month name.

%c Date and time, locale-specific.

%d Day of the month as decimal number (01–31).

%H Hours as decimal number (00–23).

%I Hours as decimal number (01–12).

%j Day of year as decimal number (001–366).

%m Month as decimal number (01–12).

%M Minute as decimal number (00–59).

%p AM/PM indicator in the locale. Used in conjunction with %I and not with %H.

%S Second as decimal number (00–61), allowing for up to two leap-seconds.

%U Week of the year as decimal number (00–53) using the first Sunday as day 1 of week 1.

%w Weekday as decimal number (0–6, Sunday is 0).

%W Week of the year as decimal number (00–53) using the first Monday as day 1 of week 1.

%x Date, locale-specific.

%X Time, locale-specific.

%y Year without century (00–99). If you use this on input, which century you get is system-specific. So don't! Often values up to 69 are prefixed by 20 and 70–99 by 19.

%Y Year with century.

%Z (output only.) Time zone as a character string (empty if not available). Note: do not use this on GNU/Linux unless the TZ environment variable is set.

Where leading zeros are shown they will be used on output but are optional on input.

ISOdatetime and ISOdate are convenience wrappers for strptime, that differ only in their defaults.

Value

The format methods and strftime return character vectors representing the time.

strptime turns character representations into an object of class "POSIXlt".

ISOdatetime and ISOdate return an object of class "POSIXct".

Note

The default formats follow the rules of the ISO 8601 international standard which expresses a day as "2001-02-03" and a time as "14:01:02" using leading zeroes as here. The ISO form uses no space to separate dates and times.

If the date string does not specify the date completely, the returned answer may be system-specific. The most common behaviour is to assume that unspecified seconds, minutes or hours are zero, and a missing year, month or day is the current one.

If the timezone specified is invalid on your system, what happens is system-specific but it will probably be ignored.

OS facilities will probably not print years before 1CE (aka 1AD) correctly.

References

International Organization for Standardization (1988, 1997, ...) *ISO 8601. Data elements and interchange formats – Information interchange – Representation of dates and times.* The 1997 version is available on-line at `ftp://ftp.qsl.net/pub/g1smd/8601v03.pdf`

See Also

DateTimeClasses for details of the date-time classes; `locales` to query or set a locale.

Your system's help pages on `strftime` and `strptime` to see how to specify their formats.

Examples

```
## locale-specific version of date()
format(Sys.time(), "%a %b %d %X %Y")
## we would include the timezone as in format(Sys.time(),
## "%a %b %d %X %Y %Z") but this crashes some GNU/Linux
## systems

## read in date info in format 'ddmmmyyyy' This will give
## NA(s) in some locales; setting the C locale as in the
## commented lines will overcome this on most systems. lct
## <- Sys.getlocale("LC_TIME"); Sys.setlocale("LC_TIME",
## "C")
x <- c("1jan1960", "2jan1960", "31mar1960", "30jul1960")
z <- strptime(x, "%d%b%Y")
## Sys.setlocale("LC_TIME", lct)
z

## read in date/time info in format 'm/d/y h:m:s'
dates <- c("02/27/92", "02/27/92", "01/14/92",
           "02/28/92", "02/01/92")
times <- c("23:03:20", "22:29:56", "01:03:30",
           "18:21:03", "16:56:26")
x <- paste(dates, times)
z <- strptime(x, "%m/%d/%y %H:%M:%S")
z
```

Sys.time *Get Current Time and Timezone*

Description

Sys.time returns the system's idea of the current time and Sys.timezone returns the current time zone.

Usage

```
Sys.time()
Sys.timezone()
```

Value

Sys.time returns an object of class "POSIXct" (see DateTimeClasses).

Sys.timezone returns an OS-specific character string, possibly an empty string.

See Also

date for the system time in a fixed-format character string.

Examples

```
Sys.time()
## locale-specific version of date()
format(Sys.time(), "%a %b %d %X %Y")

Sys.timezone()
```

time — *Sampling Times of Time Series*

Description

`time` creates the vector of times at which a time series was sampled.

`cycle` gives the positions in the cycle of each observation.

`frequency` returns the number of samples per unit time and `deltat` the time interval between observations (see `ts`).

Usage

```
time(x, ...)
## Default S3 method:
time(x, offset=0, ...)

cycle(x, ...)
frequency(x, ...)
deltat(x, ...)
```

Arguments

x	a univariate or multivariate time-series, or a vector or matrix.
offset	can be used to indicate when sampling took place in the time unit. 0 (the default) indicates the start of the unit, 0.5 the middle and 1 the end of the interval.
...	extra arguments for future methods.

Details

These are all generic functions, which will use the `tsp` attribute of `x` if it exists. `time` and `cycle` have methods for class `ts` that coerce the result to that class.

References

Becker, R. A., Chambers, J. M. and Wilks, A. R. (1988) *The New S Language.* Wadsworth & Brooks/Cole.

See Also

ts, start, tsp, window.

date for clock time, system.time for CPU usage.

Examples

```
data(presidents)
cycle(presidents)
# a simple series plot: c() makes the x and y arguments
# into vectors
plot(c(time(presidents)), c(presidents), type="l")
```

ts   *Time-Series Objects*

Description

The function `ts` is used to create time-series objects.

`as.ts` and `is.ts` coerce an object to a time-series and test whether an object is a time series.

Usage

```
ts(data = NA, start = 1, end = numeric(0), frequency = 1,
   deltat = 1, ts.eps = getOption("ts.eps"), class = ,
   names = )
as.ts(x)
is.ts(x)
```

Arguments

data	a numeric vector or matrix of the observed time-series values. A data frame will be coerced to a numeric matrix via `data.matrix`.
start	the time of the first observation. Either a single number or a vector of two integers, which specify a natural time unit and a (1-based) number of samples into the time unit. See the examples for the use of the second form.
end	the time of the last observation, specified in the same way as `start`.
frequency	the number of observations per unit of time.
deltat	the fraction of the sampling period between successive observations; e.g., 1/12 for monthly data. Only one of `frequency` or `deltat` should be provided.
ts.eps	time series comparison tolerance. Frequencies are considered equal if their absolute difference is less than `ts.eps`.
class	class to be given to the result, or none if NULL or "none". The default is "ts" for a single series, c("mts", "ts") for multiple series.

names	a character vector of names for the series in a multiple series: defaults to the colnames of data, or Series 1, Series 2, ...
x	an arbitrary R object.

Details

The function ts is used to create time-series objects. These are vector or matrices with class of "ts" (and additional attributes) which represent data which has been sampled at equispaced points in time. In the matrix case, each column of the matrix data is assumed to contain a single (univariate) time series. Time series must have an least one observation, and although they need not be numeric there is very limited support for non-numeric series.

Class "ts" has a number of methods. In particular arithmetic will attempt to align time axes, and subsetting to extract subsets of series can be used (e.g., EuStockMarkets[, "DAX"]). However, subsetting the first (or only) dimension will return a matrix or vector, as will matrix subsetting.

The value of argument frequency is used when the series is sampled an integral number of times in each unit time interval. For example, one could use a value of 7 for frequency when the data are sampled daily, and the natural time period is a week, or 12 when the data are sampled monthly and the natural time period is a year. Values of 4 and 12 are assumed in (e.g.) print methods to imply a quarterly and monthly series respectively.

as.ts will use the tsp attribute of the object if it has one to set the start and end times and frequency.

is.ts tests if an object is a time series. It is generic: you can write methods to handle specific classes of objects, see InternalMethods.

References

Becker, R. A., Chambers, J. M. and Wilks, A. R. (1988) *The New S Language.* Wadsworth & Brooks/Cole.

See Also

tsp, frequency, start, end, time, window; print.ts, the print method for time series objects; plot.ts, the plot method for time series objects. Standard package ts for many additional time-series functions.

Examples

```
# 2nd Quarter of 1959
ts(1:10, frequency = 4, start = c(1959, 2))
# print.ts(.)
print(ts(1:10, freq = 7, start = c(12, 2)), calendar=TRUE)
## Using July 1954 as start date:
gnp <- ts(cumsum(1 + round(rnorm(100), 2)),
          start = c(1954, 7), frequency = 12)
plot(gnp) # using 'plot.ts' for time-series plot

## Multivariate
z <- ts(matrix(rnorm(300), 100, 3), start=c(1961, 1),
        frequency=12)
class(z)
plot(z)
plot(z, plot.type="single", lty=1:3)

## A phase plot:
data(nhtemp)
plot(nhtemp, c(nhtemp[-1], NA), cex = .8, col="blue",
     main = "Lag plot of New Haven temperatures")
## a clearer way to do this would be
library(ts)
plot(nhtemp, lag(nhtemp, 1), cex = .8, col="blue",
     main = "Lag plot of New Haven temperatures")
```

ts-methods   *Methods for Time Series Objects*

Description

Methods for objects of class "ts", typically the result of ts.

Usage

```
## S3 method for class 'ts':
diff(x, lag=1, differences=1, ...)

## S3 method for class 'ts':
na.omit(object, ...)
```

Arguments

x	an object of class "ts" containing the values to be differenced.
lag	an integer indicating which lag to use.
differences	an integer indicating the order of the difference.
object	a univariate or multivariate time series.
...	further arguments to be passed to or from methods.

Details

The na.omit method omits initial and final segments with missing values in one or more of the series. 'Internal' missing values will lead to failure.

Value

For the na.omit method, a time series without missing values. The class of object will be preserved.

See Also

diff; na.omit, na.fail, na.contiguous.

| tsp | *Tsp Attribute of Time-Series-like Objects* |

Description

`tsp` returns the `tsp` attribute (or `NULL`). It is included for compatibility with S version 2. `tsp<-` sets the `tsp` attribute. `hasTsp` ensures `x` has a `tsp` attribute, by adding one if needed.

Usage

```
tsp(x)
tsp(x) <- value
hasTsp(x)
```

Arguments

x   a vector or matrix or univariate or multivariate time-series.

value   a numeric vector of length 3 or NULL.

Details

The `tsp` attribute was previously described here as `c(start(x), end(x), frequency(x))`, but this is incorrect. It gives the start time *in time units*, the end time and the frequency.

Assignments are checked for consistency.

Assigning `NULL` which removes the `tsp` attribute *and* any "ts" class of x.

References

Becker, R. A., Chambers, J. M. and Wilks, A. R. (1988) *The New S Language.* Wadsworth & Brooks/Cole.

See Also

`ts`, `time`, `start`.

weekdays — *Extract Parts of a POSIXt Object*

Description

Extract the weekday, month or quarter, or the Julian time (days since some origin). These are generic functions: the methods for the internal date-time classes are documented here.

Usage

```
weekdays(x, abbreviate)
## S3 method for class 'POSIXt':
weekdays(x, abbreviate = FALSE)

months(x, abbreviate)
## S3 method for class 'POSIXt':
months(x, abbreviate = FALSE)

quarters(x, abbreviate)
## S3 method for class 'POSIXt':
quarters(x, ...)

julian(x, ...)
## S3 method for class 'POSIXt':
julian(x, origin = as.POSIXct("1970-01-01", tz="GMT"), ...)
```

Arguments

x	an object inheriting from class "POSIXt".
abbreviate	logical. Should the names be abbreviated?
origin	an length-one object inheriting from class "POSIXt".
...	arguments for other methods.

Value

weekdays and months return a character vector of names in the locale in use.

quarters returns a character vector of "Q1" to "Q4".

julian returns the number of days (possibly fractional) since the origin, with the origin as a "origin" attribute.

Note

Other components such as the day of the month or the year are very easy to compute: just use `as.POSIXlt` and extract the relevant component.

See Also

`DateTimeClasses`

Examples

```
weekdays(.leap.seconds)
months(.leap.seconds)
quarters(.leap.seconds)
```

window   *Time Windows*

Description

window is a generic function which extracts the subset of the object x observed between the times `start` and `end`. If a frequency is specified, the series is then re-sampled at the new frequency.

Usage

```
window(x, ...)

## S3 method for class 'ts':
window(x, ...)

## Default S3 method:
window(x, start = NULL, end = NULL,
       frequency = NULL, deltat = NULL, extend = FALSE, ...)
```

Arguments

x	a time-series or other object.
start	the start time of the period of interest.
end	the end time of the period of interest.
frequency, deltat	
	the new frequency can be specified by either (or both if they are consistent).
extend	logical. If true, the `start` and `end` values are allowed to extend the series. If false, attempts to extend the series give a warning and are ignored.
...	further arguments passed to or from other methods.

Details

The start and end times can be specified as for `ts`. If there is no observation at the new `start` or `end`, the immediately following (`start`) or preceding (`end`) observation time is used.

Value

The value depends on the method. `window.default` will return a vector or matrix with an appropriate `tsp` attribute.

`window.ts` differs from `window.default` only in ensuring the result is a `ts` object.

If `extend = TRUE` the series will be padded with `NA` if needed.

References

Becker, R. A., Chambers, J. M. and Wilks, A. R. (1988) *The New S Language*. Wadsworth & Brooks/Cole.

See Also

`time`, `ts`.

Examples

```
data(presidents)
window(presidents, 1960, c(1969,4)) # values in the 1960's
window(presidents, deltat=1)   # All Qtr1s
window(presidents, start=c(1945,3), deltat=1)   # All Qtr3s
window(presidents, 1944, c(1979,2), extend=TRUE)
```

Chapter 6

Base package — datasets

airmiles   *Passenger Miles on Commercial US Airlines, 1937–1960*

Description

The revenue passenger miles flown by commercial airlines in the United States for each year from 1937 to 1960.

Usage

```
data(airmiles)
```

Format

A time-series of 24 observations; yearly, 1937–1960.

Source

F.A.A. Statistical Handbook of Aviation.

References

Brown, R. G. (1963) *Smoothing, Forecasting and Prediction of Discrete Time Series.* Prentice-Hall.

Examples

```
data(airmiles)
plot(airmiles, main = "airmiles data",
 xlab = "Passenger-miles flown by US commercial airlines",
 col = 4)
```

airquality   *New York Air Quality Measurements*

Description

Daily air quality measurements in New York, May to September 1973.

Usage

data(airquality)

Format

A data frame with 154 observations on 6 variables.

[,1]	Ozone	numeric	Ozone (ppb)
[,2]	Solar.R	numeric	Solar R (lang)
[,3]	Wind	numeric	Wind (mph)
[,4]	Temp	numeric	Temperature (degrees F)
[,5]	Month	numeric	Month (1–12)
[,6]	Day	numeric	Day of month (1–31)

Details

Daily readings of the following air quality values for May 1, 1973 (a Tuesday) to September 30, 1973.

- Ozone: Mean ozone in parts per billion from 1300 to 1500 hours at Roosevelt Island
- Solar.R: Solar radiation in Langleys in the frequency band 4000–7700 Angstroms from 0800 to 1200 hours at Central Park
- Wind: Average wind speed in miles per hour at 0700 and 1000 hours at LaGuardia Airport
- Temp: Maximum daily temperature in degrees Fahrenheit at La Guardia Airport.

Source

The data were obtained from the New York State Department of Conservation (ozone data) and the National Weather Service (meteorological data).

References

Chambers, J. M., Cleveland, W. S., Kleiner, B. and Tukey, P. A. (1983) *Graphical Methods for Data Analysis.* Belmont, CA: Wadsworth.

Examples

```
data(airquality)
pairs(airquality, panel = panel.smooth,
      main = "airquality data")
```

anscombe — Anscombe's Quartet of "Identical" Simple Linear Regressions

Description

Four x-y datasets which have the same traditional statistical properties (mean, variance, correlation, regression line, etc.), yet are quite different.

Usage

```
data(anscombe)
```

Format

A data frame with 11 observations on 8 variables.

x1 == x2 == x3	the integers 4:14, specially arranged
x4	values 8 and 19
y1, y2, y3, y4	numbers in (3, 12.5) with mean 7.5 and sdev 2.03

Source

Tufte, Edward R. (1989) *The Visual Display of Quantitative Information*, 13–14. Graphics Press.

References

Anscombe, Francis J. (1973) Graphs in statistical analysis. *American Statistician*, **27**, 17–21.

Examples

```
data(anscombe)
summary(anscombe)

## now some "magic" to do the 4 regressions in a loop:
ff <- y ~ x
for(i in 1:4) {
  ff[2:3] <- lapply(paste(c("y","x"), i, sep=""), as.name)
  ## or    ff[[2]] <- as.name(paste("y", i, sep=""))
  ##       ff[[3]] <- as.name(paste("x", i, sep=""))
```

```
  assign(paste("lm.",i,sep=""),
         lmi <- lm(ff, data= anscombe))
  print(anova(lmi))
}

## See how close they are (numerically!)
sapply(objects(pat="lm\.[1-4]$"), function(n) coef(get(n)))
lapply(objects(pat="lm\.[1-4]$"),
       function(n) summary(get(n))$coef)

## Now, do what you should have done in the first place:
## PLOTS
op <- par(mfrow=c(2,2), mar=.1+c(4,4,1,1), oma= c(0,0,2,0))
for(i in 1:4) {
  ff[2:3] <- lapply(paste(c("y","x"), i, sep=""), as.name)
  plot(ff, data =anscombe, col="red", pch=21, bg="orange",
       cex=1.2, xlim=c(3,19), ylim=c(3,13))
  abline(get(paste("lm.",i,sep="")), col="blue")
}
mtext("Anscombe's 4 Regression data sets", outer = TRUE,
      cex=1.5)
par(op)
```

| attenu | *The Joyner–Boore Attenuation Data* |

Description

This data gives peak accelerations measured at various observation stations for 23 earthquakes in California. The data have been used by various workers to estimate the attenuating affect of distance on ground acceleration.

Usage

data(attenu)

Format

A data frame with 182 observations on 5 variables.

[,1]	event	numeric	Event Number
[,2]	mag	numeric	Moment Magnitude
[,3]	station	factor	Station Number
[,4]	dist	numeric	Station-hypocenter distance (km)
[,5]	accel	numeric	Peak acceleration (g)

Source

Joyner, W.B., D.M. Boore and R.D. Porcella (1981). Peak horizontal acceleration and velocity from strong-motion records including records from the 1979 Imperial Valley, California earthquake. USGS Open File report 81-365. Menlo Park, Ca.

References

Boore, D. M. and Joyner, W.B.(1982) The empirical prediction of ground motion, *Bull. Seism. Soc. Am.*, **72**, S269–S268.

Bolt, B. A. and Abrahamson, N. A. (1982) New attenuation relations for peak and expected accelerations of strong ground motion, *Bull. Seism. Soc. Am.*, **72**, 2307–2321.

Bolt B. A. and Abrahamson, N. A. (1983) Reply to W. B. Joyner & D. M. Boore's "Comments on: New attenuation relations for peak and expected accelerations for peak and expected accelerations of strong ground motion", *Bull. Seism. Soc. Am.*, **73**, 1481–1483.

Brillinger, D. R. and Preisler, H. K. (1984) An exploratory analysis of the Joyner-Boore attenuation data, *Bull. Seism. Soc. Am.*, **74**, 1441–1449.

Brillinger, D. R. and Preisler, H. K. (1984) *Further analysis of the Joyner-Boore attenuation data.* Manuscript.

Examples

```
data(attenu)
## check the data class of the variables
sapply(attenu, data.class)
summary(attenu)
pairs(attenu, main = "attenu data")
coplot(accel ~ dist | as.factor(event), data = attenu,
       show = FALSE)
coplot(log(accel) ~ log(dist) | as.factor(event),
       data = attenu, panel = panel.smooth,
       show.given = FALSE)
```

attitude — The Chatterjee-Price Attitude Data

Description

From a survey of the clerical employees of a large financial organization, the data are aggregated from the questionnaires of the approximately 35 employees for each of 30 (randomly selected) departments. The numbers give the percent proportion of favourable responses to seven questions in each department.

Usage

```
data(attitude)
```

Format

A dataframe with 30 observations on 7 variables. The first column are the short names from the reference, the second one the variable names in the data frame:

Y	rating	numeric	Overall rating
X[1]	complaints	numeric	Handling of employee complaints
X[2]	privileges	numeric	Does not allow special privileges
X[3]	learning	numeric	Opportunity to learn
X[4]	raises	numeric	Raises based on performance
X[5]	critical	numeric	Too critical
X[6]	advancel	numeric	Advancement

Source

Chatterjee, S. and Price, B. (1977) *Regression Analysis by Example.* New York: Wiley. (Section 3.7, p.68ff of 2nd ed.(1991).)

Examples

```
data(attitude)
pairs(attitude, main = "attitude data")
summary(attitude)
summary(fm1 <- lm(rating ~ ., data = attitude))
opar <- par(mfrow = c(2, 2), oma = c(0, 0, 1.1, 0),
            mar = c(4.1, 4.1, 2.1, 1.1))
plot(fm1)
```

```
summary(fm2 <- lm(rating ~ complaints, data = attitude))
plot(fm2)
par(opar)
```

cars — Speed and Stopping Distances of Cars

Description

The data gives the speed of cars and the distances taken to stop. Note that the data were recorded in the 1920s.

Usage

data(cars)

Format

A data frame with 50 observations on 2 variables.

[,1]	speed	numeric	Speed (mph)
[,2]	dist	numeric	Stopping distance (ft)

Source

Ezekiel, M. (1930) *Methods of Correlation Analysis.* Wiley.

References

McNeil, D. R. (1977) *Interactive Data Analysis.* Wiley.

Examples

```
data(cars)
plot(cars, xlab = "Speed (mph)",
    ylab = "Stopping distance (ft)", las = 1)
lines(lowess(cars$speed, cars$dist, f = 2/3, iter = 3),
    col = "red")
title(main = "cars data")
plot(cars, xlab = "Speed (mph)",
    ylab = "Stopping distance (ft)", las = 1, log = "xy")
title(main = "cars data (logarithmic scales)")
lines(lowess(cars$speed, cars$dist, f = 2/3, iter = 3),
    col = "red")
summary(fm1 <- lm(log(dist) ~ log(speed), data = cars))
opar <- par(mfrow = c(2, 2), oma = c(0, 0, 1.1, 0),
        mar = c(4.1, 4.1, 2.1, 1.1))
```

```
plot(fm1)
par(opar)

## An example of polynomial regression
plot(cars, xlab = "Speed (mph)",
    ylab = "Stopping distance (ft)",
    las = 1, xlim = c(0, 25))
d <- seq(0, 25, len = 200)
for(degree in 1:4) {
  fm <- lm(dist ~ poly(speed, degree), data = cars)
  assign(paste("cars", degree, sep="."), fm)
  lines(d, predict(fm, data.frame(speed=d)), col = degree)
}
anova(cars.1, cars.2, cars.3, cars.4)
```

chickwts	Chicken Weights by Feed Type

Description

An experiment was conducted to measure and compare the effectiveness of various feed supplements on the growth rate of chickens.

Usage

```
data(chickwts)
```

Format

A data frame with 71 observations on 2 variables.

weight a numeric variable giving the chick weight.

feed a factor giving the feed type.

Details

Newly hatched chicks were randomly allocated into six groups, and each group was given a different feed supplement. Their weights in grams after six weeks are given along with feed types.

Source

Anonymous (1948) *Biometrika*, **35**, 214.

References

McNeil, D. R. (1977) *Interactive Data Analysis.* New York: Wiley.

Examples

```
data(chickwts)
boxplot(weight ~ feed, data = chickwts, col = "lightgray",
    varwidth = TRUE, notch = TRUE, main = "chickwt data",
    ylab = "Weight at six weeks (gm)")
anova(fm1 <- lm(weight ~ feed, data = chickwts))
opar <- par(mfrow = c(2, 2), oma = c(0, 0, 1.1, 0),
            mar = c(4.1, 4.1, 2.1, 1.1))
plot(fm1)
par(opar)
```

co2   *Mauna Loa Atmospheric CO2 Concentration*

Description

Atmospheric concentrations of CO_2 are expressed in parts per million (ppm) and reported in the preliminary 1997 SIO manometric mole fraction scale.

Usage

```
data(co2)
```

Format

A time series of 468 observations; monthly from 1959 to 1997.

Details

The values for February, March and April of 1964 were missing and have been obtained by interpolating linearly between the values for January and May of 1964.

Source

Keeling, C. D. and Whorf, T. P., Scripps Institution of Oceanography (SIO), University of California, La Jolla, California USA 92093-0220.

ftp://cdiac.esd.ornl.gov/pub/maunaloa-co2/maunaloa.co2.

References

Cleveland, W. S. (1993) *Visualizing Data.* New Jersey: Summit Press.

Examples

```
data(co2)
plot(co2,
  ylab = expression("Atmospheric concentration of CO"[2]),
  las = 1)
title(main = "co2 data set")
```

data *Data Sets*

Description

Loads specified data sets, or list the available data sets.

Usage

```
data(..., list = character(0), package = .packages(),
    lib.loc = NULL, verbose = getOption("verbose"),
    envir = .GlobalEnv)
```

Arguments

...	a sequence of names or literal character strings.
list	a character vector.
package	a name or character vector giving the packages to look into for data sets. By default, all packages in the search path are used, then the 'data' subdirectory (if present) of the current working directory.
lib.loc	a character vector of directory names of R libraries, or NULL. The default value of NULL corresponds to all libraries currently known. If the default is used, the loaded packages are searched before the libraries.
verbose	a logical. If TRUE, additional diagnostics are printed.
envir	the environment where the data should be loaded.

Details

Currently, four formats of data files are supported:

1. files ending '.R' or '.r' are source()d in, with the R working directory changed temporarily to the directory containing the respective file.
2. files ending '.RData' or '.rda' are load()ed.
3. files ending '.tab', '.txt' or '.TXT' are read using read.table(..., header = TRUE), and hence result in a data frame.
4. files ending '.csv' or '.CSV' are read using read.table(..., header = TRUE, sep = ";"), and also result in a data frame.

If more than one matching file name is found, the first on this list is used.

The data sets to be loaded can be specified as a sequence of names or character strings, or as the character vector list, or as both.

For each given data set, the first two types ('.R' or '.r', and '.RData' or '.rda' files) can create several variables in the load environment, which might all be named differently from the data set. The second two ('.tab', '.txt', or '.TXT', and '.csv' or '.CSV' files) will always result in the creation of a single variable with the same name as the data set.

If no data sets are specified, data lists the available data sets. It looks for a new-style data index in the 'Meta' or, if this is not found, an old-style '00Index' file in the 'data' directory of each specified package, and uses these files to prepare a listing. If there is a 'data' area but no index, available data files for loading are computed and included in the listing, and a warning is given: such packages are incomplete. The information about available data sets is returned in an object of class "packageIQR". The structure of this class is experimental. In earlier versions of R, an empty character vector was returned along with listing available data sets.

If lib.loc is not specified, the data sets are searched for amongst those packages already loaded, followed by the 'data' directory (if any) of the current working directory and then packages in the specified libraries. If lib.loc *is* specified, packages are searched for in the specified libraries, even if they are already loaded from another library.

To just look in the 'data' directory of the current working directory, set package = NULL.

Value

a character vector of all data sets specified, or information about all available data sets in an object of class "packageIQR" if none were specified.

Note

The data files can be many small files. On some file systems it is desirable to save space, and the files in the 'data' directory of an installed package can be zipped up as a zip archive 'Rdata.zip'. You will need to provide a single-column file 'filelist' of file names in that directory.

One can take advantage of the search order and the fact that a '.R' file will change directory. If raw data are stored in 'mydata.txt' then one can set up 'mydata.R' to read 'mydata.txt' and pre-process it, *base — data* 611

e.g., using `transform`. For instance one can convert numeric vectors to factors with the appropriate labels. Thus, the '.R' file can effectively contain a metadata specification for the plaintext formats.

See Also

`help` for obtaining documentation on data sets, `save` for *creating* the second ('.rda') kind of data, typically the most efficient one.

Examples

```
# list all available data sets
data()
# list the data sets in the base package
data(package = "base")
# load the data sets 'USArrests' and 'VADeaths'
data(USArrests, "VADeaths")
# give information on data set 'USArrests'
help(USArrests)
```

discoveries — Yearly Numbers of Important Discoveries

Description

The numbers of "great" inventions and scientific discoveries in each year from 1860 to 1959.

Usage

```
data(discoveries)
```

Format

A time series of 100 values.

Source

The World Almanac and Book of Facts, 1975 Edition, pages 315–318.

References

McNeil, D. R. (1977) *Interactive Data Analysis*. Wiley.

Examples

```
data(discoveries)
plot(discoveries, ylab = "Number of important discoveries",
     las = 1)
title(main = "discoveries data set")
```

esoph — Smoking, Alcohol and (O)esophageal Cancer

Description

Data from a case-control study of (o)esophageal cancer in Ile-et-Vilaine, France.

Usage

data(esoph)

Format

A data frame with records for 88 age/alcohol/tobacco combinations.

[,1]	"agegp"	Age group	1 25–34 years
			2 35–44
			3 45–54
			4 55–64
			5 65–74
			6 75+
[,2]	"alcgp"	Alcohol consumption	1 0–39 gm/day
			2 40–79
			3 80–119
			4 120+
[,3]	"tobgp"	Tobacco consumption	1 0– 9 gm/day
			2 10–19
			3 20–29
			4 30+
[,4]	"ncases"	Number of cases	
[,5]	"ncontrols"	Number of controls	

Author(s)

Thomas Lumley

Source

Breslow, N. E. and Day, N. E. (1980) *Statistical Methods in Cancer Research. 1: The Analysis of Case-Control Studies.* IARC Lyon / Oxford University Press.

Examples

```
data(esoph)
summary(esoph)
## effects of alcohol, tobacco and interaction,
## age-adjusted
model1 <-
  glm(cbind(ncases, ncontrols) ~ agegp + tobgp * alcgp,
      data = esoph, family = binomial())
anova(model1)
## Try a linear effect of alcohol and tobacco
model2 <-
  glm(cbind(ncases, ncontrols) ~ agegp + unclass(tobgp)
      + unclass(alcgp), data = esoph, family = binomial())
summary(model2)
## Re-arrange data for a mosaic plot
ttt <- table(esoph$agegp, esoph$alcgp, esoph$tobgp)
ttt[ttt == 1] <- esoph$ncases
tt1 <- table(esoph$agegp, esoph$alcgp, esoph$tobgp)
tt1[tt1 == 1] <- esoph$ncontrols
tt <- array(c(ttt, tt1), c(dim(ttt),2),
            c(dimnames(ttt), list(c("Cancer", "control"))))
mosaicplot(tt, main = "esoph data set", color = TRUE)
```

euro Conversion Rates of Euro Currencies

Description

Conversion rates between the various Euro currencies.

Usage

```
data(euro)
```

Format

euro is a named vector of length 11, euro.cross a named matrix of size 11 by 11.

Details

The data set euro contains the value of 1 Euro in all currencies participating in the European monetary union (Austrian Schilling ATS, Belgian Franc BEF, German Mark DEM, Spanish Peseta ESP, Finnish Markka FIM, French Franc FRF, Irish Punt IEP, Italian Lira ITL, Luxembourg Franc LUF, Dutch Guilder NLG and Portugese Escudo PTE). These conversion rates were fixed by the European Union on December 31, 1998. To convert old prices to Euro prices, divide by the respective rate and round to 2 digits.

The data set euro.cross contains conversion rates between the various Euro currencies, i.e., the result of outer(1 / euro, euro).

Examples

```
data(euro)
cbind(euro)

## These relations hold:
# [6 digit precision in Euro's definition]
euro == signif(euro,6)
all(euro.cross == outer(1/euro, euro))

## Convert 20 Euro to Belgian Franc
20 * euro["BEF"]
## Convert 20 Austrian Schilling to Euro
20 / euro["ATS"]
```

```
## Convert 20 Spanish Pesetas to Italian Lira
20 * euro.cross["ESP", "ITL"]

dotchart(euro,
  main = "euro data: 1 Euro in currency unit")
dotchart(1/euro,
  main = "euro data: 1 currency unit in Euros")
dotchart(log(euro, 10),
  main = "euro data: log10(1 Euro in currency unit)")
```

eurodist   *Distances Between European Cities*

Description

The data give the road distances (in km) between 21 cities in Europe. The data are taken from a table in "The Cambridge Encyclopaedia".

Usage

data(eurodist)

Format

A dist object based on 21 objects. (You must have the **mva** package loaded to have the methods for this kind of object available).

Source

Crystal, D. Ed. (1990) *The Cambridge Encyclopaedia.* Cambridge: Cambridge University Press,

faithful   *Old Faithful Geyser Data*

Description

Waiting time between eruptions and the duration of the eruption for the Old Faithful geyser in Yellowstone National Park, Wyoming, USA.

Usage

data(faithful)

Format

A data frame with 272 observations on 2 variables.

[,1]	eruptions	numeric	Eruption time in mins
[,2]	waiting	numeric	Waiting time to next eruption

Details

A closer look at faithful$eruptions reveals that these are heavily rounded times originally in seconds, where multiples of 5 are more frequent than expected under non-human measurement. For a "better" version of the eruptions times, see the example below.

There are many versions of this dataset around: Azzalini and Bowman (1990) use a more complete version.

Source

W. Härdle.

References

Härdle, W. (1991) *Smoothing Techniques with Implementation in S.* New York: Springer.

Azzalini, A. and Bowman, A. W. (1990). A look at some data on the Old Faithful geyser. *Applied Statistics* **39**, 357–365.

See Also

geyser in package **MASS** for the Azzalini-Bowman version.

Examples

```
data(faithful)
f.tit <-  "faithful data: Eruptions of Old Faithful"

ne60 <- round(e60 <- 60 * faithful$eruptions)
all.equal(e60, ne60)              # relative diff. ~ 1/10000
table(zapsmall(abs(e60 - ne60))) # 0, 0.02 or 0.04
faithful$better.eruptions <- ne60 / 60
te <- table(ne60)
te[te >= 4] # (too) many multiples of 5 !
plot(names(te), te, type="h", main = f.tit,
     xlab = "Eruption time (sec)")

plot(faithful[, -3], main = f.tit,
     xlab = "Eruption time (min)",
     ylab = "Waiting time to next eruption (min)")
lines(lowess(faithful$eruptions, faithful$waiting,
             f = 2/3, iter = 3),
      col = "red")
```

Formaldehyde	*Determination of Formaldehyde*

Description

These data are from a chemical experiment to prepare a standard curve for the determination of formaldehyde by the addition of chromatropic acid and concentrated sulphuric acid and the reading of the resulting purple color on a spectophotometer.

Usage

```
data(Formaldehyde)
```

Format

A data frame with 6 observations on 2 variables.

[,1]	carb	numeric	Carbohydrate (ml)
[,2]	optden	numeric	Optical Density

Source

Bennett, N. A. and N. L. Franklin (1954) *Statistical Analysis in Chemistry and the Chemical Industry.* New York: Wiley.

References

McNeil, D. R. (1977) *Interactive Data Analysis.* New York: Wiley.

Examples

```
data(Formaldehyde)
plot(optden ~ carb, data = Formaldehyde,
     xlab = "Carbohydrate (ml)", ylab = "Optical Density",
     main = "Formaldehyde data", col = 4, las = 1)
abline(fm1 <- lm(optden ~ carb, data = Formaldehyde))
summary(fm1)
opar <- par(mfrow = c(2,2), oma = c(0, 0, 1.1, 0))
plot(fm1)
par(opar)
```

freeny *Freeny's Revenue Data*

Description

Freeny's data on quarterly revenue and explanatory variables.

Usage

```
data(freeny)
```

Format

There are three 'freeny' data sets.

`freeny.y` is a time series with 39 observations on quarterly revenue from (1962,2Q) to (1971,4Q).

`freeny.x` is a matrix of explanatory variables. The columns are `freeny.y` lagged 1 quarter, price index, income level, and market potential.

Finally, `freeny` is a data frame with variables y, lag.quarterly. revenue, price.index, income.level, and market.potential obtained from the above two data objects.

Source

A. E. Freeny (1977) *A Portable Linear Regression Package with Test Programs.* Bell Laboratories memorandum.

References

Becker, R. A., Chambers, J. M. and Wilks, A. R. (1988) *The New S Language.* Wadsworth & Brooks/Cole.

Examples

```
data(freeny)
summary(freeny)
pairs(freeny, main = "freeny data")
summary(fm1 <- lm(y ~ ., data = freeny))
opar <- par(mfrow = c(2, 2), oma = c(0, 0, 1.1, 0),
            mar = c(4.1, 4.1, 2.1, 1.1))
plot(fm1)
par(opar)
```

HairEyeColor — *Hair and Eye Color of Statistics Students*

Description

Distribution of hair and eye color and sex in 592 statistics students.

Usage

```
data(HairEyeColor)
```

Format

A 3-dimensional array resulting from cross-tabulating 592 observations on 3 variables. The variables and their levels are as follows:

No	Name	Levels
1	Hair	Black, Brown, Red, Blond
2	Eye	Brown, Blue, Hazel, Green
3	Sex	Male, Female

Details

This data set is useful for illustrating various techniques for the analysis of contingency tables, such as the standard chi-squared test or, more generally, log-linear modelling, and graphical methods such as mosaic plots, sieve diagrams or association plots.

References

Snee, R. D. (1974), Graphical display of two-way contingency tables. *The American Statistician*, **28**, 9–12.

Friendly, M. (1992), Graphical methods for categorical data. *SAS User Group International Conference Proceedings*, **17**, 190–200. http://www.math.yorku.ca/SCS/sugi/sugi17-paper.html

Friendly, M. (1992), Mosaic displays for loglinear models. *Proceedings of the Statistical Graphics Section*, American Statistical Association, pp. 61–68. http://www.math.yorku.ca/SCS/Papers/asa92.html

See Also

`chisq.test`, `loglin`, `mosaicplot`

Examples

```
data(HairEyeColor)
## Full mosaic
mosaicplot(HairEyeColor)
## Aggregate over sex:
x <- apply(HairEyeColor, c(1, 2), sum)
x
mosaicplot(x, main = "Relation between hair and eye color")
```

infert — *Infertility after Spontaneous and Induced Abortion*

Description

This is a matched case-control study dating from before the availability of conditional logistic regression.

Usage

data(infert)

Format

1.	Education		0 = 0-5 years
			1 = 6-11 years
			2 = 12+ years
2.	age		age in years of case
3.	parity		count
4.	number of prior induced abortions		0 = 0
			1 = 1
			2 = 2 or more
5.	case status		1 = case
			0 = control
6.	number of prior spontaneous abortions		0 = 0
			1 = 1
			2 = 2 or more
7.	matched set number		1-83
8.	stratum number		1-63

Note

One case with two prior spontaneous abortions and two prior induced abortions is omitted.

Source

Trichopoulos et al. (1976) *Br. J. of Obst. and Gynaec.* **83**, 645–650.

Examples

data(infert)

```
model1 <- glm(case ~ spontaneous+induced, data=infert,
             family=binomial())
summary(model1)
## adjusted for other potential confounders:
summary(model2 <-
  glm(case ~ age+parity+education+spontaneous+induced,
      data=infert, family=binomial()))
## Really should be analysed by conditional logistic
## regression which is in the survival package
if(require(survival)){
  model3 <-
    clogit(case~spontaneous+induced+strata(stratum),
           data=infert)
  summary(model3)
  detach() # survival (conflicts)
}
```

InsectSprays — *Effectiveness of Insect Sprays*

Description

The counts of insects in agricultural experimental units treated with different insecticides.

Usage

```
data(InsectSprays)
```

Format

A data frame with 72 observations on 2 variables.

[,1]	count	numeric	Insect count
[,2]	spray	factor	The type of spray

Source

Beall, G., (1942) The Transformation of data from entomological field experiments, *Biometrika*, **29**, 243–262.

References

McNeil, D. (1977) *Interactive Data Analysis*. New York: Wiley.

Examples

```
data(InsectSprays)
boxplot(count ~ spray, data = InsectSprays,
        xlab = "Type of spray", ylab = "Insect count",
        main = "InsectSprays data", varwidth = TRUE,
        col = "lightgray")
fm1 <- aov(count ~ spray, data = InsectSprays)
summary(fm1)
opar <- par(mfrow = c(2,2), oma = c(0, 0, 1.1, 0))
plot(fm1)
fm2 <- aov(sqrt(count) ~ spray, data = InsectSprays)
summary(fm2)
plot(fm2)
par(opar)
```

| iris | *Edgar Anderson's Iris Data* |

Description

This famous (Fisher's or Anderson's) iris data set gives the measurements in centimeters of the variables sepal length and width and petal length and width, respectively, for 50 flowers from each of 3 species of iris. The species are *Iris setosa, versicolor,* and *virginica.*

Usage

```
data(iris)
data(iris3)
```

Format

iris is a data frame with 150 cases (rows) and 5 variables (columns) named Sepal.Length, Sepal.Width, Petal.Length, Petal.Width, and Species.

iris3 gives the same data arranged as a 3-dimensional array of size 50 by 4 by 3, as represented by S-PLUS. The first dimension gives the case number within the species subsample, the second the measurements with names Sepal L., Sepal W., Petal L., and Petal W., and the third the species.

Source

Fisher, R. A. (1936) The use of multiple measurements in taxonomic problems. *Annals of Eugenics,* **7**, Part II, 179–188.

The data were collected by Anderson, Edgar (1935). The irises of the Gaspe Peninsula, *Bulletin of the American Iris Society,* **59**, 2–5.

References

Becker, R. A., Chambers, J. M. and Wilks, A. R. (1988) *The New S Language.* Wadsworth & Brooks/Cole. (has iris3 as iris.)

See Also

matplot some examples of which use iris.

Examples

```
data(iris3)
dni3 <- dimnames(iris3)
ii <- data.frame(matrix(aperm(iris3, c(1,3,2)), ncol=4,
                dimnames=list(NULL, sub(" L.",".Length",
                sub(" W.",".Width", dni3[[2]])))),
                Species = gl(3,50,lab=sub("S","s",
                sub("V","v",dni3[[3]]))))
data(iris)
all.equal(ii, iris) # TRUE
```

islands *Areas of the World's Major Landmasses*

Description

The areas in thousands of square miles of the landmasses which exceed 10,000 square miles.

Usage

```
data(islands)
```

Format

A named vector of length 48.

Source

The World Almanac and Book of Facts, 1975, page 406.

References

McNeil, D. R. (1977) *Interactive Data Analysis.* Wiley.

Examples

```
data(islands)
dotchart(log(islands, 10),
    main = "islands data: log10(area) (log10(sq. miles))")
dotchart(log(islands[order(islands)], 10),
    main = "islands data: log10(area) (log10(sq. miles))")
```

LifeCycleSavings   *Intercountry Life-Cycle Savings Data*

Description

Data on the savings ratio 1960–1970.

Usage

```
data(LifeCycleSavings)
```

Format

A data frame with 50 observations on 5 variables.

[,1]	sr	numeric	aggregate personal savings
[,2]	pop15	numeric	% of population under 15
[,3]	pop75	numeric	% of population over 75
[,4]	dpi	numeric	real per-capita disposable income
[,5]	ddpi	numeric	% growth rate of dpi

Details

Under the life-cycle savings hypothesis as developed by Franco Modigliani, the savings ratio (aggregate personal saving divided by disposable income) is explained by per-capita disposable income, the percentage rate of change in per-capita disposable income, and two demographic variables: the percentage of population less than 15 years old and the percentage of the population over 75 years old. The data are averaged over the decade 1960–1970 to remove the business cycle or other short-term fluctuations.

Source

The data were obtained from Belsley, Kuh and Welsch (1980). They in turn obtained the data from Sterling (1977).

References

Sterling, Arnie (1977) Unpublished BS Thesis. Massachusetts Institute of Technology.

Belsley, D. A., Kuh. E. and Welsch, R. E. (1980) *Regression Diagnostics*. New York: Wiley.

Examples

```
data(LifeCycleSavings)
pairs(LifeCycleSavings, panel = panel.smooth,
      main = "LifeCycleSavings data")
fm1 <- lm(sr ~ pop15 + pop75 + dpi + ddpi,
          data = LifeCycleSavings)
summary(fm1)
```

longley *Longley's Economic Regression Data*

Description

A macroeconomic data set which provides a well-known example for a highly collinear regression.

Usage

data(longley)

Format

A data frame with 7 economical variables, observed yearly from 1947 to 1962 ($n = 16$).

GNP.deflator: GNP implicit price deflator (1954 = 100)

GNP: Gross National Product.

Unemployed: number of unemployed.

Armed.Forces: number of people in the armed forces.

Population: 'noninstitutionalized' population $\geq$ 14 years of age.

Year: the year (time).

Employed: number of people employed.

The regression lm(Employed ~ .) is known to be highly collinear.

Source

J. W. Longley (1967) An appraisal of least-squares programs from the point of view of the user. *Journal of the American Statistical Association*, **62**, 819–841.

References

Becker, R. A., Chambers, J. M. and Wilks, A. R. (1988) *The New S Language.* Wadsworth & Brooks/Cole.

Examples

```
## give the data set in the form it is used in S-PLUS:
data(longley)
longley.x <- data.matrix(longley[, 1:6])
longley.y <- longley[, "Employed"]
pairs(longley, main = "longley data")
summary(fm1 <- lm(Employed ~ ., data = longley))
opar <- par(mfrow = c(2, 2), oma = c(0, 0, 1.1, 0),
            mar = c(4.1, 4.1, 2.1, 1.1))
plot(fm1)
par(opar)
```

morley — Michaelson-Morley Speed of Light Data

Description

The classical data of Michaelson and Morley on the speed of light. The data consists of five experiments, each consisting of 20 consecutive 'runs'. The response is the speed of light measurement, suitably coded.

Usage

```
data(morley)
```

Format

A data frame contains the following components:

Expt The experiment number, from 1 to 5.

Run The run number within each experiment.

Speed Speed-of-light measurement.

Details

The data is here viewed as a randomized block experiment with 'experiment' and 'run' as the factors. 'run' may also be considered a quantitative variate to account for linear (or polynomial) changes in the measurement over the course of a single experiment.

Source

A. J. Weekes (1986) *A Genstat Primer.* London: Edward Arnold.

Examples

```
data(morley)
morley$Expt <- factor(morley$Expt)
morley$Run <- factor(morley$Run)
attach(morley)
plot(Expt, Speed, main = "Speed of Light Data",
     xlab = "Experiment No.")
fm <- aov(Speed ~ Run + Expt, data = morley)
summary(fm)
```

```
fm0 <- update(fm, . ~ . - Run)
anova(fm0, fm)
detach(morley)
```

mtcars — *Motor Trend Car Road Tests*

Description

The data was extracted from the 1974 *Motor Trend* US magazine, and comprises fuel consumption and 10 aspects of automobile design and performance for 32 automobiles (1973–74 models).

Usage

```
data(mtcars)
```

Format

A data frame with 32 observations on 11 variables.

[, 1]	mpg	Miles/(US) gallon
[, 2]	cyl	Number of cylinders
[, 3]	disp	Displacement (cu.in.)
[, 4]	hp	Gross horsepower
[, 5]	drat	Rear axle ratio
[, 6]	wt	Weight (lb/1000)
[, 7]	qsec	1/4 mile time
[, 8]	vs	V/S
[, 9]	am	Transmission (0 = automatic, 1 = manual)
[,10]	gear	Number of forward gears
[,11]	carb	Number of carburettors

Source

Henderson and Velleman (1981), Building multiple regression models interactively. *Biometrics*, **37**, 391–411.

Examples

```
data(mtcars)
pairs(mtcars, main = "mtcars data")
coplot(mpg ~ disp | as.factor(cyl), data = mtcars,
       panel = panel.smooth, rows = 1)
```

nhtemp   *Average Yearly Temperatures in New Haven*

Description

The mean annual temperature in degrees Fahrenheit in New Haven, Connecticut, from 1912 to 1971.

Usage

```
data(nhtemp)
```

Format

A time series of 60 observations.

Source

Vaux, J. E. and Brinker, N. B. (1972) *Cycles*, **1972**, 117–121.

References

McNeil, D. R. (1977) *Interactive Data Analysis*. New York: Wiley.

Examples

```
data(nhtemp)
plot(nhtemp, main = "nhtemp data",
  ylab = "Mean annual temperature in New Haven (deg F)")
```

OrchardSprays *Potency of Orchard Sprays*

Description

An experiment was conducted to assess the potency of various constituents of orchard sprays in repelling honeybees, using a Latin square design.

Usage

```
data(OrchardSprays)
```

Format

A data frame with 64 observations on 4 variables.

[,1]	rowpos	numeric	Row of the design
[,2]	colpos	numeric	Column of the design
[,3]	treatment	factor	Treatment level
[,4]	decrease	numeric	Response

Details

Individual cells of dry comb were filled with measured amounts of lime sulphur emulsion in sucrose solution. Seven different concentrations of lime sulphur ranging from a concentration of 1/100 to 1/1,562,500 in successive factors of 1/5 were used as well as a solution containing no lime sulphur.

The responses for the different solutions were obtained by releasing 100 bees into the chamber for two hours, and then measuring the decrease in volume of the solutions in the various cells.

An 8 × 8 Latin square design was used and the treatments were coded as follows:

A	highest level of lime sulphur
B	next highest level of lime sulphur
.	
.	
.	
G	lowest level of lime sulphur
H	no lime sulphur

Source

Finney, D. J. (1947) *Probit Analysis.* Cambridge.

References

McNeil, D. R. (1977) *Interactive Data Analysis.* New York: Wiley.

Examples

```
data(OrchardSprays)
pairs(OrchardSprays, main = "OrchardSprays data")
```

phones   *The World's Telephones*

Description

The number of telephones in various regions of the world (in thousands).

Usage

```
data(phones)
```

Format

A matrix with 7 rows and 8 columns. The columns of the matrix give the figures for a given region, and the rows the figures for a year.

The regions are: North America, Europe, Asia, South America, Oceania, Africa, Central America.

The years are: 1951, 1956, 1957, 1958, 1959, 1960, 1961.

Source

AT&T (1961) *The World's Telephones*.

References

McNeil, D. R. (1977) *Interactive Data Analysis*. New York: Wiley.

Examples

```
data(phones)
matplot(rownames(phones), phones, type = "b", log = "y",
        xlab = "Year",
        ylab = "Number of telephones (1000's)")
legend(1951.5, 80000, colnames(phones), col = 1:6,
       lty = 1:5, pch = rep(21, 7))
title(main = "phones data: log scale for response")
```

`PlantGrowth`   *Results from an Experiment on Plant Growth*

Description

Results from an experiment to compare yields (as measured by dried weight of plants) obtained under a control and two different treatment conditions.

Usage

```
data(PlantGrowth)
```

Format

A data frame of 30 cases on 2 variables.

[, 1]	weight	numeric
[, 2]	group	factor

The levels of group are 'ctrl', 'trt1', and 'trt2'.

Source

Dobson, A. J. (1983) *An Introduction to Statistical Modelling.* London: Chapman and Hall.

Examples

```
## One factor ANOVA example from Dobson's book, cf. Table
## 7.4:
data(PlantGrowth)
boxplot(weight ~ group, data = PlantGrowth,
        main = "PlantGrowth data",
        ylab = "Dried weight of plants", col = "lightgray",
        notch = TRUE, varwidth = TRUE)
anova(lm(weight ~ group, data = PlantGrowth))
```

precip — Annual Precipitation in US Cities

Description

The average amount of precipitation (rainfall) in inches for each of 70 United States (and Puerto Rico) cities.

Usage

```
data(precip)
```

Format

A named vector of length 70.

Source

Statistical Abstracts of the United States, 1975.

References

McNeil, D. R. (1977) *Interactive Data Analysis.* New York: Wiley.

Examples

```
data(precip)
dotchart(precip[order(precip)], main = "precip data")
title(sub = "Average annual precipitation (in.)")
```

presidents — Quarterly Approval Ratings of US Presidents

Description

The (approximately) quarterly approval rating for the President of the United states from the first quarter of 1945 to the last quarter of 1974.

Usage

```
data(presidents)
```

Format

A time series of 120 values.

Details

The data are actually a fudged version of the approval ratings. See McNeil's book for details.

Source

The Gallup Organisation.

References

McNeil, D. R. (1977) *Interactive Data Analysis.* New York: Wiley.

Examples

```
data(presidents)
plot(presidents, las = 1, ylab = "Approval rating (%)",
    main = "presidents data")
```

pressure	*Vapor Pressure of Mercury as a Function of Temperature*

Description

Data on the relation between temperature in degrees Celsius and vapor pressure of mercury in millimeters (of mercury).

Usage

data(pressure)

Format

A data frame with 19 observations on 2 variables.

[, 1]	temperature	numeric	temperature (deg C)
[, 2]	pressure	numeric	pressure (mm)

Source

Weast, R. C., ed. (1973) *Handbook of Chemistry and Physics.* CRC Press.

References

McNeil, D. R. (1977) *Interactive Data Analysis.* New York: Wiley.

Examples

```
data(pressure)
plot(pressure, xlab = "Temperature (deg C)",
     ylab = "Pressure (mm of Hg)",
     main = "pressure data: Vapor Pressure of Mercury")
plot(pressure, xlab = "Temperature (deg C)",   log = "y",
     ylab = "Pressure (mm of Hg)",
     main = "pressure data: Vapor Pressure of Mercury")
```

quakes   *Locations of Earthquakes off Fiji*

Description

The data set give the locations of 1000 seismic events of MB > 4.0 The events occurred in a cube near Fiji since 1964.

Usage

    data(quakes)

Format

A data frame with 1000 observations on 5 variables.

[,1]	lat	numeric	Latitude of event
[,2]	long	numeric	Longitude
[,3]	depth	numeric	Depth (km)
[,4]	mag	numeric	Richter Magnitude
[,5]	stations	numeric	Number of stations reporting

Details

There are two clear planes of seismic activity. One is a major plate junction; the other is the Tonga trench off New Zealand. These data constitute a subsample from a larger dataset of containing 5000 observations.

Source

This is one of the Harvard PRIM-H project data sets. They in turn obtained it from Dr. John Woodhouse, Dept. of Geophysics, Harvard University.

Examples

    data(quakes)
    pairs(quakes, main = "Fiji Earthquakes, N = 1000",
          cex.main=1.2, pch=".")

randu *Random Numbers from Congruential Generator RANDU*

Description

400 triples of successive random numbers were taken from the VAX FORTRAN function RANDU running under VMS 1.5.

Usage

data(randu)

Format

A data frame with 400 observations on 3 variables named x, y and z which give the first, second and third random number in the triple.

Details

In three dimensional displays it is evident that the triples fall on 15 parallel planes in 3-space. This can be shown theoretically to be true for all triples from the RANDU generator.

These particular 400 triples start 5 apart in the sequence, that is they are ((U[5i+1], U[5i+2], U[5i+3]), i= 0, ..., 399), and they are rounded to 6 decimal places.

Under VMS versions 2.0 and higher, this problem has been fixed.

Source

David Donoho

Examples

```
## We could re-generate the dataset by the following R code
seed <- as.double(1)
RANDU <- function() {
    seed <<- ((2^16 + 3) * seed) %% (2^31)
    seed/(2^31)
}
for(i in 1:400) {
    U <- c(RANDU(), RANDU(), RANDU(), RANDU(), RANDU())
    print(round(U[1:3], 6))
}
```

rivers   *Lengths of Major North American Rivers*

Description

This data set gives the lengths (in miles) of 141 "major" rivers in North America, as compiled by the US Geological Survey.

Usage

data(rivers)

Format

A vector containing 141 observations.

Source

World Almanac and Book of Facts, 1975, page 406.

References

McNeil, D. R. (1977) *Interactive Data Analysis.* New York: Wiley.

sleep — *Student's Sleep Data*

Description

Data which show the effect of two soporific drugs (increase in hours of sleep) on groups consisting of 10 patients each.

Usage

```
data(sleep)
```

Format

A data frame with 20 observations on 2 variables.

[, 1]	extra	numeric	increase in hours of sleep
[, 2]	group	factor	patient group

Source

Student (1908) The probable error of the mean. *Biometrika*, **6**, 20.

References

Scheffé, Henry (1959) *The Analysis of Variance.* New York, NY: Wiley.

Examples

```
data(sleep)
## ANOVA
anova(lm(extra ~ group, data = sleep))
```

stackloss   *Brownlee's Stack Loss Plant Data*

Description

Operational data of a plant for the oxidation of ammonia to nitric acid.

Usage

```
data(stackloss)
```

Format

stackloss is a data frame with 21 observations on 4 variables.

[,1]	Air Flow	Flow of cooling air
[,2]	Water Temp	Cooling Water Inlet Temperature
[,3]	Acid Conc.	Concentration of acid [per 1000, minus 500]
[,4]	stack.loss	Stack loss

For compatibility with S-PLUS, the data sets stack.x, a matrix with the first three (independent) variables of the data frame, and stack.loss, the numeric vector giving the fourth (dependent) variable, are provided as well.

Details

"Obtained from 21 days of operation of a plant for the oxidation of ammonia (NH_3) to nitric acid (HNO_3). The nitric oxides produced are absorbed in a countercurrent absorption tower". (Brownlee, cited by Dodge, slightly reformatted by MM.)

Air Flow represents the rate of operation of the plant. Water Temp is the temperature of cooling water circulated through coils in the absorption tower. Acid Conc. is the concentration of the acid circulating, minus 50, times 10: that is, 89 corresponds to 58.9 per cent acid. stack.loss (the dependent variable) is 10 times the percentage of the ingoing ammonia to the plant that escapes from the absorption column unabsorbed; that is, an (inverse) measure of the over-all efficiency of the plant.

Source

Brownlee, K. A. (1960, 2nd ed. 1965) *Statistical Theory and Methodology in Science and Engineering.* New York: Wiley. pp. 491–500.

References

Becker, R. A., Chambers, J. M. and Wilks, A. R. (1988) *The New S Language.* Wadsworth & Brooks/Cole.

Dodge, Y. (1996) The guinea pig of multiple regression. In: *Robust Statistics, Data Analysis, and Computer Intensive Methods; In Honor of Peter Huber's 60th Birthday, 1996, Lecture Notes in Statistics* **109**, Springer-Verlag, New York.

Examples

```
data(stackloss)
summary(lm.stack <- lm(stack.loss ~ stack.x))
```

state	US State Facts and Figures

Description

Data sets related to the 50 states of the United States of America.

Usage

data(state)

Details

R currently contains the following "state" data sets. Note that all data are arranged according to alphabetical order of the state names.

state.abb: character vector of 2-letter abbreviations for the state names.

state.area: numeric vector of state areas (in square miles).

state.center: list with components named x and y giving the approximate geographic center of each state in negative longitude and latitude. Alaska and Hawaii are placed just off the West Coast.

state.division: factor giving state divisions (New England, Middle Atlantic, South Atlantic, East South Central, West South Central, East North Central, West North Central, Mountain, and Pacific).

state.name: character vector giving the full state names.

state.region: factor giving the region (Northeast, South, North Central, West) that each state belongs to.

state.x77: matrix with 50 rows and 8 columns giving the following statistics in the respective columns.

> Population: population estimate as of July 1, 1975
>
> Income: per capita income (1974)
>
> Illiteracy: illiteracy (1970, percent of population)
>
> Life Exp: life expectancy in years (1969–71)
>
> Murder: murder and non-negligent manslaughter rate per 100,000 population (1976)
>
> HS Grad: percent high-school graduates (1970)
>
> Frost: mean number of days with minimum temperature below freezing (1931–1960) in capital or large city
>
> Area: land area in square miles

Source

U.S. Department of Commerce, Bureau of the Census (1977) *Statistical Abstract of the United States.*

U.S. Department of Commerce, Bureau of the Census (1977) *County and City Data Book.*

References

Becker, R. A., Chambers, J. M. and Wilks, A. R. (1988) *The New S Language.* Wadsworth & Brooks/Cole.

sunspots   *Monthly Sunspot Numbers, 1749–1983*

Description

Monthly mean relative sunspot numbers from 1749 to 1983. Collected at Swiss Federal Observatory, Zurich until 1960, then Tokyo Astronomical Observatory.

Usage

```
data(sunspots)
```

Format

A time series of monthly data from 1749 to 1983.

Source

Andrews, D. F. and Herzberg, A. M. (1985) *Data: A Collection of Problems from Many Fields for the Student and Research Worker*. New York: Springer-Verlag.

See Also

sunspot.month (package **ts**) has a longer (and a bit different) series.

Examples

```
data(sunspots)
plot(sunspots, main = "sunspots data", xlab = "Year",
     ylab = "Monthly sunspot numbers")
```

swiss — Swiss Fertility and Socioeconomic Indicators (1888) Data

Description

Standardized fertility measure and socio-economic indicators for each of 47 French-speaking provinces of Switzerland at about 1888.

Usage

data(swiss)

Format

A data frame with 47 observations on 6 variables, *each* of which is in percent, i.e., in $[0, 100]$.

[,1]	Fertility	I_g, "common standardized fertility measure"
[,2]	Agriculture	% of males involved in agricultural occupation
[,3]	Examination	% "draftees" with highest mark on army exam
[,4]	Education	% "draftees" educated beyond primary school
[,5]	Catholic	% catholic (as opposed to "protestant")
[,6]	Infant.Mortality	live births who live less than 1 year.

All variables but 'Fertility' give proportions of the population.

Details

(paraphrasing Mosteller and Tukey):

Switzerland, in 1888, was entering a period known as the "demographic transition"; i.e., its fertility was beginning to fall from the high level typical of underdeveloped countries.

The data collected are for 47 French-speaking "provinces" at about 1888.

Here, all variables are scaled to $[0, 100]$, where in the original, all but "Catholic" were scaled to $[0, 1]$.

Source

Project "16P5", pages 549–551 in

Mosteller, F. and Tukey, J. W. (1977) *Data Analysis and Regression: A Second Course in Statistics.* Addison-Wesley, Reading Mass.

indicating their source as "Data used by permission of Franice van de Walle. Office of Population Research, Princeton University, 1976. Unpublished data assembled under NICHD contract number No 1-HD-O-2077."

References

Becker, R. A., Chambers, J. M. and Wilks, A. R. (1988) *The New S Language.* Wadsworth & Brooks/Cole.

Examples

```
data(swiss)
pairs(swiss, panel = panel.smooth, main = "swiss data",
      col = 3 + (swiss$Catholic > 50))
summary(lm(Fertility ~ . , data = swiss))
```

Titanic *Survival of passengers on the Titanic*

Description

This data set provides information on the fate of passengers on the fatal maiden voyage of the ocean liner 'Titanic', summarized according to economic status (class), sex, age and survival.

Usage

```
data(Titanic)
```

Format

A 4-dimensional array resulting from cross-tabulating 2201 observations on 4 variables. The variables and their levels are as follows:

No	Name	Levels
1	Class	1st, 2nd, 3rd, Crew
2	Sex	Male, Female
3	Age	Child, Adult
4	Survived	No, Yes

Details

The sinking of the Titanic is a famous event, and new books are still being published about it. Many well-known facts—from the proportions of first-class passengers to the "women and children first" policy, and the fact that that policy was not entirely successful in saving the women and children in the third class—are reflected in the survival rates for various classes of passenger.

These data were originally collected by the British Board of Trade in their investigation of the sinking. Note that there is not complete agreement among primary sources as to the exact numbers on board, rescued, or lost.

Source

Dawson, Robert J. MacG. (1995), The 'Unusual Episode' Data Revisited. *Journal of Statistics Education*, **3**. http://www.amstat.org/publications/jse/v3n3/datasets.dawson.html

The source provides a data set recording class, sex, age, and survival status for each person on board of the Titanic, and is based on data originally collected by the British Board of Trade and reprinted in:

British Board of Trade (1990), *Report on the Loss of the 'Titanic' (S.S.)*. British Board of Trade Inquiry Report (reprint). Gloucester, UK: Allan Sutton Publishing.

Examples

```
data(Titanic)
mosaicplot(Titanic, main = "Survival on the Titanic")
## Higher survival rates in children?
apply(Titanic, c(3, 4), sum)
## Higher survival rates in females?
apply(Titanic, c(2, 4), sum)
## Use loglm() in package 'MASS' for further analysis ...
```

ToothGrowth   *The Effect of Vitamin C on Tooth Growth in Guinea Pigs*

Description

The response is the length of odontoblasts (teeth) in each of 10 guinea pigs at each of three dose levels of Vitamin C (0.5, 1, and 2 mg) with each of two delivery methods (orange juice or ascorbic acid).

Usage

```
data(ToothGrowth)
```

Format

A data frame with 60 observations on 3 variables.

[,1]	len	numeric	Tooth length
[,2]	supp	factor	Supplement type (VC or OJ).
[,3]	dose	numeric	Dose in milligrams.

Source

C. I. Bliss (1952) *The Statistics of Bioassay.* Academic Press.

References

McNeil, D. R. (1977) *Interactive Data Analysis.* New York: Wiley.

Examples

```
data(ToothGrowth)
coplot(len ~ dose | supp, data = ToothGrowth,
       panel = panel.smooth,
       xlab = "length vs dose, given type of supplement")
```

trees Girth, Height and Volume for Black Cherry Trees

Description

This data set provides measurements of the girth, height and volume of timber in 31 felled black cherry trees. Note that girth is the diameter of the tree (in inches) measured at 4 ft 6 in above the ground.

Usage

```
data(trees)
```

Format

A data frame with 31 observations on 3 variables.

[,1]	Girth	numeric	Tree diameter in inches
[,2]	Height	numeric	Height in ft
[,3]	Volume	numeric	Volume of timber in cubic ft

Source

Ryan, T. A., Joiner, B. L. and Ryan, B. F. (1976) *The Minitab Student Handbook.* Duxbury Press.

References

Atkinson, A. C. (1985) *Plots, Transformations and Regression.* Oxford University Press.

Examples

```
data(trees)
pairs(trees, panel = panel.smooth, main = "trees data")
plot(Volume ~ Girth, data = trees, log = "xy")
coplot(log(Volume) ~ log(Girth) | Height, data = trees,
       panel = panel.smooth)
summary(fm1 <- lm(log(Volume) ~ log(Girth), data=trees))
summary(fm2 <- update(fm1, ~ . + log(Height), data=trees))
step(fm2)
## i.e., Volume ~= c * Height * Girth^2 seems reasonable
```

UCBAdmissions — *Student Admissions at UC Berkeley*

Description

Aggregate data on applicants to graduate school at Berkeley for the six largest departments in 1973 classified by admission and sex.

Usage

```
data(UCBAdmissions)
```

Format

A 3-dimensional array resulting from cross-tabulating 4526 observations on 3 variables. The variables and their levels are as follows:

No	Name	Levels
1	Admit	Admitted, Rejected
2	Gender	Male, Female
3	Dept	A, B, C, D, E, F

Details

This data set is frequently used for illustrating Simpson's paradox, see Bickel et al. (1975). At issue is whether the data show evidence of sex bias in admission practices. There were 2691 male applicants, of whom 1198 (44.5%) were admitted, compared with 1835 female applicants of whom 557 (30.4%) were admitted. This gives a sample odds ratio of 1.83, indicating that males were almost twice as likely to be admitted. In fact, graphical methods (as in the example below) or log-linear modelling show that the apparent association between admission and sex stems from differences in the tendency of males and females to apply to the individual departments (females used to apply "more" to departments with higher rejection rates).

This data set can also be used for illustrating methods for graphical display of categorical data, such as the general-purpose mosaic plot or the "fourfold display" for 2-by-2-by-k tables. See the home page of Michael Friendly (http://www.math.yorku.ca/SCS/friendly.html) for further information.

References

Bickel, P. J., Hammel, E. A., and O'Connell, J. W. (1975) Sex bias in graduate admissions: Data from Berkeley. *Science*, **187**, 398–403.

Examples

```
data(UCBAdmissions)
## Data aggregated over departments
apply(UCBAdmissions, c(1, 2), sum)
mosaicplot(apply(UCBAdmissions, c(1, 2), sum),
           main = "Student admissions at UC Berkeley")
## Data for individual departments
opar <- par(mfrow = c(2, 3), oma = c(0, 0, 2, 0))
for(i in 1:6)
  mosaicplot(UCBAdmissions[,,i],
    xlab = "Admit", ylab = "Sex",
    main = paste("Department", LETTERS[i]))
mtext(expression(bold("Student admissions at Berkeley")),
      outer = TRUE, cex = 1.5)
par(opar)
```

USArrests   *Violent Crime Rates by US State*

Description

This data set contains statistics, in arrests per 100,000 residents for assault, murder, and rape in each of the 50 US states in 1973. Also given is the percent of the population living in urban areas.

Usage

```
data(USArrests)
```

Format

A data frame with 50 observations on 4 variables.

[,1]	Murder	numeric	Murder arrests (per 100,000)
[,2]	Assault	numeric	Assault arrests (per 100,000)
[,3]	UrbanPop	numeric	Percent urban population
[,4]	Rape	numeric	Rape arrests (per 100,000)

Source

World Almanac and Book of facts 1975. (Crime rates).

Statistical Abstracts of the United States 1975. (Urban rates).

References

McNeil, D. R. (1977) *Interactive Data Analysis.* New York: Wiley.

See Also

The state data sets.

Examples

```
data(USArrests)
pairs(USArrests, panel = panel.smooth,
      main = "USArrests data")
```

USJudgeRatings — Lawyers' Ratings of State Judges in the US Superior Court

Description

Lawyers' ratings of state judges in the US Superior Court.

Usage

```
data(USJudgeRatings)
```

Format

A data frame containing 43 observations on 12 numeric variables.

[,1]	CONT	Number of contacts of lawyer with judge.
[,2]	INTG	Judicial integrity.
[,3]	DMNR	Demeanor.
[,4]	DILG	Diligence.
[,5]	CFMG	Case flow managing.
[,6]	DECI	Prompt decisions.
[,7]	PREP	Preparation for trial.
[,8]	FAMI	Familiarity with law.
[,9]	ORAL	Sound oral rulings.
[,10]	WRIT	Sound written rulings.
[,11]	PHYS	Physical ability.
[,12]	RTEN	Worthy of retention.

Source

New Haven Register, 14 January, 1977 (from John Hartigan).

Examples

```
data(USJudgeRatings)
pairs(USJudgeRatings, main = "USJudgeRatings data")
```

USPersonalExpenditure — *Personal Expenditure Data*

Description

This data set consists of United States personal expenditures (in billions of dollars) in the categories; food and tobacco, household operation, medical and health, personal care, and private education for the years 1940, 1945, 1950, and 1960.

Usage

```
data(USPersonalExpenditure)
```

Format

A matrix with 5 rows and 5 columns.

Source

The World Almanac and Book of Facts, 1962, page 756.

References

Tukey, J. W. (1977) *Exploratory Data Analysis.* Addison-Wesley.

McNeil, D. R. (1977) *Interactive Data Analysis.* Wiley.

Examples

```
data(USPersonalExpenditure)
USPersonalExpenditure
eda::medpolish(log10(USPersonalExpenditure))
```

uspop *Populations Recorded by the US Census*

Description

This data set gives the population of the United States (in millions) as recorded by the decennial census for the period 1790–1970.

Usage

data(uspop)

Format

A time series of 19 values.

Source

McNeil, D. R. (1977) *Interactive Data Analysis.* New York: Wiley.

Examples

```
data(uspop)
plot(uspop, log = "y", main = "uspop data", xlab = "Year",
    ylab = "U.S. Population (millions)")
```

VADeaths   *Death Rates in Virginia (1940)*

Description

Death rates per 100 in Virginia in 1940.

Usage

```
data(VADeaths)
```

Format

A matrix with 5 rows and 5 columns.

Details

The death rates are cross-classified by age group (rows) and population group (columns). The age groups are: 50–54, 55–59, 60–64, 65–69, 70–74 and the population groups are Rural/Male, Rural/Female, Urban/Male and Urban/Female.

This provides a rather nice 3-way analysis of variance example.

Source

Moyneau, L., Gilliam, S. K., and Florant, L. C.(1947) Differences in Virginia death rates by color, sex, age, and rural or urban residence. *American Sociological Review*, **12**, 525–535.

References

McNeil, D. R. (1977) *Interactive Data Analysis.* Wiley.

Examples

```
data(VADeaths)
n <- length(dr <- c(VADeaths))
nam <- names(VADeaths)
d.VAD <- data.frame(
  Drate = dr,
  age = rep(ordered(rownames(VADeaths)),length=n),
  gender= gl(2,5,n, labels= c("M", "F")),
  site =  gl(2,10,   labels= c("rural", "urban")))
```

```
coplot(Drate ~ as.numeric(age) | gender * site,
       data = d.VAD, panel = panel.smooth,
       xlab = "VADeaths data - Given: gender")
summary(aov.VAD <- aov(Drate ~ .^2, data = d.VAD))
opar <- par(mfrow = c(2,2), oma = c(0, 0, 1.1, 0))
plot(aov.VAD)
par(opar)
```

volcano	*Topographic Information on Auckland's Maunga Whau*

Volcano

Description

Maunga Whau (Mt Eden) is one of about 50 volcanos in the Auckland volcanic field. This data set gives topographic information for Maunga Whau on a 10m by 10m grid.

Usage

```
data(volcano)
```

Format

A matrix with 87 rows and 61 columns, rows corresponding to grid lines running east to west and columns to grid lines running south to north.

Source

Digitized from a topographic map by Ross Ihaka. These data should not be regarded as accurate.

See Also

`filled.contour` for a nice plot.

Examples

```
data(volcano)
filled.contour(volcano, color = terrain.colors, asp = 1)
title(main = "volcano data: filled contour map")
```

warpbreaks — The Number of Breaks in Yarn during Weaving

Description

This data set gives the number of warp breaks per loom, where a loom corresponds to a fixed length of yarn.

Usage

```
data(warpbreaks)
```

Format

A data frame with 54 observations on 3 variables.

[,1]	breaks	numeric	The number of breaks
[,2]	wool	factor	The type of wool (A or B)
[,3]	tension	factor	The level of tension (L, M, H)

There are measurements on 9 looms for each of the six types of warp (AL, AM, AH, BL, BM, BH).

Source

Tippett, L. H. C. (1950) *Technological Applications of Statistics*. Wiley. Page 106.

References

Tukey, J. W. (1977) *Exploratory Data Analysis*. Addison-Wesley.

McNeil, D. R. (1977) *Interactive Data Analysis*. Wiley.

See Also

xtabs for ways to display these data as a table.

Examples

```
data(warpbreaks)
summary(warpbreaks)
opar <- par(mfrow = c(1,2), oma = c(0, 0, 1.1, 0))
plot(breaks ~ tension, data=warpbreaks, col="lightgray",
     varwidth=TRUE, subset = wool == "A", main="Wool A")
```

```
plot(breaks ~ tension, data=warpbreaks, col="lightgray",
     varwidth=TRUE, subset = wool == "B", main="Wool B")
mtext("warpbreaks data", side = 3, outer = TRUE)
par(opar)
summary(fm1 <- lm(breaks ~ wool*tension, data=warpbreaks))
anova(fm1)
```

women   *Average Heights and Weights for American Women*

Description

This data set gives the average heights and weights for American women aged 30–39.

Usage

```
data(women)
```

Format

A data frame with 15 observations on 2 variables.

[,1]	height	numeric	Height (in)
[,2]	weight	numeric	Weight (lbs)

Details

The data set appears to have been taken from the American Society of Actuaries *Build and Blood Pressure Study* for some (unknown to us) earlier year.

The World Almanac notes: "The figures represent weights in ordinary indoor clothing and shoes, and heights with shoes".

Source

The World Almanac and Book of Facts, 1975.

References

McNeil, D. R. (1977) *Interactive Data Analysis.* Wiley.

Examples

```
data(women)
plot(women, xlab = "Height (in)", ylab = "Weight (lb)",
     main = "women data: American women aged 30-39")
```

Other Books From The Publisher

Network Theory publishes books about free software under free documentation licenses. Our current catalogue includes the following titles:

- **Comparing and Merging Files with GNU diff and patch** by David MacKenzie, Paul Eggert, and Richard Stallman (ISBN 0-9541617-5-0) $19.95 (£12.95)

- **Version Management with CVS** by Per Cederqvist et al. (ISBN 0-9541617-1-8) $29.95 (£19.95)

- **GNU Bash Reference Manual** by Chet Ramey and Brian Fox (ISBN 0-9541617-7-7) $29.95 (£19.95)

- **An Introduction to R** by W.N. Venables, D.M. Smith and the R Development Core Team (ISBN 0-9541617-4-2) $19.95 (£12.95)

- **GNU Octave Manual** by John W. Eaton (ISBN 0-9541617-2-6) $29.99 (£19.99)

- **GNU Scientific Library Reference Manual - Second Edition** by M. Galassi, J. Davies, J. Theiler, B. Gough, G. Jungman, M. Booth, F. Rossi (ISBN 0-9541617-3-4) $39.99 (£24.99)

- **An Introduction to Python** by Guido van Rossum and Fred L. Drake, Jr. (ISBN 0-9541617-6-9) $19.95 (£12.95)

- **Python Language Reference Manual** by Guido van Rossum and Fred L. Drake, Jr. (ISBN 0-9541617-8-5) $19.95 (£12.95)

All titles are available for order from bookstores worldwide. Sales of the manuals fund the development of more free software and documentation. For details visit the website http://www.network-theory.co.uk/

Index

bold numbers denote primary references to sections.
roman numbers denote references to entries within a section.
italic numbers denote references to an entry in the text of another section.

*Topic **NA**
    naprint, 495
    naresid, 496
*Topic **algebra**
    backsolve, 252
    chol, 257
    chol2inv, 260
    colSums, 262
    crossprod, 266
    eigen, 271
    matrix, 290
    qr, 298
    QR.Auxiliaries, 301
    solve, 308
    svd, 319
*Topic **aplot**
    abline, 6
    arrows, 8
    axis, 12
    box, 24
    bxp, 31
    contour, 40
    coplot, 44
    filled.contour, 61
    frame, 67
    grid, 71
    Hershey, 74
    image, 86
    Japanese, 92
    legend, 98
    lines, 104
    matplot, 108
    mtext, 116
    persp, 139
    plot.window, 170
    plot.xy, 172
    plotmath, 173
    points, 180
    polygon, 183
    rect, 200
    rug, 204
    screen, 206
    segments, 209
    symbols, 223
    text, 228
    title, 231
*Topic **arith**
    all.equal, 245
    approxfun, 247
    Arithmetic, 250
    cumsum, 267
    diff, 565
    Extremes, 274
    findInterval, 278
    gl, 280
    matmult, 289
    ppoints, 192
    prod, 297
    range, 303
    Round, 305

sign, 307
sort, 310
sum, 318
tabulate, 321
*Topic **array**
   backsolve, 252
   chol, 257
   chol2inv, 260
   colSums, 262
   contrast, 417
   crossprod, 266
   eigen, 271
   expand.grid, 429
   lm.fit, 463
   matmult, 289
   matplot, 108
   matrix, 290
   qr, 298
   QR.Auxiliaries, 301
   svd, 319
*Topic **category**
   gl, 280
   loglin, 475
   plot.table, 166
*Topic **character**
   strwidth, 218
*Topic **chron**
   as.POSIX*, 558
   axis.POSIXct, 15
   cut.POSIXt, 560
   DateTimeClasses, 562
   difftime, 567
   hist.POSIXt, 569
   rep, 572
   round.POSIXt, 575
   seq.POSIXt, 576
   strptime, 579
   Sys.time, 583
   weekdays, 591
*Topic **color**
   col2rgb, 36
   colors, 39
   gray, 70

hsv, 82
palette, 123
Palettes, 125
rgb, 203
*Topic **datasets**
   airmiles, 596
   airquality, 597
   anscombe, 599
   attenu, 601
   attitude, 603
   cars, 605
   chickwts, 607
   co2, 608
   data, 609
   discoveries, 612
   esoph, 613
   euro, 615
   eurodist, 617
   faithful, 618
   Formaldehyde, 620
   freeny, 621
   HairEyeColor, 622
   infert, 624
   InsectSprays, 626
   iris, 627
   islands, 629
   LifeCycleSavings, 630
   longley, 632
   morley, 634
   mtcars, 636
   nhtemp, 637
   OrchardSprays, 638
   phones, 640
   PlantGrowth, 641
   precip, 642
   presidents, 643
   pressure, 644
   quakes, 645
   randu, 646
   rivers, 647
   sleep, 648
   stackloss, 649
   state, 651

INDEX

sunspots, 653
swiss, 654
Titanic, 656
ToothGrowth, 658
trees, 659
UCBAdmissions, 660
USArrests, 662
USJudgeRatings, 663
USPersonalExpenditure, 664
uspop, 665
VADeaths, 666
volcano, 668
warpbreaks, 669
women, 671

∗Topic **design**
contrast, 417
contrasts, 419
TukeyHSD, 547

∗Topic **device**
dev.xxx, 50
dev2, 52
Devices, 57
Gnome, 68
gtk, 73
pdf, 137
pictex, 143
png, 177
postscript, 186
quartz, 198
screen, 206
x11, 234
xfig, 236

∗Topic **distribution**
bandwidth, 324
Beta, 326
Binomial, 328
birthday, 330
Cauchy, 332
Chisquare, 334
density, 337
Exponential, 342
FDist, 344

GammaDist, 346
Geometric, 348
hist, 78
Hypergeometric, 350
Logistic, 352
Lognormal, 354
Multinomial, 356
NegBinomial, 358
Normal, 361
Poisson, 363
ppoints, 192
qqnorm, 196
r2dtable, 365
Random, 367
Random.user, 372
sample, 374
SignRank, 376
TDist, 378
Tukey, 380
Uniform, 382
Weibull, 384
Wilcoxon, 386

∗Topic **documentation**
data, 609

∗Topic **dplot**
approxfun, 247
axTicks, 17
boxplot.stats, 29
col2rgb, 36
colors, 39
convolve, 264
fft, 276
hist, 78
hist.POSIXt, 569
hsv, 82
jitter, 93
layout, 95
n2mfrow, 119
Palettes, 125
panel.smooth, 127
par, 128
plot.density, 154
ppoints, 192

pretty, 194
screen, 206
splinefun, 315
strwidth, 218
units, 233
xy.coords, 238
xyz.coords, 240
∗Topic **environment**
layout, 95
par, 128
∗Topic **hplot**
assocplot, 10
barplot, 19
boxplot, 25
chull, 34
contour, 40
coplot, 44
curve, 48
dotchart, 59
filled.contour, 61
fourfoldplot, 64
hist, 78
hist.POSIXt, 569
image, 86
interaction.plot, 89
matplot, 108
mosaicplot, 112
pairs, 120
panel.smooth, 127
persp, 139
pie, 145
plot, 147
plot.data.frame, 149
plot.default, 150
plot.design, 155
plot.factor, 158
plot.formula, 159
plot.histogram, 161
plot.lm, 163
plot.table, 166
plot.ts, 168
qqnorm, 196
stars, 211

stripchart, 216
sunflowerplot, 220
symbols, 223
termplot, 226
∗Topic **iplot**
dev.xxx, 50
frame, 67
identify, 84
layout, 95
locator, 106
par, 128
plot.histogram, 161
recordPlot, 199
∗Topic **logic**
all.equal, 245
∗Topic **manip**
cut.POSIXt, 560
expand.model.frame, 430
model.extract, 487
rep, 572
seq.POSIXt, 576
sort, 310
∗Topic **math**
abs, 244
Bessel, 254
convolve, 264
deriv, 268
fft, 276
Hyperbolic, 281
integrate, 282
kappa, 285
log, 287
nextn, 292
poly, 293
polyroot, 295
Special, 313
splinefun, 315
Trig, 322
∗Topic **methods**
plot.data.frame, 149
∗Topic **models**
add1, 390
AIC, 393

INDEX

alias, 395
anova, 398
anova.glm, 399
anova.lm, 401
aov, 403
AsIs, 405
C, 407
case/variable.names, 409
coef, 411
confint, 412
deviance, 421
df.residual, 422
dummy.coef, 423
eff.aovlist, 425
effects, 427
expand.grid, 429
extractAIC, 432
factor.scope, 434
family, 436
fitted, 439
formula, 440
glm, 443
glm.control, 449
glm.summaries, 451
is.empty.model, 457
labels, 458
lm.summaries, 468
logLik, 470
logLik.glm, 472
logLik.lm, 473
loglin, 475
make.link, 483
makepredictcall, 484
manova, 486
model.extract, 487
model.frame, 489
model.matrix, 491
model.tables, 493
naprint, 495
naresid, 496
offset, 501
power, 512
predict.glm, 513

preplot, 193
profile, 517
proj, 518
relevel, 521
replications, 522
residuals, 524
se.contrast, 525
stat.anova, 527
step, 529
summary.aov, 532
summary.glm, 534
summary.lm, 537
summary.manova, 540
terms, 542
terms.formula, 543
terms.object, 545
TukeyHSD, 547
update, 551
update.formula, 553
vcov, 554
∗Topic **multivariate**
    stars, 211
    symbols, 223
∗Topic **nonlinear**
    deriv, 268
    nlm, 497
    optim, 502
    vcov, 554
∗Topic **nonparametric**
    sunflowerplot, 220
∗Topic **optimize**
    constrOptim, 414
    glm.control, 449
    nlm, 497
    optim, 502
    optimize, 509
    uniroot, 549
∗Topic **print**
    labels, 458
∗Topic **programming**
    model.extract, 487
∗Topic **regression**
    anova, 398

anova.glm, 399
anova.lm, 401
aov, 403
case/variable.names, 409
coef, 411
contrast, 417
contrasts, 419
df.residual, 422
effects, 427
expand.model.frame, 430
fitted, 439
glm, 443
glm.summaries, 451
influence.measures, 453
lm, 459
lm.fit, 463
lm.influence, 465
lm.summaries, 468
ls.diag, 478
ls.print, 480
lsfit, 481
plot.lm, 163
predict.glm, 513
predict.lm, 515
qr, 298
residuals, 524
stat.anova, 527
summary.aov, 532
summary.glm, 534
summary.lm, 537
termplot, 226
weighted.residuals, 555
*Topic smooth
bandwidth, 324
density, 337
sunflowerplot, 220
*Topic sysdata
colors, 39
palette, 123
Random, 367
Random.user, 372
*Topic ts
diff, 565

plot.ts, 168
print.ts, 571
start, 578
time, 584
ts, 586
ts-methods, 589
tsp, 590
window, 593
*Topic univar
Extremes, 274
range, 303
sort, 310
*Topic utilities
all.equal, 245
as.POSIX*, 558
axis.POSIXct, 15
DateTimeClasses, 562
dev2bitmap, 55
difftime, 567
findInterval, 278
integrate, 282
jitter, 93
n2mfrow, 119
relevel, 521
strptime, 579
Sys.time, 583
* *(Arithmetic)*, 250
*.difftime *(difftime)*, 567
+ *(Arithmetic)*, 250
+.POSIXt *(DateTimeClasses)*,
    562
- *(Arithmetic)*, 250
-.POSIXt *(DateTimeClasses)*,
    562
.Device *(dev.xxx)*, 50
.Devices *(dev.xxx)*, 50
.Pars *(par)*, 128
.PostScript.Options
    (postscript), 186
.Random.seed *(Random)*, 367
.Device, *198*
.Machine, *510*
.Random.seed, *362*, *383*

INDEX

.leap.seconds
  (DateTimeClasses),
  562
.ps.prolog *(postscript)*, 186
/ *(Arithmetic)*, 250
/.difftime *(difftime)*, 567
==, *246*
[, 558
[.AsIs *(AsIs)*, 405
[.POSIXct *(DateTimeClasses)*,
  562
[.POSIXlt *(DateTimeClasses)*,
  562
[.data.frame, *489*
[.difftime *(difftime)*, 567
[<-.POSIXct
  (DateTimeClasses),
  562
[<-.POSIXlt
  (DateTimeClasses),
  562
[[.POSIXct
  (DateTimeClasses),
  562
%*%, *266*
%*% *(matmult)*, 289
%/% *(Arithmetic)*, 250
%% *(Arithmetic)*, 250
%o%, *266*
^ *(Arithmetic)*, 250
~ *(formula)*, 440

abline, **6**, 71, *184*
abs, **244**, *307*
acos, *281*
acos *(Trig)*, 322
acosh *(Hyperbolic)*, 281
adapt, *283*
add.scope *(factor.scope)*,
  434
add1, **390**, *398*, *433*, *434*, *530*,
  531
AIC, **393**, *432*, *433*

airmiles, **596**
airquality, **597**
alias, **395**, *404*
all, *246*
all.vars, *442*
all.equal, **245**
all.equal.POSIXct
  (DateTimeClasses),
  562
anova, *392*, **398**, *400*, *402*, *445*,
  447, *460*, *478*, *527*
anova.glm, *445*, *447*, *452*, *527*
anova.lm, *462*, *469*, *527*
anova.glm, **399**
anova.glmlist *(anova.glm)*,
  399
anova.lm, **401**
anova.lmlist *(anova.lm)*, 401
anova.mlm *(anova.lm)*, 401
anscombe, **599**
aov, *155*, *391*, *392*, **403**, *418*,
  419, *424*, *425*, *427*,
  432, *434*, *459*, *462*,
  468, *486*, *494*, *518*,
  519, *529*, *533*, *541*,
  546–548
apply, *262*, *263*
approx, *279*, *316*
approx *(approxfun)*, 247
approxfun, **247**, *316*
Arithmetic, *244*, **250**, *288*,
  289, *314*, *512*
arrows, **8**, *209*
as.data.frame, *470*
as.integer, *306*
as.POSIXct, **564**, *580*
as.POSIXlt, **564**, *592*
as.character.POSIXt
  (strptime), 579
as.data.frame, *405*
as.data.frame.logLik
  (logLik), 470

as.data.frame.POSIXct
    (DateTimeClasses),
    562
as.data.frame.POSIXlt
    (DateTimeClasses),
    562
as.difftime (difftime), 567
as.formula (formula), 440
as.matrix (matrix), 290
as.matrix.POSIXlt
    (DateTimeClasses),
    562
as.POSIX*, 558
as.POSIXct (as.POSIX*), 558
as.POSIXlt (as.POSIX*), 558
as.qr (qr), 298
as.ts (ts), 586
asin, *281*
asin (Trig), 322
asinh (Hyperbolic), 281
AsIs, 405
assocplot, 10, *114*
atan, *281*
atan (Trig), 322
atan2 (Trig), 322
atanh (Hyperbolic), 281
attenu, 601
attitude, **603**
attr.all.equal (all.equal),
    245
attributes, *246*, *275*
axis, 12, **15**, **17**, *18*, *21*, *32*, *74*,
    128, *173*, *175*, *204*
axis.POSIXct, *569*
axis.POSIXct, **15**
axTicks, *13*, **17**, *71*, *133*

backsolve, 252, *253*, *309*
bandwidth, **324**
bandwidth.nrd, *325*
barplot, **19**, *100*, *158*, *201*
bcv, *325*
Bessel, 254, *314*

bessel (Bessel), 254
besselI (Bessel), 254
besselJ (Bessel), 254
besselK (Bessel), 254
besselY (Bessel), 254
Beta, **326**
beta, *255*, *327*
beta (Special), 313
Binomial, **328**
binomial, *447*
binomial (family), 436
birthday, **330**
bitmap, *57*, *178*
bitmap (dev2bitmap), 55
box, **24**, *32*, *62*, *166*, *184*, *201*,
    213
boxplot, **25**, *29*–*31*, *158*, *160*,
    216
boxplot.formula, *26*
boxplot.stats, *27*
boxplot.stats, *29*
bquote, *175*
bs, *484*
bw.nrd, *337*, *339*
bw.bcv (bandwidth), 324
bw.nrd (bandwidth), 324
bw.nrd0 (bandwidth), 324
bw.SJ (bandwidth), 324
bw.ucv (bandwidth), 324
bxp, *26*, *27*, *30*, **31**

C, *407*, *418*, *419*
c, *564*
c.POSIXct (DateTimeClasses),
    562
c.POSIXlt (DateTimeClasses),
    562
call, *98*, *268*
capabilities, *58*, *178*, *272*,
    320
cars, *484*, **605**
case.names
    (case/variable.names),

INDEX

409
case/variable.names, **409**
cat, *449*
Cauchy, **332**
cbind.ts *(ts)*, 586
ceiling *(Round)*, 305
character, *180*, *229*, *231*
check.options, *187*, *190*
chickwts, **607**
chisq.test, *11*, *622*
Chisquare, **334**, *345*
chol, *252*, **257**, *260*, *272*
chol2inv, *258*, **260**
choose *(Special)*, 313
chull, **34**
class, *78*, *160*, *396*, *460*
close.screen *(screen)*, 206
cm *(units)*, 233
cm.colors *(Palettes)*, 125
co.intervals *(coplot)*, 44
co2, **608**
coef, **411**, *428*, *452*, *469*
coefficients, *398*, *439*, *445*, *462*, *524*
coefficients *(coef)*, 411
col2rgb, **36**, *39*, *123*, *126*, *203*
colMeans *(colSums)*, 262
colors, *36*, **39**, *123*, *126*, *134*, *135*, *172*
colours *(colors)*, 39
colSums, **262**
Comparison, *310*
complex, *295*
confint, **412**
confint.nls, *412*
constrOptim, **414**, *506*
contour, **40**, *63*, *74*, *77*, *87*, *92*, *141*
contr.helmert, *419*
contr.poly, *294*, *419*
contr.sum, *407*, *419*
contr.treatment, *419*, *521*

contr.helmert *(contrast)*, 417
contr.poly *(contrast)*, 417
contr.sum *(contrast)*, 417
contr.treatment *(contrast)*, 417
contrast, **417**
contrasts, *407*, *418*, **419**, *491*, *525*, *526*
contrasts<- *(contrasts)*, 419
convolve, **264**, *277*, *292*
cooks.distance, *164*, *466*
cooks.distance *(influence.measures)*, 453
coplot, **44**, *127*, *442*
cos, *281*
cos *(Trig)*, 322
cosh *(Hyperbolic)*, 281
covratio, *466*
covratio *(influence.measures)*, 453
coxph, *227*, *546*
crossprod, **266**
cummax *(cumsum)*, 267
cummin *(cumsum)*, 267
cumprod, *297*
cumprod *(cumsum)*, 267
cumsum, **267**, *297*
curve, **48**
cut, *87*, *560*
cut.POSIXt, *564*
cut.POSIXt, **560**
cycle *(time)*, 584

D *(deriv)*, 268
data, **609**
data.frame, *148*, *149*, *290*, *406*, *489*, *490*
data.matrix, *291*
data.frame, *405*
date, *583*, *585*

DateTimeClasses, 15, 559, **562**, *568*, *575*, *577*, 582, 583, *592*
dbeta, *347*
dbeta *(Beta)*, 326
dbinom, *359*, *363*, *364*
dbinom *(Binomial)*, 328
dcauchy *(Cauchy)*, 332
dchisq, *345*, *347*
dchisq *(Chisquare)*, 334
deltat *(time)*, 584
density, *81*, *148*, *154*, *162*, *222*, *324*, *325*, **337**
deparse, *224*
deriv, **268**, *497*, *499*
deriv3 *(deriv)*, 268
det, *272*, *300*
dev.cur, *54*, *58*
dev.print, *58*, *178*
dev.control *(dev2)*, 52
dev.copy *(dev2)*, 52
dev.copy2eps *(dev2)*, 52
dev.cur *(dev.xxx)*, 50
dev.interactive *(Devices)*, 57
dev.list *(dev.xxx)*, 50
dev.next *(dev.xxx)*, 50
dev.off *(dev.xxx)*, 50
dev.prev *(dev.xxx)*, 50
dev.print *(dev2)*, 52
dev.set *(dev.xxx)*, 50
dev.xxx, **50**
dev2, **52**
dev2bitmap, **55**, *58*
deviance, **421**, **422**, **433**, **452**, *469*
device *(Devices)*, 57
Devices, *51*, **57**, *68*, *73*, *138*, *144*, *178*, *190*, *198*, *207*, *235*, *237*
dexp, *385*
dexp *(Exponential)*, 342
df, *379*

df *(FDist)*, 344
df.residual, *421*, *452*, *469*
df.residual, **422**
dfbeta *(influence.measures)*, 453
dfbetas, *466*
dfbetas *(influence.measures)*, 453
dffits, *466*
dffits *(influence.measures)*, 453
dgamma, *327*, *335*, *343*
dgamma *(GammaDist)*, 346
dgeom, *359*
dgeom *(Geometric)*, 348
dhyper *(Hypergeometric)*, 350
diag, *289*
diff, **565**, *589*
diff.ts, *566*
diff.ts *(ts-methods)*, 589
diffinv, *566*
difftime, *563*, *564*, **567**, *576*
digamma *(Special)*, 313
dim, *275*, *290*
dimnames, *166*, *290*
discoveries, **612**
dlnorm, *362*
dlnorm *(Lognormal)*, 354
dlogis *(Logistic)*, 352
dmultinom *(Multinomial)*, 356
dnbinom, *329*, *349*, *364*
dnbinom *(NegBinomial)*, 358
dnorm, *355*
dnorm *(Normal)*, 361
dotchart, *21*, *59*, *146*
double, *240*, *274*
dpois, *329*, *359*
dpois *(Poisson)*, 363
drop, *289*
drop.scope *(factor.scope)*, 434

INDEX

drop1, *398, 400, 402, 433, 434,
    530, 531*
drop1 *(add1)*, 390
dsignrank, *387*
dsignrank *(SignRank)*, 376
dt, *333, 345*
dt *(TDist)*, 378
dummy.coef, **423**
dunif *(Uniform)*, 382
dweibull, *343*
dweibull *(Weibull)*, 384
dwilcox, *377*
dwilcox *(Wilcoxon)*, 386

ecdf, *279*
eff.aovlist, **425**
effects, *398,* **427,** *447, 452,
    461, 462, 469*
eigen, **271,** *300, 320*
end, *587*
end *(start)*, 578
environment, 609
erase.screen *(screen)*, 206
Error *(aov)*, 403
esoph, *447,* **613**
euro, **615**
eurodist, **617**
exp, *343*
exp *(log)*, 287
expand.model.frame, *490*
expand.grid, **429**
expand.model.frame, **430**
expm1 *(log)*, 287
Exponential, **342**
expression, *98, 218, 229, 231,
    268, 269*
extractAIC, *392, 394, 421,
    432, 530*
extractAIC.glm, *530*
Extremes, **274**

factor, *45, 155, 158, 280, 321,
    445, 491, 521*

factor.scope, **434**
faithful, **618**
family, **436,** *443, 446, 472,
    483, 512*
family.glm *(glm.summaries)*,
    451
family.lm *(lm.summaries)*,
    468
FDist, **344**
fft, *264, 265,* **276,** *292, 338*
filled.contour, *42, 87, 668*
filled.contour, **61**
filter, *265*
findInterval, **278**
fitted, **439,** *452, 469*
fitted.values, *398, 411, 447,
    462, 524*
fivenum, *30*
floor *(Round)*, 305
Formaldehyde, **620**
format, *290*
format.POSIXct, *559*
format.POSIXlt, *559*
format.POSIXct *(strptime)*,
    579
format.POSIXlt *(strptime)*,
    579
formula, *155, 159, 268, 405,
    406,* **440,** *490, 542,
    545, 546*
formula.lm *(lm.summaries)*,
    468
forwardsolve *(backsolve)*,
    252
fourfoldplot, **64**
frame, **67**
freeny, **621**
frequency, *587*
frequency *(time)*, 584
function, *45, 148*

Gamma, *472*
Gamma *(family)*, 436

gamma, *254, 255, 347*
gamma *(Special)*, 313
gammaCody *(Bessel)*, 254
GammaDist, **346**
gaussian, *472*
gaussian *(family)*, 436
Geometric, **348**
gl, **280**
glm, *163, 227, 391, 399, 400, 411, 418, 419, 421, 422, 436, 437, 439, 440, 442,* **443**, *449, 451, 452, 454, 457, 462, 469, 472, 483, 501, 514, 524, 529, 530, 534, 535, 542, 554, 555*
glm.control, *444*
glm.fit, *449*
glm.control, **449**
glm.summaries, **451**
GNOME, *57*
GNOME *(Gnome)*, 68
Gnome, **68**
gnome *(Gnome)*, 68
graphics.off, *58*
graphics.off *(dev.xxx)*, 50
gray, *39,* **70**, *82, 123, 126, 135, 203*
grey *(gray)*, 70
grid, **71**
GTK, *57*
GTK *(gtk)*, 73
gtk, **73**

HairEyeColor, **622**
hasTsp *(tsp)*, 590
hat, *164, 466, 478*
hat *(influence.measures)*, 453
hatvalues *(influence.measures)*, 453

heat.colors, *39, 86, 87*
heat.colors *(Palettes)*, 125
heatmap, *87*
help, *611*
Hershey, *41,* **74**, *92, 229*
hist, *21,* **78**, *161, 162, 170, 201, 339, 569*
hist.default, *569*
hist.POSIXt, **569**
hsv, *39, 70,* **82**, *87, 123, 125, 126, 131, 203*
Hyperbolic, **281**
Hypergeometric, **350**

I, *441*
I *(AsIs)*, 405
identical, *245*
identify, **84**, *107*
image, *42, 56, 58, 63,* **86**, *141, 170*
Inf, *250*
infert, *447*, **624**
influence, *164, 454–456, 469, 555*
influence *(lm.influence)*, 465
influence.measures, *465, 466*
influence.measures, **453**
InsectSprays, **626**
Insurance, *501*
integer, *274, 369*
integrate, **282**
interaction.plot, *156*
interaction.plot, **89**
InternalMethods, 290, 587
interpSpline, *316*
inverse.gaussian, *472*
inverse.gaussian *(family)*, 436
invisible, *123*
iris, **627**
iris3 *(iris)*, 627
is.empty.model, **457**

is.matrix *(matrix)*, 290
is.mts *(ts)*, 586
is.na.POSIXlt
    (DateTimeClasses),
    562
is.qr *(qr)*, 298
is.ts *(ts)*, 586
is.unsorted *(sort)*, 310
islands, **629**
ISOdate *(strptime)*, 579
ISOdatetime *(strptime)*, 579

Japanese, 77, **92**
jitter, **93**, *205*, *221*
jpeg, *56–58*
jpeg *(png)*, 177
julian *(weekdays)*, 591

kappa, **285**

La.chol *(chol)*, 257
La.chol2inv *(chol2inv)*, 260
La.eigen *(eigen)*, 271
La.svd *(svd)*, 319
labels, **458**
layout, *51*, **95**, *119*, *132*, *135*, *207*
lbeta *(Special)*, 313
lchoose *(Special)*, 313
lcm *(layout)*, 95
legend, *90*, **98**, *201*
levelplot, *62*, *63*
lgamma *(Special)*, 313
LifeCycleSavings, **630**
lines, *7*, *49*, *71*, **104**, *109*, *127*, *128*, *141*, *148*, *161*, *168*, *169*, *172*, *181*, *184*, *209*, *226*, *239*
lines.formula
    (plot.formula), 159
lines.histogram
    (plot.histogram),
    161

lines.ts *(plot.ts)*, 168
list, *299*
lm, *163*, *227*, *391*, *392*, *401*, *402*, *404*, *409*, *411*, *418*, *419*, *421*, *422*, *427*, *432*, *434*, *439*, *440*, *442*, *447*, *454*, *457*, **459**, *463–466*, *468*, *469*, *473*, *480*, *482*, *516*, *519*, *524*, *529*, *537*, *539*, *542*, *555*
lm.fit, *460*, *462*
lm.influence, *164*, *454–456*, *462*, *478*, *480*, *555*
lm.wfit, *462*
lm.fit, **463**
lm.influence, **465**
lm.summaries, **468**
lm.wfit *(lm.fit)*, 463
load, *609*
locales, *582*
locator, *85*, *98*, **106**
log, *244*, **287**
log10 *(log)*, 287
log1p *(log)*, 287
log2 *(log)*, 287
logb *(log)*, 287
Logistic, **352**
logLik, *393*, *394*, **470**
logLik.glm, *471*
logLik.gls, *471*
logLik.lm, *471*, *472*
logLik.lme, *471*
logLik.glm, **472**
logLik.lm, **473**
loglin, *113*, *114*, **475**, *622*
Lognormal, **354**
longley, **632**
lowess, *127*, *239*
ls.diag, *480*, *482*
ls.print, *478*, *482*
ls.diag, **478**
ls.print, **480**

lsfit, *300*, *301*, *478*, *480*, **481**

make.link, *512*
make.link, **483**
makepredictcall, **484**
makepredictcall.poly *(poly)*, 293
manova, **486**, *541*
Math.difftime *(difftime)*, 567
Math.POSIX1t *(DateTimeClasses)*, 562
Math.POSIXt *(DateTimeClasses)*, 562
matlines *(matplot)*, 108
matmult, **289**
matplot, **108**, *627*
matpoints *(matplot)*, 108
matrix, *46*, *109*, *289*, **290**
max, *248*, *303*
max *(Extremes)*, 274
mean, *248*, *262*, *379*
mean.POSIXct *(DateTimeClasses)*, 562
mean.POSIX1t *(DateTimeClasses)*, 562
Methods, *303*
methods, *245*, *451*, *468*, *534*
min, *248*, *303*
min *(Extremes)*, 274
Mod, *246*
mode, *246*
model.extract, *492*
model.frame, *430*, *441*, *484*, *487*, *491*, *492*, *501*
model.frame.default, *484*
model.matrix, *460*, *490*, *543*, *553*
model.offset, *501*

model.tables, *403*, *404*, *424*, *519*, *523*, *526*, *533*, *548*
model.tables.aovlist, *425*
model.extract, **487**
model.frame, **489**
model.matrix, **491**
model.offset *(model.extract)*, 487
model.response *(model.extract)*, 487
model.tables, **493**
model.weights *(model.extract)*, 487
months *(weekdays)*, 591
morley, **634**
mosaicplot, *11*, *65*, **112**, *166*, *622*
mosaicplot.default, *113*
mosaicplot.formula, *113*
mtcars, **636**
mtext, *74*, **116**, *128*, *173*, *175*, *229*, *232*
Multinomial, **356**
mvfft *(fft)*, 276

n2mfrow, **119**
NA, *25*, *29*, *36*, *71*, *196*, *200*, *238*, *240*, *250*, *303*, *565*
na.action, *465*
na.contiguous, *589*
na.exclude, *465*, *466*
na.fail, *430*, *444*, *459*, *489*, *589*
na.omit, *430*, *444*, *459*, *489*, *589*
na.omit.ts *(ts-methods)*, 589
names, *275*
NaN, *29*, *250*
napredict, *439*
napredict *(naresid)*, 496
naprint, **495**
naresid, **496**, *524*
nchar, *218*

INDEX

nclass.FD, *80*
nclass.scott, *80*
nclass.Sturges, *80, 81*
NegBinomial, **358**
nextn, *265, 277,* **292**
nhtemp, **637**
nlm, *269,* **497,** *506, 510, 550*
nls, *269, 499*
Normal, **361**
ns, *484*
numeric, *303*

offset, *460, 487,* **501,** *545*
Ops, *567*
Ops.difftime *(difftime)*, 567
Ops.POSIXt
  (DateTimeClasses),
  562
Ops.ts *(ts)*, 586
optim, *269, 414, 415, 499,* **502**
optimise *(optimize)*, 509
optimize, *499, 505, 506,* **509,** *550*
options, *53, 58, 85, 106, 135, 419, 444, 449, 459, 489*
OrchardSprays, **638**
order, *311*

pairs, *46,* **120,** *127, 149*
palette, *36, 39, 63,* **123,** *126, 134, 172*
Palettes, **125**
panel.smooth, *46, 163*
panel.smooth, **127**
par, *7, 8, 13, 17, 18, 21, 24, 41, 59, 70, 87, 90, 96, 104, 105, 107, 109, 113, 116, 117, 119, 127,* **128,** *141, 147, 148, 151, 155, 158, 159, 163, 168, 170, 181, 183, 184, 200, 201, 203, 207, 209, 212, 213, 221, 227, 229, 231, 233*
pbeta *(Beta)*, 326
pbinom *(Binomial)*, 328
pbirthday *(birthday)*, 330
pcauchy *(Cauchy)*, 332
pchisq, *378, 380*
pchisq *(Chisquare)*, 334
pdf, *56, 57,* **137**
pentagamma *(Special)*, 313
periodicSpline, *316*
persp, **139**
pexp *(Exponential)*, 342
pf *(FDist)*, 344
pgamma, *359*
pgamma *(GammaDist)*, 346
pgeom *(Geometric)*, 348
phones, **640**
phyper *(Hypergeometric)*, 350
pictex, *57,* **143**
pie, **145**
PlantGrowth, **641**
plnorm *(Lognormal)*, 354
plogis *(Logistic)*, 352
plot, *21, 71, 87, 100, 104, 105, 108, 109, 117, 128,* **147,** *149, 150, 152, 155, 158, 166, 168, 170, 172, 180, 181, 221, 238*
plot.default, *15, 32, 41, 62, 67, 109, 135, 148, 149, 155, 158, 160, 166, 168, 170, 172, 213, 220, 239*
plot.density, *339*
plot.factor, *160, 166*
plot.formula, *148, 158*
plot.histogram, *78, 79*
plot.lm, *227*
plot.new, *170*
plot.ts, *587*

plot.window, *21*, *60*, *62*, *67*,
    151, *152*
plot.xy, *105*, *170*, *181*
plot.data.frame, 149
plot.default, *133*, **150**
plot.density, 154
plot.design, 155
plot.factor, 158
plot.formula, 159
plot.function *(curve)*, 48
plot.histogram, 161
plot.lm, 163
plot.mlm *(plot.lm)*, 163
plot.new, *132*
plot.new *(frame)*, 67
plot.POSIXct *(axis.POSIXct)*,
    15
plot.POSIXlt *(axis.POSIXct)*,
    15
plot.table, 166
plot.ts, 168
plot.TukeyHSD *(TukeyHSD)*,
    547
plot.window, 170
plot.xy, 172
plotmath, *74*, 99, *117*, *143*,
    173, *229*, *232*
pmax *(Extremes)*, 274
pmin *(Extremes)*, 274
pnbinom *(NegBinomial)*, 358
png, *56–58*, **177**
pnorm, *381*
pnorm *(Normal)*, 361
points, *32*, *46*, *71*, *99*, *105*,
    108, *109*, *127*, *128*,
    141, *148*, *151*, *163*,
    172, **180**, *227*, *239*
points.default, *172*
points.formula
    (plot.formula), 159
Poisson, **363**
poisson *(family)*, 436
poly, **293**, *484*

polygon, *34*, *145*, **183**, *201*, *209*
polym *(poly)*, 293
polyroot, **295**, *550*
POSIXct *(DateTimeClasses)*,
    562
POSIXlt *(DateTimeClasses)*,
    562
POSIXt, *565*
POSIXt *(DateTimeClasses)*,
    562
postscript, *51*, *52*, *55–58*, *135*,
    137, *138*, *144*, *180*,
    186
power, *436*, *437*, **512**
power.t.test, *512*
ppoints, **192**, *197*
ppois *(Poisson)*, 363
precip, **642**
predict, *430*, *462*, *496*, *516*
predict.glm, *227*, *447*
predict.lm, *461*, *462*
predict.glm, **513**
predict.lm, **515**
predict.mlm *(predict.lm)*,
    515
predict.poly *(poly)*, 293
preplot, **193**
presidents, **643**
pressure, **644**
pretty, *13*, *18*, **194**
print, **396**, *404*, *571*
print.ts, *587*
print.anova *(anova)*, 398
print.aov *(aov)*, 403
print.aovlist *(aov)*, 403
print.AsIs *(AsIs)*, 405
print.density *(density)*, 337
print.difftime *(difftime)*,
    567
print.dummy.coef
    (dummy.coef), 423
print.family *(family)*, 436
print.formula *(formula)*, 440

print.glm *(glm)*, 443
print.infl
    (influence.measures),
    453
print.integrate *(integrate)*,
    282
print.lm *(lm)*, 459
print.logLik *(logLik)*, 470
print.mtable *(alias)*, 395
print.packageIQR *(data)*, 609
print.POSIXct
    (DateTimeClasses),
    562
print.POSIXlt
    (DateTimeClasses),
    562
print.recordedplot
    (recordPlot), 199
print.summary.aov
    (summary.aov), 532
print.summary.aovlist
    (summary.aov), 532
print.summary.glm
    (summary.glm), 534
print.summary.lm
    (summary.lm), 537
print.summary.manova
    (summary.manova),
    540
print.tables.aov
    (model.tables), 493
print.terms *(terms)*, 542
print.ts, **571**
print.TukeyHSD *(TukeyHSD)*,
    547
prod, **297**
profile, **517**
profile.glm, *517*
profile.nls, *517*
proj, *404*, *494*, **518**
ps.options, *236*, *237*
ps.options *(postscript)*, 186
psignrank *(SignRank)*, 376

pt *(TDist)*, 378
ptukey *(Tukey)*, 380
punif *(Uniform)*, 382
pweibull *(Weibull)*, 384
pwilcox *(Wilcoxon)*, 386

qbeta *(Beta)*, 326
qbinom *(Binomial)*, 328
qbirthday *(birthday)*, 330
qcauchy *(Cauchy)*, 332
qchisq *(Chisquare)*, 334
qexp *(Exponential)*, 342
qf *(FDist)*, 344
qgamma *(GammaDist)*, 346
qgeom *(Geometric)*, 348
qhyper *(Hypergeometric)*, 350
qlnorm *(Lognormal)*, 354
qlogis *(Logistic)*, 352
qnbinom *(NegBinomial)*, 358
qnorm, *369*, *381*
qnorm *(Normal)*, 361
qpois *(Poisson)*, 363
qqline *(qqnorm)*, 196
qqnorm, *192*, **196**
qqplot, *192*
qqplot *(qqnorm)*, 196
qr, *252*, *258*, *272*, *285*, *286*,
    298, *301*, *302*, *320*,
    463, *464*
qr.Q, *300*
qr.qy, *302*
qr.R, *300*
qr.solve, *309*
qr.X, *300*
QR.Auxiliaries, **301**
qr.Q *(QR.Auxiliaries)*, 301
qr.R *(QR.Auxiliaries)*, 301
qr.X *(QR.Auxiliaries)*, 301
qsignrank *(SignRank)*, 376
qt *(TDist)*, 378
qtukey, *548*
qtukey *(Tukey)*, 380
quakes, **645**

quantile, *29*
quarters *(weekdays)*, 591
quartz, **198**
quasi *(family)*, 436
quasibinomial *(family)*, 436
quasipoisson *(family)*, 436
qunif *(Uniform)*, 382
quote, *175*
qweibull *(Weibull)*, 384
qwilcox *(Wilcoxon)*, 386

R CMD BATCH, *57*
r2dtable, **365**
rainbow, *39, 70, 82, 86, 87, 123, 135, 203*
rainbow *(Palettes)*, 125
Random, **367**
Random.user, *369*, **372**
randu, **646**
range, *46, 275*, **303**
range.default, *303*
rank, *311*
rbeta *(Beta)*, 326
rbinom, *357*
rbinom *(Binomial)*, 328
rcauchy *(Cauchy)*, 332
rchisq *(Chisquare)*, 334
read.table, *609*
recordPlot, **199**
rect, *24, 161, 184*, **200**
relevel, **521**
rep, *238, 240*, **572**
replayPlot *(recordPlot)*, 199
replications, *494*, **522**
resid, *496*
resid *(residuals)*, 524
residuals, *227, 398, 411, 427, 439, 447, 452, 462, 469*, **524**, *555*
residuals.glm, *469*, **535**
residuals.glm *(glm.summaries)*, 451

residuals.lm *(lm.summaries)*, 468
rexp *(Exponential)*, 342
rf *(FDist)*, 344
rgamma *(GammaDist)*, 346
rgb, *36, 39, 70, 82, 126, 135,* **203**
rgeom *(Geometric)*, 348
rhyper *(Hypergeometric)*, 350
rivers, **647**
rlnorm *(Lognormal)*, 354
rlogis *(Logistic)*, 352
rmultinom *(Multinomial)*, 356
rnbinom *(NegBinomial)*, 358
RNG *(Random)*, 367
RNGkind, *372*
RNGkind *(Random)*, 367
RNGversion *(Random)*, 367
rnorm, *371, 383*
rnorm *(Normal)*, 361
Round, **305**
round, *567*
round *(Round)*, 305
round.POSIXt, *564*
round.difftime *(difftime)*, 567
round.POSIXt, **575**
rowMeans *(colSums)*, 262
rowsum, *263*
rowSums *(colSums)*, 262
rpois *(Poisson)*, 363
Rprof, *517*
rsignrank *(SignRank)*, 376
rstandard *(influence.measures)*, 453
rstudent, *468*
rstudent *(influence.measures)*, 453
rt *(TDist)*, 378
rug, *94*, **204**, **226**
runif, *362, 371*

INDEX   693

runif *(Uniform)*, 382
rweibull *(Weibull)*, 384
rwilcox *(Wilcoxon)*, 386

SafePrediction, *514*, *516*
SafePrediction
    (makepredictcall),
    484
sample, **374**
save, *611*
scale, *484*
screen, **206**
sd, *379*
se.contrast, *494*
se.contrast.aovlist, *425*
se.contrast, **525**
segments, *7*, *9*, *184*, *201*, **209**
seq, *573*, *576*
seq.POSIXt, *560*, *564*, *569*
seq.POSIXt, **576**
sequence, *573*
set.seed *(Random)*, 367
sign, **307**
signif *(Round)*, 305
SignRank, **376**
sin, *244*, *281*
sin *(Trig)*, 322
sinh *(Hyperbolic)*, 281
sleep, **648**
smooth.spline, *316*
solve, *252*, *260*, **308**
solve.qr, *300*, *309*
solve.qr *(qr)*, 298
sort, **310**
source, *609*
Special, *244*, *251*, **313**
spline, *248*
spline *(splinefun)*, 315
splinefun, *49*, *248*, **315**
split.screen, *135*
split.screen, *132*
split.screen *(screen)*, 206
sqrt, *251*, *288*, *314*

sqrt *(abs)*, 244
stack.loss *(stackloss)*, 649
stack.x *(stackloss)*, 649
stackloss, **649**
stars, **211**, *224*
start, *578*, *585*, *587*, *590*
stat.anova, *399*
stat.anova, **527**
state, **651**, *662*
stem, *81*, *162*
step, *392*, *432*, *433*, **529**
stepAIC, *531*
str.logLik *(logLik)*, 470
str.POSIXt
    (DateTimeClasses),
    562
strftime *(strptime)*, 579
strheight *(strwidth)*, 218
stripchart, *27*, *149*, **216**
strptime, *15*, *558*, *559*, *564*,
    569, **579**
strwidth, *99*, *130*, **218**
substitute, *98*, *175*
sum, *262*, *297*, **318**
summary, *398*, *404*, *445*, *447*,
    466, *533*, *535*, *539*
summary.aov, *404*
summary.glm, *445*, *447*, *452*
summary.lm, *462*, *466*, *469*, *478*
summary.manova, *486*
summary.aov, **532**
summary.aovlist
    (summary.aov), 532
Summary.difftime *(difftime)*,
    567
summary.glm, **534**
summary.infl
    (influence.measures),
    453
summary.lm, **537**
summary.manova, **540**
summary.mlm *(summary.lm)*,
    537

Summary.POSIXct
    (DateTimeClasses),
    562
summary.POSIXct
    (DateTimeClasses),
    562
Summary.POSIXlt
    (DateTimeClasses),
    562
summary.POSIXlt
    (DateTimeClasses),
    562
sunflowerplot, **220**, *224*
sunspot.month, *653*
sunspots, **653**
survreg, *227*
svd, *258*, *272*, *286*, *300*, **319**
swiss, **654**
symbols, **223**
symnum, *534*, *537*
Syntax, *251*
Sys.time, *564*
Sys.time, **583**
Sys.timezone *(Sys.time)*, 583
system.time, *585*

table, *166*, *321*, *477*
tabulate, **321**
tan, *281*
tan *(Trig)*, 322
tanh *(Hyperbolic)*, 281
TDist, **378**
termplot, *164*, **226**
terms, *155*, *442*, *446*, *461*, *492*,
    542, *544–546*, *553*
terms.formula, *542*, *545*
terms.object, *542*, *544*
terms.formula, **543**
terms.object, 545
terrain.colors, *86*, *87*, *123*
terrain.colors *(Palettes)*,
    125
tetragamma *(Special)*, 313

text, *41*, *74*, *77*, *92*, *100*, *116*,
    117, *128*, *129*, *168*,
    169, *173*, *175*, *189*,
    218, **228**, *232*
time, *238*, *578*, **584**, *587*, *590*,
    594
Titanic, **656**
title, *21*, *32*, *41*, *60*, *62*, *74*,
    109, *117*, *128*, *139*,
    147, *155*, *163*, *175*,
    229, **231**
ToothGrowth, **658**
topo.colors, *39*, *86*, *87*
topo.colors *(Palettes)*, 125
trees, **659**
Trig, *288*, **322**
trigamma *(Special)*, 313
TRUE, *169*
truehist, *81*
trunc *(Round)*, 305
trunc.POSIXt, *564*
trunc.POSIXt *(round.POSIXt)*,
    575
ts, *169*, *565*, *571*, *578*, *584*, *585*,
    586, *589*, *590*, *593*,
    594
ts-methods, **589**
tsp, *578*, *584*, *585*, *587*, **590**,
    594
tsp<- *(tsp)*, 590
Tukey, **380**
TukeyHSD, *404*, *494*, *533*, **547**
typeof, *250*

UCBAdmissions, **660**
ucv, *325*
Uniform, **382**
uniroot, *295*, *499*, *510*, **549**
units, **233**
update, **551**
update.formula, *530*, *551*
update.formula, **553**
USArrests, **662**

INDEX

USJudgeRatings, **663**
USPersonalExpenditure, **664**
uspop, **665**

VADeaths, **666**
variable.names
    (case/variable.names),
    409
vcov, **554**
volcano, **668**

warning, *250*
warpbreaks, **669**
weekdays, **591**
weekdays.POSIXt, *564*
Weibull, **384**
weighted.residuals, *469*
weighted.residuals, **555**
weights, *555*
weights *(lm.summaries)*, 468
weights.glm *(glm)*, 443
which.min, *275*
width.SJ, *325*
Wilcoxon, **386**
window, *585*, *587*, **593**
women, **671**

X11, *57*, *58*, *177*, *178*
X11 *(x11)*, 234
x11, *68*, *73*, *135*, **234**
xfig, *57*, **236**
xinch *(units)*, 233
xtabs, *669*
xy.coords, *34*, *98*, *99*, *104*, *150*,
    152, *170*, *172*, *181*,
    184, *220*, *223*, *228*,
    241, *247*, *315*
xy.coords, **238**
xyinch *(units)*, 233
xyz.coords, **240**

yinch *(units)*, 233

zapsmall *(Round)*, 305

Printed in the United States
52017LVS00003B/10